Everyday Information

Everyday Information

The Evolution of Information Seeking in America

Edited by William Aspray and Barbara M. Hayes

The MIT Press
Cambridge, Massachusetts
London, England

KH

For information about special quantity discounts, please email special_sales@ mitpress.mit.edu.

This book was set in Stone Sans and Stone Serif by Toppan Best-set Premedia Limited. Printed and bound in the United States of America.

Library of Congress Cataloging-in-Publication Data

Everyday information : the evolution of information seeking in America / edited by William Aspray and Barbara M. Hayes.
   p.   cm.
Includes bibliographical references and index.
ISBN 978-0-262-01501-1 (alk. paper)—ISBN 978-0-262-51561-0 (pbk. : alk. paper)
1. Information behavior—United States—Case studies. I. Aspray, William. II. Hayes, Barbara M., 1955–
ZA3075.E94 2011
025.5′24—dc22

2010020602

10 9 8 7 6 5 4 3 2 1

10/17/11

Contents

1 Introduction 1
William Aspray and Barbara M. Hayes

2 One Hundred Years of Car Buying 9
William Aspray

3 Informed Giving: Information on Philanthropy in Everyday Life 71
Barbara M. Hayes

4 Airline Travel: A History of Information-Seeking Behavior by Leisure and Business Passengers 121
Rachel D. Little, Cecilia D. Williams, and Jeffrey R. Yost

5 Genealogy as a Hobby 157
James W. Cortada

6 Sports Fans and Their Information-Gathering Habits: How Media Technologies Have Brought Fans Closer to Their Teams over Time 185
Jameson Otto, Sara Metz, and Nathan Ensmenger

7 Information in the Hobby of Gourmet Cooking: Four Contexts 217
Jenna Hartel

8 The Transformation of Public Information in the United States 249
Gary Chapman and Angela Newell

9 Active Readership: The Case of the American Comics Reader 277
George Royer, Beth Nettels, and William Aspray

10 Information Exchange and Relationship Maintenance among American Youth: The Case of Text Messaging 305
Arturo Longoria, Gesse Stark-Smith, and Barbara M. Hayes

11 Conclusion 329
William Aspray and Barbara M. Hayes

Editors and Contributors 341
Index 347

1 Introduction

William Aspray and Barbara M. Hayes

Consider a suburban American family on a weekday morning: parents Kim and Sean, children Jamie and Jordan. Kim powers up the laptop to check the family stock portfolio while preparing lunches. She considers a solicitation from a local charity and browses vacation packages online, comparing them to those advertised in the morning paper. Sean retrieves an email answer to a genealogical query about a distant ancestor. He scours car ads in the paper and ponders whether the family should take advantage of the federal rebates on new, high-efficiency automobiles. With coffee in hand, both adults call Kim's parents on the East Coast to see how they are doing after a bout with the flu and then listen to the local weather and traffic reports. While gathering his books together, Jamie checks the details of yesterday's Yankee game on his cell phone and emails friends in a fantasy baseball league to discuss Derek Jeter's performance. Jordan texts several friends to determine what they are wearing to the school play. All of this activity takes place before breakfast! From the moment they wake to the end of the day, each member of this family seeks and uses information to engage in society and construct the myriad of small decisions that will collectively define their everyday lives.

The Problem This Book Addresses

Many of the everyday activities in American life are information activities. Americans gather information from many sources and in many ways. They seek information about major purchases, local sports events, and community activities. They seek political information and information about their heritage and genealogy. They seek information about the best places to vacation and the best schools for their children. They seek information from government sources about a diverse range of topics including health care, social services, visas, recycling, public museums, taxes, government forms, and voter registration. In these and many other ways, everyday life in America is filled with information activity. There is a strong belief, largely unexplored, that the Internet has either made American life more information-rich or changed the rules of information-seeking behavior. However, it is clear that Americans receive their information from many sources in addition to the Internet— from family and friends, experts or other trusted members of their community, professional and social organizations, sales people and consumer advocates, print sources and television, and so on. These kinds of information-seeking behaviors went on long before the advent of the Internet, although the Internet may have reshaped that behavior in various ways.

This book is organized around nine specific everyday activities in American life that involve the purchase of goods and services (cars, airline travel), participation in hobbies (sports, genealogy, comics reading, and cooking), being informed about sectors of society (government information, philanthropy), and communication and relationship maintenance (text messaging). In order to add coherence to the book, the chapters all follow a single rubric. Each is a case study of a particular activity, identifying the information-gathering questions that are being asked as part of this activity and the sources used to answer them. Each chapter discusses not only present day information-seeking behavior but also the information-seeking behaviors of Americans in the past and the evolution of those behaviors. Each chapter considers which forces have caused the questions and the sources to change over time. The book concludes with a synthetic essay in which the editors identify some of the common themes in information-seeking behavior in everyday American life and how it has changed over the past century, drawing examples from the nine case studies.

Almost every chapter in the book draws on literature and methods from multiple academic disciplines. Each chapter relies, to some degree, on some type of history, whether it is cultural or technological, because history enables the author or authors to examine change in questions and sources over time. Some chapters rely on other methodologies such as ethnography (of individuals who pursue gourmet cooking as a hobby), psychology (of reasons for philanthropic giving), economics (of the relationship between optimal price and time spent searching for a car), literary theory and cultural studies (on reading comics), and communication theory (on text messaging as a means of youth to stay connected).

Topics Covered

Chapter 2, by William Aspray, examines car buying in America as an information issue since the creation of the Ford Model T and the formation of General Motors in 1908. He identifies fifty questions car buyers would typically ask, in categories such as general understanding of the car-buying process, automobile technology, makes and models, dealers and manufacturers, and financial considerations. The chapter explains how questions buyers ask and the information sources they consult have been shaped by forces endogenous to the car industry such as the strategy and structure of automobile manufacturing firms, available technologies, foreign competition, and interconnected industries such as motels and fast food restaurants; as well as by exogenous forces such as economic depression, war, suburbanization, the entry of women into the workforce, the environmental movement, and oil shortages.

Chapter 3, by Barbara M. Hayes, examines the evolution of an increasingly rich culture of information around philanthropy in everyday American life. The chapter examines how individual motives shape information-seeking behavior related to charitable giving. It explores the ways in which exogenous forces such as religion, war, and economic crisis have, over time, shaped the nature and distribution of information prompting Americans to give. Finally, the chapter examines endogenous factors such as the professionalization of philanthropic workers, the efforts of "citizen philanthropists," and the use of modern digital media, all of which have contributed to a rich information environment designed to engage every American in a lifetime of philanthropic activity.

Chapter 4, by Rachel D. Little, Cecilia D. Williams, and Jeffrey R. Yost, traces airline travel by leisure and business passengers since the first passenger flight in the United States in 1913. The chapter authors consider how exogenous factors such as fuel prices, the general state of the economy, and increases in global commerce, terrorism, and government deregulation have all affected travel and the information issues related to it. They also consider endogenous factors in the airline industry such as technological innovation (including jet engines and pressurized cabins), frequent flyer programs, computerized reservation systems, and the disintermediation of ticketing through the rise of online airline reservation systems such as Orbitz and Expedia and the fall in importance of travel agents.

Chapter 5, by James W. Cortada, traces genealogy as a hobby since the end of the American Civil War. This chapter considers diverse practices ranging from recording family ancestry in the family Bible to use of online genealogical databases. Forces for change studied by Cortada include the increasing geographic and social mobility of American households, government sources of information such as the decadal population census and passport records, powerful new information technologies for carrying out genealogical research, and key events such as the American Bicentennial and the television miniseries *Roots* that stimulated new interest in genealogy across America.

Chapter 6, by Jameson Otto, Sara Metz, and Nathan Ensmenger, traces information issues—in particular, the obsession for performance statistics—among sports fans and participants in fantasy sports leagues. The chapter authors focus on a series of changes in the media (newspaper, radio, television, Internet) in providing fans with information about their favorite teams and players. They also consider the impact of such factors as local pride, gambling, high salaries for athletes, performance-enhancing drugs and other forms of cheating, and sports business practices such as free agency.

Chapter 7, by Jenna Hartel, considers information in the hobby of gourmet cooking. Based on an ethnographic study, mainly carried out in Los Angeles, of gourmet-cooking social groups and classes, online forums, public lectures, and farmers markets, Hartel examines information issues associated with living a gourmet lifestyle, expressing culinary expertise, staying informed and inspired about gourmet cooking, and trying a new recipe. She sets this work in a historical context by studying early television shows about cooking and the rise of popular magazines on gourmet cooking.

Chapter 8, by Gary Chapman and Angela Newell, discusses access to government information by individuals. It considers the shaping role of government legislation such as the Freedom of Information Act, the creation of the National Archives, the political movement promoting open access to government, the Enlightenment idea of the informed citizen and its adoption by America's "Founding Fathers," the growth of a free press, assaults on free access to information in times of war and terrorism, the development of digital archives, the use of interactive tools in the military, Web 2.0 and the rise of eGovernment, and the need for rapid dissemination of geospatial data in times of natural disasters.

Chapter 9, by George Royer, Beth Nettels, and William Aspray, explores information issues related to the practice of reading. More specifically, this chapter examines the readership of comics, detailing the transition of the comics reader from passive consumer to active participant in shaping both the future of the medium and a participatory reading culture. The chapter traces comics reading in America since comics originated in 1895, and considers such factors as immigration, newspaper circulation battles, world wars and patriotism, government regulation and comic book industry self-censorship, evolution in the form of comics (strips, books, interactives), direct marketing, the rise of comic book shops, technological advances, and the emergence of readers as creators.

Chapter 10, by Arturo Longoria, Gesse Stark-Smith, and Barbara M. Hayes, considers the use of the medium of text messaging for information exchange and relationship management among youth. New technologies, such as test messaging, Twitter, and Facebook, are ways for young people to construct their own social worlds and navigate through these worlds independently. This can be contrasted with earlier times, when young people used the landline telephone, letters, radios, and other media to communicate with one another; and to meet in such spaces as arcades, restaurants and diners, parks, and clubhouses.

Chapter 11, by the editors, draws from the nine case studies presented in chapters 2 through 10 a set of conclusions about information-seeking behavior in everyday American life, and how the collective questions raised and sources consulted have changed over time. In particular, chapter 11 evaluates the role that the Internet has played in shaping daily information-seeking behavior, and whether the Internet therefore deserves a privileged place in the study of information in everyday American life. This chapter also discusses how the research methodology developed in this book is related to other information-seeking behavior methodologies.

Situating This Book in the Scholarly Literature

We came to this topic through our work on the social and economic influences of the Internet and in particular through the literature on the Internet's role in everyday life. The earliest social study of the Internet investigated what happened online, divorced from the work or home or "third place" setting (e.g., coffeehouse, library, or social club) in which the Internet activity occurred. Starting less than ten years ago, a few scholars began to investigate Internet activity in the home setting. These scholars considered such issues as where the computer was placed in the home, who used it, what it was used for, and how it fit into the social dynamic of other (offline) activities in the home. Only a few, particular everyday home activities, such as staying connected with friends in another part of the world or seeking health information, received much attention. The breakout book in this field was an edited volume giving various national perspectives (Wellman and Haythornthwaite 2002). This was followed by two small but well-crafted ethnographic studies on the Internet and everyday life—in Canada (Bakardjieva 2005) and Australia (Lally 2002). As far as we know, this kind of study of the Internet in American everyday life has not yet been done. This manuscript advances this line of scholarship not only by covering the American scene of the Internet in everyday life, but also by considering the Internet as only one of many means of gathering information in the pursuit of everyday life.

There are various other literatures that touch on themes similar to those in this book, although none of them is exactly on the same topic. For example, there is a rapidly expanding literature on the social study of the Internet. It includes works examining how youth with ready access to the Internet behave, such as Palfrey and Gasser 2008 and Solove 2008.

From the extensive body of literature in information science on information-seeking behavior, the most relevant perhaps is the work of Karen Fisher and others at the University of Washington (e.g., Fisher, Landry, and Naumer 2007). She and her colleagues focus more on developing abstract social science concepts—such as the notion of information grounds—than on specific everyday activities, and they do not generally employ cultural and technological history in their studies.

There is also a body of scholarship on everyday life, ranging from critical theory (e.g., de Certeau 2002; Lefevre 2008) to American cultural history (e.g., Larkin 1989; Schlereth 1992) to fact books (e.g., McCutcheon 1993). However, none of these books focus on information issues. Some of the most astute historical scholarship on the use of information and of information and communication technologies over several centuries of American history has been written by Daniel Headrick (2000). However, his work focuses on corporations, governments, and nations, not on the home.

This book illuminates information aspects of everyday activities in American life from the nineteenth century to the present. It considers the forces—both those internal to these activities and others more global such as war, economic depression, workforce changes, and social movements—that change the information dimensions of everyday life. This book can be used as a primary text in a course on information in everyday life, or as one of several texts in information science courses on information-seeking behavior and in American cultural studies courses. It could also be used profitably in other courses in communications studies, library and information science, American studies, and American history.

References

Bakardjieva, Maria. 2005. *Internet Society: The Internet in Everyday Life*. London and Thousand Oaks, CA: Sage Publications.

de Certeau, Michel. 2002. *The Practice of Everyday Life*. Berkeley, CA: University of California Press.

Fisher, Karen E., Carol F. Landry, and Charles Naumer. 2007. Social Spaces, Casual Interactions, Meaningful Exchanges: "Information Ground" Characteristics Based on the College Student Experience. *Information Research* 12 (2) (January).

Haythornthwaite, Carolyn, and Barry Wellman. 2002. *The Internet and Everyday Life*. Malden, MA, and Oxford: Wiley-Blackwell.

Headrick, Daniel. 2000. *When Information Came of Age*. New York and Oxford: Oxford University Press.

Lally, Elaine. 2002. *At Home with Computers*. Oxford: Berg.

Larkin, Jack. 1989. *The Reshaping of Everyday Life, 1790–1840*. New York: Harper Perennial.

Lefevre, Henri. 2008. *Critique of Everyday Life*. London: Verso.

McCutcheon, Marc. 1993. *The Writer's Guide to Everyday Life in the 1800s*. Cincinnati, OH: Writer's Digest.

Palfrey, John, and Urs Gasser. 2008. *Born Digital: Understanding the First Generation of Digital Natives*. New York: Basic Books.

Schlereth, Thomas. 1992. *Victorian America: Transformations in Everyday Life, 1876–1915*. New York: Harper Perennial.

Solove, Daniel. 2008. *The Future of Reputation: Gossip, Rumor, and Privacy on the Internet*. New Haven: Yale University Press.

2 One Hundred Years of Car Buying

William Aspray

Last year a Yale University physicist calculated that since Chevy offered 46 models, 32 engines, 20 transmissions, 21 colors (plus nine two-tone combinations) and more than 400 accessories and options, the number of different cars that a Chevrolet customer conceivably could order was greater than the number of atoms in the universe. This seemingly would put General Motors one notch higher than God in the chain of command.
—Hal Higdon, *New York Times Magazine*, 1966, as quoted in Foster 2003, 146–147

I visited showrooms, I spoke with car owners around the country, I kicked tires. I underlined passages in *Consumer Reports*. Finally, I spent a week in Detroit, talking to executives and assembly-line workers alike, watching simulated head-on crashes and riding in test cars at 120 mph. In essence, I did what every car buyer dreams of doing before purchasing a car.
—Matthew Stevenson, *The Saturday Evening Post*, July/August 1982, 66

Introduction

America is deeply rooted in car culture. Except for people who live in large cities with densely laid public transit systems, Americans are utterly dependent on their cars. In 2002, 200 million cars were operated on American roads and driven over three trillion miles.[1] The car purchase is important to the American family not only because the family members otherwise do not have the mobility to shop, attend school or church, or see friends, but also because this is typically the second most expensive durable purchase an American family makes, after the purchase of a home. The transaction is fraught with stress because of the widely varying costs to purchase, operate, and maintain a vehicle; together with the low confidence consumers have in car manufacturers and dealers to provide them with accurate information. The transaction is also complex; this chapter identifies more than fifty information issues that are commonly raised in an American family's quest to purchase a new or used car.

The act of buying a car is an excellent case study for understanding how information needs in American life have changed over time. The Ford Model T was introduced just over a hundred years ago—in 1908, the same year in which the company General Motors (GM) was founded. Henry Ford, the creator of the Model T, famously stated that an individual could buy one of his cars in any color, so long as it was black. Within a few years of the introduction of the Model T, however, one did have choices. There were choices of body style—including some that are not available today, such as a tractor body for your Ford; and beginning in 1923, with General Motor's eggshell-blue Oakland model, one had a choice of colors.[2] But the information issues of buying a car then generally were simpler than they are today. Until the mid-1920s, it was not common for consumers to buy their cars on credit. Cars then were not so technologically advanced or complicated as they are today, and neither safety nor environmental issues were foremost concerns. The information sources for learning about cars in the early twentieth century were also more limited. For example, no car enthusiast magazine was published in the United States between the two world wars, and there was no consumer magazine to rate cars until the publication of *Consumers Union Report* in 1936 (renamed *Consumer Reports* in 1942).

This chapter considers three elements involved in the process of car buying in America since the Model T's introduction a century ago.[3] First, what types of information have prospective new or used car purchasers sought in the course of making their purchase? Second, what sources of information has the typical car buyer consulted when buying a car? Third, what endogenous and exogenous drivers have caused the information types and sources to change over time?

These change drivers have been complex. Historian Sally Clarke has described how "private and public entities—the courts, insurance underwriters, engineering societies, state motor vehicle administrations, the Justice Department, the Federal Trade Commission (FTC), and the Board of Governors of the Federal Reserve System—regulated relations between buyers and sellers" (Clarke 2007, 1–2). Other major external forces at play have included economic depressions, war, suburbanization, an expanding workforce, growing wealth, environmental and safety concerns, the rise of the media, and foreign competition.

By extending this chapter's case study, the first presented in this book, our intention is to examine the information questions, information sources, and exogenous and endogenous shaping forces in such detail that readers clearly understand the organizational rubric used throughout the book. The first part of this chapter reports on information issues and sources for today's car buyers; examines in detail two particularly important means—advertising and dealerships—by which prospective buyers learn about cars; and uses the example of Cord automobiles in the 1930s to illustrate the relationship between salesman and buyer. The second part of the chapter offers a historical overview of the American automobile and explains how these developments have influenced information issues for car buyers. The final part presents methodological considerations and conclusions, followed by an extensive bibliography of sources.

Car Buying as an Information Issue Today

The decision to buy a car is a complex one, partly because of the large number of choices available (see the Higdon epigraph that opens this chapter); and partly because of how much Americans expect from their cars:

What is expected of the American automobile is demanded of few machines. In the course of a year, it is supposed to be safe and easy to drive; log about 10,000 miles; haul the Little League to games; tow boats down to the landing; take the kids to college; provide comfort and cool air on a long vacation; ride practically mainte-nance free, except for the odd quart of oil; keep its finish without being kept in a garage; deliver 50-pound bags of fertilizer to the lawn; pass trucks on the interstate; fit into small, downtown parking spaces; have a sound system like that of Carnegie Hall; pollute nothing; maintain a high trade-in value; stand up to the abuse of the Dukes of Hazzard; provide a romantic setting for teenagers; and give the family dog a chance to take in some local errands. (Stevenson 1982, 66)

In this section, we identify more than fifty issues that a person seeking to buy a car might typically consider. (The complete list of these issues is given in table 2.1.) Probably no car buyer considers every one of these issues as part of the purchase decision, but each of these issues is routinely consid-ered by many Americans when purchasing a car. The issues are grouped into five categories: general understanding of the car-buying process, ques-tions about technology, information about makes and models, information about dealers and manufacturers, and financial considerations.

Even before a potential buyer begins to search for a new car, he (or she) is likely to know a great deal about the car-buying process just from having lived in America. This might include information about the general nature of the distribution system (the various cars' manufacturers distribute

Table 2.1
Information issues for car buyers

1920s	1950s	Today	Issue
General understanding of car buying			
x	x	x	General understanding of cars and car-buying issues
	x	x	Features and pricing on used cars
?	x	x	Time of year for closeouts and special sales events
?	x	x	Size of dealership inventories, scarcity of model, ability to find the right combination of features, and the ability to bargain
		x	Info on types of cars for particular lifestyles (e.g., minivans for soccer moms, ease of use of infant seats, ease of access for the aged, etc.)
?	?	x	Opportunities for aftermarket customization
Technology issues			
x	x	x	New auto technology
x		x	Alternative propulsion technologies (gas, diesel, hybrid, electric, etc.)
Features of makes and models			
x	x	x	Features and pricing on new makes and models
x	x	x	Side-by-side comparisons of favored models
			Attributes identified by car buyers to discriminate between cars (the first sixteen come from Gupta and Ratchford 1992):
?		x	• size
x	x	x	• price
x	x	x	• sportiness
		x	• gas mileage
		x	• foreign/domestic
x	x	x	• acceleration
x	x	x	• handling
?	?	x	• comfort
x	x	x	• luxury
	?	x	• reliability
		x	• safety
x	x	x	• style
?	?	x	• quality
?	?	x	• roominess

Table 2.1
(continued)

1920s	1950s	Today	Issue
		x	• trunk space
x	?	x	• engine power
x		x	• quietness
x			• ventilation
		x	Reliability of a particular model over time
		x	Secret warranties, manufacturer rebates, and other special discounts
?	?	x	Maintenance and service requirements and cost
		x	Safety features and ratings
	?	x	Where in life cycle of a model (first year, last year, etc.)
?		x	Gas mileage and fuel type
x			Ability to drive on unpaved roads
Issues about dealers and manufacturers			
?		x	Reliability of car manufacturer overall
		x	Comparison shopping for price across dealers for a given new car
		x	Info about local dealers—complaints, awards, hours, loaner policy, etc.
		x	Domestic/foreign content and domestic dealer network
Financial considerations			
x	x	x	Whether to pay cash or finance the car
		x	Advantages and disadvantages of leasing vs. buying
		x	How to negotiate a purchase or lease or loan
?	?	x	New vs. used (financial implications)
?	?	x	How an auto purchase fits into one's overall family financial situation
		x	Identifying total costs of ownership
?	?	x	Monthly payments
	?	x	Where to go for a loan
	?	x	Trade in to dealer or sell to private party
	?	x	Value of your trade-in
		x	Value of dealer add-ons (window tinting, rust proofing, paint-chip clear coat, etc.)
x			Personal auto vs. public transit

Note: The bulleted items are from Gupta and Ratchford 1992.

exclusively through a network of geographically dispersed dealers), the names of popular manufacturers, the names of popular models, and possibly something about the big vendors in used cars (CarMax in many cities nationwide, or aggressive local marketers of used cars such as Rocky's Auto in Denver). But most serious car buyers will want to know much more—for example, details of the practice of changing models annually; the timing of year-end discounts; the range of car body types and their value in serving particular uses such as hauling kids to sporting events or providing the elderly or infirm easy access; the amount of domestic versus foreign content in cars if one wants to "buy American"; and opportunities for after-market customization of one's car.

The automobile is a complex technological artifact, comprising thousands of parts. Most people do not understand the mechanics of the different systems that make up a car, nor do they generally have to. The internal workings have become more opaque to the typical driver over time as electronics have become a more prevalent element of automotive systems. A steady stream of innovation has made cars safer, accelerate and handle better, consume less fuel, and ride better. These innovations are often introduced by a high-end manufacturer (such as Mercedes, Lexus, or Cadillac), and trickle down over the next few years to less expensive makes when the cost drops and the innovation is well received by consumers. Given a constant stream of new technologies, there is always some particular technological innovation available on some makes of cars but not others. Many consumers feel the need to understand what this technology is, how it would affect their driving experience, what it costs, and which cars offer it.

Safety features have been one of the biggest drivers of this stream of innovation. Consider the anti-lock braking system (ABS), which improves stopping ability. It was invented in 1929, but it was not introduced into airplanes, racing cars, and experimental cars such as the Ford Zodiac until the 1950s. ABS was introduced into high-end American automobiles such as Lincoln and Cadillac in the early 1970s, vastly improved by Mercedes and the engineering company Bosch in the late 1970s, then slowly spread into many car models, including typical family sedans and even economy models, in the 1980s and 1990s.

In times when fuel prices have risen, such as during the oil crisis of 1973 and the world competition for oil in 2008, consumer interest rose in alternative technologies to the gasoline-fueled combustion engine, such as diesel engines, electric cars, and hybrid vehicles. While there has always been a small group of people interested in these alternative technologies for environmental or technological reasons, widespread interest in them in America has been limited to times of high gas prices. As prices at the pump increase, consumers become more sensitive to the relative fuel economies of gasoline-fueled models.

Perhaps the greatest amount of scrutiny among American car buyers concerns comparison of makes and models. Social scientists Gupta and Ratchford (1992) identified sixteen attributes used by buyers to discriminate in a side-by-side comparison of cars: size, price, sportiness, gas mileage, foreign/domestic, acceleration, handling, comfort, luxury, reliability, safety, style, quality, roominess, trunk space, and engine power. One can add to this list: fuel type required, frequency of maintenance, warranty, and where the model is in its life cycle.

This last factor, life cycle, may have a different value to different customers. Early adopters may want the latest offerings. Others may be keen to have a particular feature and so might buy the first mid-cost family sedan available with anti-lock brakes. Some buyers might avoid a model in its first year, with the expectation that the manufacturer will tweak the design in subsequent years to fix the initial "bugs." Some people won't buy a model in its final year of production to avoid buying a product that will soon become obsolete, while others will savor the opportunity of end-of-run cost discounts on a relatively bug-free design.

Buyers often have many questions about auto manufacturers and especially about their dealers. Manufacturers differ in the number of defects found in their products. In the early 1960s, the average American car was sold with twenty-four defects and one in five cars was subject to a recall. By contrast, a decade later many of the Japanese manufacturers won the loyalty of American buyers for producing cars with almost no defects. Several of the manufacturers gained competitive advantage through standards they set for their dealers on length and comprehensiveness of warranties, no-haggle pricing (e.g., Saturn), and other aspects that affect the quality of the buying experience.

The buyer's contact typically is with the dealer and not with the manufacturer—not only for the purchase, but also for ongoing maintenance and repairs—and so buyers tend to have more questions about dealers than about manufacturers. Is the dealer's business model focused on high-discount, high-volume sales or on service and the "personal touch"? Other issues might be size of a dealership's inventory, the scarcity of a particular model, the ability to find the right combination of features, and the ability to bargain on the asking price. Buyers might want to compare local dealers for price and availability. They might want to know about each dealer's quality (awards, customer complaints) and business practices (service department hours, availability and cost of loaner cars while one's car is being serviced, or the likelihood of after-sale advocacy support from the sales person).

The purchase of an automobile is one of the most important financial transactions a family makes. Not only is a car expensive to buy, maintain, operate, and repair, it also is an investment that depreciates rapidly in value. The buyer must consider a number of financial issues: how much one's current car is worth when traded to the car dealer or sold independently, whether to buy or lease, whether to pay cash or finance, how depreciation works and whether to buy a new or used car, whether to accept a loan from the automobile manufacturer or get a loan elsewhere,[4] whether the manufacturer is offering dealers secret warranties or rebates to help move cars that had been slow to sell or have some problematic mechanical or electrical feature, what the total cost of ownership will be (including purchase price, maintenance and repair costs, insurance costs, and resale value), and what the monthly lease or loan payment will be for various down payments and loan length.

Thus, the savvy purchaser will consider dozens of issues when buying a new or used automobile. Most of these issues require specific information to answer them in a reasonable way. Clearly, the task of buying an automobile is an important information issue of everyday American life.

Information Sources Available to Car Buyers Today

While the informational demands on purchasing an automobile are high, the sources available to the car buyer are extensive. (See table 2.2) Historically, the most pervasive sources of information are advertising and the dealerships, which get treated in their own sections later in the chapter.

Perhaps the single most widely consulted source of information by prospective automobile buyers is *Consumer Reports*. It was the first consumer magazine to evaluate cars, and started doing so in its initial year of publication.[5] *Consumer Reports'* origin harks back to Consumers Union, an organization owned by consumers and operating a laboratory in New York City, which was founded and began publishing reviews of consumer products in 1936. Lawrence Crooks, an independently wealthy car aficionado, managed the auto reviews section from 1936 until 1966, personally buying many of the early cars to be reviewed. Reviews in the early years were often about how to maintain and repair cars and tires—understandable during the Depression when people could not afford new cars and throughout World War II when new car manufacturing was suspended. Circulation grew from 100,000 in 1946 to 400,000 in 1950. *Consumer Reports* published its first annual auto issue in April 1953, reviewing fifty cars. In 1962, the magazine published an influential report on the wide variation in costs of

Table 2.2
Information sources available to car buyers

1920s	1950s	Today	Source
			Printed and online literature
	x	x	• consumer guides (e.g., *Consumer Reports*)
x	x	x	• mass-market periodicals (e.g., *The Saturday Evening Post*)
			• auto magazines
	x	x	For the general audience (e.g. Edmunds)
	x	x	For hobbyists (e.g., *Road & Track*)
		x	• personal finance magazines (e.g., *Kiplinger's*)
x	x	x	Manufacturer brochures
		x	Radio/TV/Internet car shows
x	x	x	Auto shows
			Advertising
	x	x	• radio
	x	x	• television
x	x	x	• magazine
	x	x	• direct mail advertising
		x	• targeted Internet
			Showroom experience
?	x	x	• salesperson information
?	?	x	• financial specialist information
x	x	x	• walk around
	x	x	• test drive
			Buyer buddies
x	x	x	• family
x	x	x	• friends
x	x	x	• mechanic

auto insurance and called successfully for reform. In 1966, the magazine reviewed its first Japanese import (the Toyota Corona, which it reviewed favorably). After publishing his landmark book *Unsafe at Any Speed*, pioneering consumer advocate Ralph Nader, the U.S. auto industry's most formidable critic, was invited to serve on the magazine's board, which he did from 1967 to 1975. In 1983, the magazine started an auto-pricing service. Four years later, the company opened a large test track and auto testing facility in Connecticut. That same year, its reports became available online through the Prodigy and CompuServe services, and five years later through America Online. By 1992, magazine circulation reached five million. In 1997, Consumer Reports Online (later called ConsumerReports. org) was formed, and five years after that it was reaching a million subscribers.

Consumer Reports is only one of many magazines serving the car-buying public. Other consumer-product-rating magazines competing with *Consumer Reports* include *Consumers Digest, Consumers' Research*, and *Which?Car*. Weekly news magazines and general-purpose popular magazines have also been a common source for car buyers: *Time, Newsweek, US News and World Report, New Republic, National Review, The Nation, Businessweek, Collier's Weekly, The Saturday Evening Post, Look*, and *Life*. Two later sections of this chapter discuss what prospective car buyers could learn from these magazines. Automobile enthusiast magazines have also been important.[6] The most influential of these today are *Car & Driver, Motor Trend*, and *Road & Track* for the general public, *Autosport* and *Motor Sport* for the serious racing buff, and *Robb Report* for those interested in luxury vehicles. [7] The popular consumer finance magazines have been a useful source too: among them *Kiplinger's Personal Finance* (formerly *The Kiplinger Magazine* and *Changing Times*), *Money*, and to some degree *Forbes* and *Fortune*.

To get a sense of what the potential car buyer can learn from a personal finance magazine, we reviewed the car-related articles appearing in *Money* magazine during the five-year period from 1973 to 1977.[8] As one might expect, the majority of the articles are directly about money issues relating to car buying: good buys on used cars when there is a generous supply in the market, how to obtain low financing costs, rebates on select models and model types, the benefits of paying off a loan quickly, the financial advantages of diesel over gasoline engines, understanding the total cost of ownership, the financial advantages of buying leftover new cars or late-model used cars, how to find cars with the best fuel economy, the financial value of long-term car loans in times of high inflation, and which technological changes result in cost cuts for car buyers.

Money magazine also contains news items of interest to car buyers. In the period surveyed, these included stories about changes in the nature of car insurance, irritations buyers might have with new seat belts designed to meet federal regulations, the rapid rise in the cost of auto parts and auto insurance because of monopolistic practices in the auto replacement-parts industry, a class action suit against General Motors for putting "inferior" Chevrolet engines in new Oldsmobiles, the rising cost of Japanese cars because of the increasing value of the yen against the dollar, movements by American makers to address fuel concerns by shrinking their cars' exterior dimensions while increasing interior dimensions, and the growing popularity of trucks as personal vehicles.

Money also addressed some specialty issues: which cars are best for buyers concerned about the environment, how light-colored cars are more visible and hence safer, how the value of convertibles increased after American companies stopped manufacturing them, and the advantages and disadvantages of luxury sedans over traditional sedans. Finally, *Money* ran occasional buyer-beware articles on such topics as misleading advertising, the problems with servicing cars that are no longer offered for sale as new models, and insurance problems for families with teenage drivers.

Television,[9] radio, and the Internet also offer programming on car buying. Examples include shows specifically about automobiles such as *MotorWeek* (television and Internet) and *Car Talk* (radio and Internet) as well as personal finance shows that occasionally offer car advice, such as *The Motley Fool* (television and Internet), *The Dave Ramsey Show* (television and radio), and *The Suze Orman Show* (television).[10]

Three other common sources of information for prospective car buyers are product literature from automobile manufacturers; visits to auto shows; and, as they are known in the academic literature, "buyer buddies" such as parents, friends, and local garage mechanics.[11]

The Special Role of Advertising[12]

"Car ads are all full of mansions, horses, surf, mountains, sunsets, chiseled chins, chic women, and caviar—anything but facts."
—Art director of an ad agency, mid-1950s, as quoted in Nelson 1970

The most pervasive source of information about cars is advertising. In her book-length diatribe, Katie Alvord gives statistics that indicate the magnitude of automobile advertising in American society (not including direct mail advertising):

It's estimated people in the western world see 3,000 to 16,000 ads and commercial images every day, or roughly three to 17 of them every waking minute. Close to a fifth of those relate to cars. U.S. car and car-product sellers are the number one advertisers on TV and in magazines, number two in newspapers. Car and car-product ads account for about a quarter of newspaper ads, close to 20 percent of TV ads, 15 percent of magazine ads, and 10 percent of radio ads. Car ads fill billboards, sports stadiums, web pages, and even show up on transit buses.

Every year, more is spent on car and car-accessory ads than on any other product category. U.S. ad spending for autos, auto accessories, equipment, and supplies topped $14 billion in 1998 . . .three of the top six [advertisers] in the U.S. . . . were auto companies. (Alvord 2000, 45)

Although we discuss radio and television advertising briefly at the end of this section, our primary focus is on magazine advertising. The craft of advertising and the technology of automobiles came to maturity at about the same time, with their first growth toward professionalism occurring in the 1920s. With the introduction in the 1920s of the model year and "dynamic obsolescence" practice (as GM's principal car stylist Harley Earl called it), within a decade the automobile industry became a leading advertiser in newspapers and magazines. Magazine advertising of automobiles rose from $3.5 million in 1921 to $9.3 million in 1927.

Ads are intended, of course, as a device to persuade, not to give the prospective car buyer objective facts about purchase choices. One of the reasons for this lack of hard facts and direct comparisons between cars was a general proscription in the American advertising industry since the 1920s against knocking down the competition directly in one's ads. One notable exception was Plymouth, which, from its founding in 1928 through the 1950s, tended to offer factual, copy-heavy, and comparative ads. Later, in the 1960s, comparative copy did become more widespread. Even then, although it directly compared features, it did not knock down the competition generally. These comparative ads were most effectively used to advertise low-end cars being sold on the basis of features available for a given cost.

The advertising industry became substantially more professional and effective over the past century as it honed its psychological edge and enhanced its aesthetic production. Color photography began to appear in car ads in 1932 and increased in the 1940s. However, "elongator" illustrations continued to be shown alongside photographs into the 1970s. The goal was to be faithful to the details but not the proportions of the cars, so as to make the cars look longer, wider, and lower. From 1933 onward, these illustrations were particularly popular with companies advertising mid-market cars because they made them resemble more expensive cars. In the 1950s, one would find ads combining both photos and illustrations. By the 1960s, color photography became commonplace, notably including atmospheric photography for personal cars, such as the Ford Thunderbird. One typical Thunderbird ad showed a couple and their car experiencing a "Thunderbird Interlude" from daily life in a beautiful soft-focus scene set in a copse of trees along a lake. Advertising messages and photographic conventions were relatively stable between the mid-1960s and 1990: with "common man" scenes for inexpensive cars and neutral or upper-class scenes for mid-market and expensive cars.

Over time, there have been changes in the advertising copy similar to those in the imagery. One common practice for many years has been the touting of some special gadget in the car's name. Before the war and into the 1950s, manufacturers gave these features special names (e.g., Cruise-O-Matic drive). By 1961, this naming practice was regarded as old-fashioned and discontinued, but identifying special features of the automobile on display continued in the ad copy.

Looking back on the American auto advertisements of the last eighty years, one can identify general trends. Until the beginning of World War I in 1914, one can see the influence of Art Nouveau styling in the advertisements. Ads for upscale cars such as Chrysler in the 1920s assumed it was reaching upper-class readers who would recognize the allusions in its ads to art movements such as Bauhaus, Cubism, and Art Deco. In contrast, the ads for less upscale cars tended to be simple and functional.

A new realism entered into advertising in 1929 and persisted through the 1930s, perhaps as a result of the stock market crash and the ensuing Depression. Car sales dropped 40 percent from 1929 to 1930. Cars continued to be portrayed in an elongated fashion but were placed against realistic backgrounds. Ads sought to portray cars as important accessories in real life, not as things one should aspire to. The copy centered on durability and value for money, and often included testimonials from satisfied users. Claims for safety were sometimes made, and the courts gave advertisers wide latitude to exaggerate these claims.[13] Low-price cars were displayed in small-town and rural settings, while expensive cars were shown in recognizably sophisticated settings. This realism remained a dominant theme in low- and medium-priced cars for many years.

Where once engineers had determined the style of cars on purely functional grounds, in the 1920s General Motors established the practice of using stylists to design cars so that they would sell on aesthetic as well as technical grounds.[14] The principal styling movement of the 1930s was streamlining, which embodied notions of efficiency and progress. Selling through styling involved reassuring buyers about their purchase; for example, copy for Lincoln-Zephyr ads in the 1930s argued for the practicality while the images in the ad emphasized the style.

As the different car models became more similar in terms of their body styles and technological underpinnings in the 1930s, testimonials came to be used to sell a particular brand. However, the testimonials of the 1930s (unlike those of the 1920s and 1940s) were most often from famous individuals, not contented ordinary customers. Generally, these ads provided little factual information to bolster the claims made by the manufacturers concerning the comfortable ride or engine efficiency.

During the early 1940s, in the midst of World War II, and continuing throughout the Cold War years of the 1950s, the ad copy was filled with patriotic themes and wild optimism. After the war, ads for both low- and high-end cars began to be filled with fantasy. The copy began to promote particular technological and design features of the cars as a way to escape from the humdrum of everyday life.

The 1950s witnessed a rapid succession of advertising themes for selling cars. For example, in Chevrolet ads at the beginning of the decade the focus was on practicality of the vehicles. But performance was Chevy's big selling point in ads of the mid-1950s. The Automobile Manufacturers Association decided in 1958 that the industry should deemphasize performance for performance's sake in ads, so Chevrolet returned to a strategy it had used in the 1930s of emphasizing performance as an attribute of vehicle comfort. By the late 1950s, many of the low-end cars began to emulate the marketing strategies of the high-end cars, emphasizing both power and comfort. Advertisements focused on performance did not appear again until the mid-1960s, and then only in the niche of specialty performance cars such as the Pontiac GTO.

If the 1930s was the decade of streamlined design, the 1950s was the decade of aerospace styling, ranging from the Cadillac tail fins to the Oldsmobile "rocket engine." By the late 1950s, some critics became concerned that this aerospace-inspired styling, which had been developed independently of functional considerations, was beginning to undermine functionality; and this concern led to a retreat from this design and advertising strategy.

The manufacturers and their advertising agencies generally believed that buyers were more concerned with style, performance, and comfort than the underlying technical details. A study of Chicago-area car owners conducted by Packard in the 1950s "found that only a minority of the population, mostly men in the lower class, [had] any real interest in the technical aspects of cars" (Stevenson 2008, 241). Other buyers expressed their own personal taste, for example, as conservatives or as individualists, through the purchase of particular brands associated with a particular reputation, or perhaps by buying bright colors or special features—but not by paying much attention to the technical details. Not surprisingly, the ad copy typically said little about a car's technical specifications.

Women had been driving cars independently in significant numbers since the 1920s, when the self-starter device became widely available. However, there were few car ads directly targeting women until the 1950s, when many middle-class families had sufficient income to afford a second car, often driven primarily by the wife. The ads displayed stereotyped men's and women's interests in cars, as stated succinctly in this Mercury ad in 1946: "Women judge a car mostly on its beauty, comfort, safety, ease of handling, and its perfection of detail. To men—power, economy, and how it's put together are most important" (Stevenson 2008, 146). The advertising of cars to women began to change in the 1960s, as part of the larger social revolution in America. Women increasingly appeared in car advertisements not as adornment but as part of everyday worklife in the home or office.

The mid-1960s witnessed a brief period of uniformity in how cars were advertised, including "a photograph of the car against a realistic—if socially optimized—background, a slogan, and a few paragraphs of copy at the foot of the page. Technical and status differentiation were increasingly established within the copy itself, rather than by illustration" (Stevenson 2008, 27). In 1967, the different segments of the new car market began to diverge in their ways of advertising. Size and niche came to matter more than brand in car advertising in the late 1960s and afterward.

The 1980s and 1990s were marked by a mature industry that returned to approaches perfected over the preceding sixty years. A survey by Stevenson of 1993 car ads showed a return to old themes, customized to the current settings: concern about foreign competition, emphasis on features and functionalism, focus on value-conscious buyers and independent women buyers, some amount of escapism, and plays to safety, snobbery, and quality—virtually every theme from six decades of car advertising was to be found. Most common were ads that focused on the car's functional values, including ads offering direct comparisons with the competition or presenting user testimonials.

Stevenson's review of car ads from 2005 and 2006 found that there were fewer ads about basic quality, economy, or reliability; brands with low sales quoted lots of statistics in an effort to benchmark their quality; suburban family images were used in advertising vans and crossover vehicles, much like the 1950s advertising of station wagons; ads employed backdrops reminiscent of those used since the 1930s, including restaurants, sports venues, highways, and classical-looking architecture; some luxury brands (notably Acura) employed substantial copy in order to give detailed discussion of features; more ads engaged environmental themes (especially but not only for hybrid vehicles); large, gas-guzzling SUVs by and large featured escape themes; and advertisers emphasized bright colors and music features (e.g., discussing XM radio equipment) in their appeals to younger buyers.

Although this section has focused on magazine advertising, car manufacturers supplemented their print advertising campaigns with radio and television. Radio programming became more popular in the 1930s and 1940s, and automakers began sponsoring programs. For example, Plymouth sponsored the *Major Bowes' Original Amateur Hour*, on which Plymouth products were personally endorsed by Bowes. General Motors sponsored a weekly show on the NBC network, *General Motors Symphony of the Air*. In the 1950s, television supplanted radio as the preferred form of car advertising. For example, DeSoto-Plymouth dealers sponsored Groucho Marx's *You Bet Your Life*, which appeared on both radio and television; and the *Dinah Shore Show* in the 1950s made America familiar with the ditty "See the USA in Your Chevrolet." There were also car ads on television, as the scholar Karal Ann Marling has noted: "There were several kinds of TV car commercials: the Motorama, the mini-drama, the pseudo-documentary, the ersatz 'lecture' by an expert. . . . What all these types have in common is an obsession with design, and specifically with a set of artistic principles that is presumed the audience understands and appreciates" (Marling 2002, 360).

Whether through magazines, radio, or television, advertising kept the new cars in front of the American populace on an everyday basis. The ads were intended to create an impression with viewers that cars are safe or fun or will enhance the owner's enjoyment or prestige. Little evidence was given for these claims. The ads gave people an overview of the car industry and its offerings, but offered little hard information that prospective buyers could use in selecting a vehicle if image was not their foremost consideration.

The Special Role of Dealerships

The dealership is another important element in the information system for the prospective car buyer. The buyer gets to look at the cars, take test drives, procure company literature, and ask questions of salesmen, finance officers, and service representatives. Because new cars have had to be purchased exclusively through dealers for most of automotive history, the relationship of the prospective buyer with the dealer is an important one—although one accompanied by dread for many buyers who do not like the high-pressure tactics and the information asymmetry inherent between the dealer's employees and the customer.

Before settling on the dealer system employed today, auto manufacturers tried various methods of product distribution: mail order, consignment, traveling salesmen, visits to local farms in the winter when the farmers were less busy, and even sales in department stores. The first dealership opened in the United States in 1895 (in Detroit). In the early days, there was no cost to obtain a dealership; one simply had to convince the manufacturer of one's usefulness to the company. Most dealers had originally been in the garage business, and many of the first dealership premises were converted bicycle or car repair garages.

After three failed attempts to organize, going back to 1905, the National Association of Auto Dealers (NADA) was formed in 1917. The initial impetus for forming the trade association was to lobby against turning the car manufacturers to war production during World War I, but the association also lobbied successfully to reduce the luxury tax on cars from 5 percent to 3 percent. In 1922 NADA began to publish the Official Used Car Guide, with up-to-date prices for used cars. This guide has been endorsed by the federal government and remains today the principal source for used car pricing.

Even before 1945 the dealers were unhappy with certain practices of the auto manufacturers, and this created ongoing tension. Prior to World War II, manufacturers assigned dealers a specific geographic territory and the dealers were fined or could lose their franchise if they sold outside their territory. Nevertheless, the manufacturers reserved the right to open additional dealerships within an existing assigned territory, which they often did when demand grew. Also, although dealers received cars at a 10 to 25 percent discount off the retail price, the manufacturer strongly discouraged any discounting to customers.

One of the Ford company's great advantages in selling cars, however, was its dealer network. In the first decade of the twentieth century, brands were not particularly well established and most dealers would sell any brand of car they could acquire. Henry Ford began franchising dealers in 1903, even before the Ford Motor Company was legally incorporated. Because the Ford company required dealers to prepay for cars they ordered, the cars generally arrived on time and at favorable prices compared to those of competitors. Most of Ford's competitors either did not have a dealer network or did not have close relations with their dealers. Buyers of other car brands often ordered directly from the manufacturer, bypassing the dealer. Ford set high standards for its authorized dealers because it wanted to establish good customer relations in part by offering quality repair and service work. The company required its dealers to run their business full time—not part time as did some dealers of other car brands— and to staff their own well-equipped showrooms and garages with trained service men, instead of having the men perform this work out of their own homes. Most of the Ford dealerships were franchised, but the company retained ownership of key dealerships in major cities. By 1913 Ford had grown to 7,000 dealerships, giving it by far the most complete national coverage of any U.S. automobile brand. And to further expand sales, Ford instructed its dealers to make home sales calls rather than wait for potential customers to walk into the showroom.

Dealerships for all auto manufacturers became more important in the 1920s and 1930s as the demand for new cars became saturated and the used car trade-in became a more important part of the business. In the 1930s dealers grew unhappy with the auto manufacturers because they were forced to take a larger number of new cars than they wanted or particular models that were hard to sell; pressured about sale prices of used cars; forced to give overly generous discounts on end-of-year sales to move remaining stock; required to use manufacturer-sponsored loan programs;[15] given no say in whether the manufacturer opened new dealerships in their sales territory; and threatened by the manufacturers with loss of their franchise license on short notice. In response, the dealers pressured Congress to investigate. The Federal Trade Commission carried out a study in 1938, as a result of which Ford and Chrysler signed consent decrees and General Motors was convicted of antitrust activity. However, these actions did not resolve the uneasy tension between the dealers and the manufacturers. After World War II, the dealers convinced Congress to pass a Good Faith act that limited some of the most objectionable of these practices, but the law was weakened in order to gain passage and never gave the dealers much satisfaction.

There was a pent-up demand for cars after World War II. In the seller's golden era of the late 1940s and early 1950s, some dealers and their salesmen adopted unethical and other questionable practices. For example, some dealers required the customer to pay a bribe to get a car delivered in a reasonable amount of time. As car historian James Flink describes the practices of this era:

Customers were subjected to the "plain pack" (inflated charges for dealer preparation of the car); the "top pack" (an inflated trade-in allowance added to the price of the new car); the "finance pack" (exorbitant rates of interest on installment sales, usually involving a kickback to the dealer from the finance agency); the "switch" (luring a customer into a salesroom with an advertised bargain, then getting him to accept a worse deal on another car); the "bush" (hiking an initially quoted price during the course of the sale by upping the figures on a conditional sales contract signed by the customer); and the "highball" (reneging on an initially high trade-in offer after the customer committed himself to buying a new car). (Flink 1988, 281–282)

Overall, the era from 1945 to 1970 was a great time to own a dealership or be a car salesman. In these years, just as in the period from 1900 to 1920, demand for automobiles outstripped supply, so sales were relatively easy to make. The number of American families able to afford a car increased, from approximately 55 percent in 1945 to 80 percent in 1965. The manufacturers became more concerned about building and sustaining brand loyalty, and in the 1940s they disallowed the dealers from selling multiple makes, which had been a common practice before the war.

Manufacturers used contractual arrangements to control the dealers' rules of operation. The contract specified procedures for selling vehicles, instructions on how to organize sales and service operations, arrangements for rebates at the end of model years (especially in years when a particular model was being replaced or undergoing major revision), details on how to treat purchasers and handle complaints, and architectural plans for the dealership's building. The manufacturers provided the dealers with literature and other media to distribute to customers. These included not only literature about specific models for sale, but also automobile-themed coloring books for children, and a stereoscopic viewer that could be used by potential buyers at the dealership to view the different colors and body styles in three dimensions. From the 1930s, Chevrolet had a model program, run by its legendary head of sales William Holler, to improve the quality of the sales experience at Chevrolet dealerships and to educate dealers on how to be good citizens in their local communities through charitable and service activities. Not surprisingly, Chevrolet had high customer loyalty from the 1930s through the 1960s.

The first dealerships were housed in garages, sheds, and small stores. Although the first purpose-built buildings for car dealerships (often in Art Deco style) were constructed in the 1920s and 1930s, it was only in the 1940s and 1950s that manufacturers and dealers gave careful attention to both the placement of the dealership in a high-traffic block on the main commercial street in town and the architecture of the dealership's building. General Motors sponsored an architectural design competition, approved by the American Institute of Architects, which led to an influential guide-book on car dealership design published by General Motors in 1948. GM and other manufacturers paid especial attention to the design of the dealer buildings, especially to the store front.

The new dealership designs of the late 1940s were intended to be merchandise and service oriented and to showcase new cars through a well-organized store front. A store front is one of the dealer's best means of advertising its products and services. The term "store front" includes not only the exterior elevation of the building and property that borders on a street, but also the view that extends into the showroom and to the signs that are placed in the front, side, and rear of the building. Dealers are interested in any elevation that will make traffic, stop, look, and enter the dealership. (Genat 2004, 44–45)

This placement and design of dealerships continued until the 1980s, when many of these buildings were replaced by much larger structures in large auto malls, often located on the outskirts of town, where land was cheaper.

Many of the dealership practices familiar today became established in the period between 1945 and 1970. Until the 1970s, dealerships were overwhelmingly a white, male world. The first African American to own a dealership in the United States was Ed Davis, who gained ownership of a Studebaker dealership in 1940, after having sold cars for a number of years before the war. Studebaker went out of business in 1956 and Davis lost his dealership, but in 1963 he became the first African American to own a dealership of one of the Big Three manufacturers, with Chrysler. Today, women and minorities remain underrepresented in selling cars and owning dealerships.

Each year, the week when new cars were introduced was a special time for the dealerships. Gala events were organized to attract high foot traffic. The opening event lost its importance in the early 1970s, when the oil crisis, sticker shock, and a different schedule by the Japanese automakers for introducing new cars made the event less effective. Today, cars are introduced not through gala events at the dealerships but instead through major auto shows.

Manufacturers took responsibility for the national advertising, including brochures describing both individual cars and the entire product line; while the dealership was responsible for the local advertising. Manufacturers' brochures used only illustrations and no photographs until the mid-1960s. Before 1945, it was not common practice for an individual to be offered a test drive during a visit to the dealership. The practice of giving salespeople their own new cars to drive (so-called demonstrators) was introduced as a means to promote test drives. Demonstrators went back to the early years of the automobile (for example, they are mentioned in a 1907 court case *Morley v. Consolidated Manufacturing Co.*), but they did not catch on widely until after World War II. Their use fell off in the 1980s, when the car business became more competitive and margins were smaller. Another big change for the dealers was the passage of the Automobile Information Disclosure Act in 1958, which made it a requirement under penalty of law that all new cars display a window sticker listing make, model, serial number, and suggested retail price.

Retail distribution of cars changed during the postwar years. Rules prohibiting an individual businessperson from owning multiple dealerships under one roof were lifted as a result of government hearings in 1956. Another reform caused manufacturers to offer five-year instead of thirty-day dealership contracts, and the government implemented an arbitration process to resolve issues between dealers and carmakers. While these new rules were particularly favorable to foreign (especially European) manufacturers trying to get a foothold in the U.S. auto market, they were also valuable to buyers, who could more readily comparison shop and might have an easier time with both their trade-in and switching from one brand to another.

In the late 1980s and 1990s the auto industry introduced new retail distribution practices. These included automobile supermarkets, mall showrooms and information centers, warehouse club purchase programs, and one-price, no-haggle pricing (Saturn for new cars; CarMax and later AutoNation and Driver's Mart for used cars). Ford chairman Jacques Nasser experimented in 1999 and 2000 with The Auto Collection, which consolidated several Ford dealers in a geographic region into a single megacompany, giving the previously independent dealers an investment position in a superdealership sales organization. This experiment was tried in several cities, including San Diego, Tulsa, and Salt Lake City. The dealers disliked losing their status as independent businessmen, and The Auto Collection practice was reversed by Ford's next chairman, Bill Ford.

Ford was the first American car company to experiment in a major way with the Internet. This occurred during Nasser's tenure as chairman, which coincided with the dot-com boom. In 1999 and 2000, Ford introduced three interactive Web sites for consumers and expanded its own Web pages. It also entered into cooperative ventures with a number of high-tech firms, including Microsoft, Yahoo!, Oracle, Hewlett Packard, UUNet, and Vehix.com. This direction was also reversed when Bill Ford took over the chairmanship from Nasser in 2001.

Perhaps because of the generally high quality of cars since the 1980s, the availability of information on the Internet, or the busy lives of potential buyers, dealers have noticed a change in the attitudes of buyers that includes their being less reliant on information from the dealer's sales staff:

Back in the '50s and '60s, we used to do the walk-around and show the customers all the features of the car. . . . Nowadays, we don't get to do the walk-around much anymore. The customers are now in a hurry to buy, and they don't want to hear the stories about the vehicle and think about it. People now have an idea of what they want, and they want to know the price right away before they even look at the car. People were interested in details back in the '50s and '60s. For instance, if someone came in to buy a new Impala, they wanted to know what the car had to offer. They wanted to know about the engine—the displacement and so forth—and they wanted to see about the trunk space, the head room, and shoulder room. Most of them today don't even care about the engine. They know it's going to be a V-8 or a V-6, but they don't care like they used to. (Owner of Culberson-Stowers Chevrolet in Pampas, Texas, as quoted in Robinson 2003, 151)

The Case of Cord

Let us conclude our discussion of dealerships with an illustration of the way in which dealers marketed two Cord automobiles during the 1930s. We will first consider the Cord L-29, also known as the Cord Front Drive. Between 1929 and 1932, slightly more than five thousand of these cars were sold. The Cord was intended as a mid-level car, priced above the budget-minded Auburn and below the luxury line Duesenberg—all products of the Auburn Automotive Company of Auburn, Indiana. The L-29 was the first production car sold with front-wheel drive and was based on successful racing car technology of the 1920s. The front-wheel drive helped the designers build a car that was low, with sweeping lines. Frank Lloyd Wright was reported to have said the car "looked becoming [next] to the houses I design" (http://www.conceptcarz.com/vehicle/z7428/Cord-L-29 .aspx). Its competition was the small Pierce-Arrow and Packard automobiles, the large Buick, Chrysler, and Willys automobiles, the LaSalle, and the Hupmobile. It was highly regarded, winning some thirty awards at car shows, but the stock market crash of 1929 and the resulting Depression doomed the car almost from its beginning.

There is unusually detailed information available about the L-29's marketing and purchasers' reactions to the car. In 1930 T. M. S. Gibson, Auburn Automotive Company's general sales manager, prepared a one-hundred-page report for use by Cord dealers and salesmen responsible for selling the new car.[16] He seemingly left no detail to chance. The report included information about cars in the $3,000 price range (which included the Cord), the advertising campaign, pricing, and warranty policies for the company and for this particular car, the history and underlying scientific principles of front-wheel drive, the technical issues related to front-wheel drive as well as its advantages (improved performance, operating ease, safety, riding ease, silence, efficiency, durability, accessibility for maintaining and repairing the car, and convenience), and miscellaneous additional information including the L-29's particular significance to the future of the Auburn Automobile Company.

The L-29 advertising plan gives a sense of the report's detail. The car was advertised in the *The Saturday Evening Post*, *Ladies Home Journal*, and *Colliers* to reach mid-class buyers and in *Vanity Fair* and *Harpers Bazaar* to reach "ultra quality buyers." It was also advertised in every major American newspaper. The advertising plan discussed direct mail, electric signs, service signs, product literature, window displays, and posters. It also provided specifications about catalogues, booklets, and folders; sales talks, sales manuals, and question-and-answer information; pricing (of both cars and accessories), used car allowance, and financing; and how to build a list of prospects for direct mail and telephone campaigns. Also listed were specifications for use in the various types of advertisements: photographs, line drawings, artists' conceptions and retouchings, performance and comparative charts, blueprints and engineering graphs, and film pictures.

In 1931 the Auburn Automobile Company published a 150-page book, *What Owners Say*, which gave testimonials by owners of the L-29. A copy was displayed in each dealership for prospective buyers. The testimonials often discussed the car's front-wheel drive, unsurprisingly since this was the first production car built with this feature. Other common responses addressed appearance, comfort, ease of handling, and to a lesser degree performance—characteristics one might expect to hear about today for a mid-priced sports car. Responses that would be less likely today included ease of shifting for women and lack of vibration. Such hard-to-find evidence speaks to what buyers (or at least owners) valued.[17]

The next car manufactured by Cord was the 810. It was introduced in 1936 but ceased production only one year later, with the demise of the company, after about three thousand units were manufactured. The car was a sensation when it was introduced at the New York, Chicago, and Los Angeles auto shows, but the company could not manufacture enough to keep up with demand. It sat very low and had an extraordinarily long front end because of an innovative design that involved the transmission extending in front of the engine. The streamlined design was emphasized by the absence of running boards, minimal chrome, pontoon fenders, manually retractable headlights, and an antenna running under the car rather than sticking up from the hood. Powered by a strong, eight-cylinder engine, the 810 set a speed record of 107 mph for a continuous twenty-four-hour test on the Great Salt Lake in Nevada. Historian Sally Clarke described the car as "perhaps the most aesthetically modern U.S. automobile of the Depression decade" (Clarke 2007, 198)[18] Extensive handwork drove the car's price well beyond the cost of sporty luxury cars offered by General Motors.

In 1935, the company produced two brochures intended for the sales staff to help with the selling of the 810. One was a nine-page guide for dealerships on how to give a test drive of the car.[19] These guidelines are illustrative:

• Have a single trained person drive on the demonstration ride and have the salesman go along as well to do the selling.
• "When driving in 4th ratio and approaching a slower moving vehicle which is to be passed pre-select 3rd ratio some distance back, then when nearly ready to pass a stroke of the clutch will engage 3rd ratio and make excellent acceleration. The change is so silent the passenger will be unaware of what has occurred, but sense the outstanding acceleration."
• "On a smooth, clear stretch at a moderate speed, show that only the pressure of two finger tips is required to steer the Cord."
• "Point out the absence of wind noise as the car glides along in 4th. Perfect streamlining is responsible for reducing wind resistance and contributes to economy as well as eliminating annoying wind noise."
• Have the prospective customer read a newspaper and write a short phrase to show how smoothly the car rides.

But the demonstrator was cautioned against frightening the prospective buyer with "drastic or extreme" tests while emphasizing "its tremendous safety and performance."

Also accompanying the report, another nine-page brochure instructed the salesman on how to effectively show the car to a customer in the showroom.[20] These notes provided the salesman with "a systematic, planned presentation of the car" and identified the fundamental facts that each salesman must be familiar with. The report offered a diagram of the car, choreographing step by step the detailed movement around the car of the salesman and buyers, including a list of facts to be discussed at each stage of the sales talk (see figure 2.1).

It is clear that the dealership has represented, at least since the 1920s, a critically important information source for a prospective car buyer. The buyer can visually inspect and sit in the various new cars, obtain manufacturer literature on the products, go for test drives, and learn about the features of a particular car and its underlying technology. Although more informative than advertising, the dealership as information source shares the same limitation as advertising in that the communication, whether oral or written, is intended to persuade the prospective buyer to purchase a car from this particular dealer at a price and with features that are in the dealer's—not the customer's—best interest.

TWENTY MINUTES OF CORD FACTS

2. AROUND THE REAR
   Width and Lowness
   Tail Lamps
   Gas Filler
   License Holder
   Body Construction
   Reduced Unsprung Weight

3. REAR INTERIOR
   Quality Finish
   Color & Upholstery Treatment
   Headroom
   No Tunnel
   Steel Floor
   Upholstery Material

4. FRONT OF CAR
   Oil and Water Filler
   Headlamps
   Front Unit
   Frame
   Front Suspension
   Front View

Figure 2.1
"Twenty Minutes with the Cord Prospect," Auburn Automobile Co. Courtesy of the
Auburn-Cord-Duesenberg Museum Archives, Auburn, Indiana.

5. FROM DRIVER'S SEAT
   Lowness
   Comfort
   Vision
   Horn Ring
   Instruments
   Unobstructed
      Compartment
   Remote Control
   Cruising Ratio
   Parking Brake
   Steering Wheel
   Steering System
   Powered Steering
      Front Drive Advantages

APPEAL FOR DEMONSTRATION

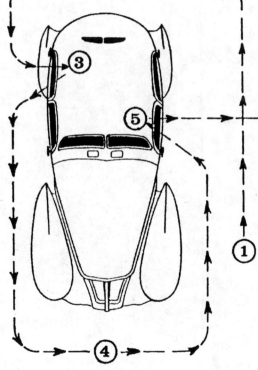

TO DEMONSTRATOR

1. GENERAL VIEW
   Lowness
   Distinctiveness
   Streamlining
   Hood & Louvers
   Retractable Headlamps
   Pontoon Fenders
   No Running Boards

Figure 2.1
(continued)

Cars before 1945

As background for understanding the information issues facing American car buyers prior to and during World War II, this section reviews America's early automotive history, with an emphasis on forces that shaped both the auto industry and car-buying practices.

The first functional gas-powered vehicles were built in Germany in the 1880s. America's first car was built in 1893 by the Duryea brothers of Springfield, Massachusetts. By 1900, there were about thirty manufacturers of cars in the United States, but altogether they had produced no more than three thousand vehicles. These cars tended to sell for three to six times the annual American family income and thus mainly were purchased by wealthy individuals as toys; a little later on doctors began purchasing cars as a kind of business tool that increased their ability to travel to patients. It cost approximately thirty cents per mile to own and operate a car in 1900, a price that dropped to 2.7 cents per mile forty years later. In 1902, to increase consumer confidence, the National Association of Automobile Manufacturers enacted a sixty-day warranty on new cars sold by its members, increased to ninety days in 1913.

As more affordable cars began to be built, automobile ownership grew rapidly in the United States during the first decade of the twentieth century, and in 1904 the United States eclipsed France as the world's largest manufacturer of automobiles. The fast rise of the United States in the automotive field is partly attributable to American might in industrial production and partly to the high and relatively evenly distributed income levels in America compared to Europe. Almost five hundred manufacturers entered the U.S. car business between 1900 and 1908, and almost three hundred of them remained in business in 1908, the year in which Henry Ford introduced the Model T and General Motors was founded.

Henry Ford completed his first car in 1896 and created the Ford Motor Company in 1903. Unlike many of his rivals, he was interested in building a car for the masses; and in order to do so he made important innovations in the assembly line method of production and later in the five-dollar-a-day wages to attract and retain workers to carry out the boring, repetitive, and physically demanding work on his assembly line. Ford's personal reputation and that of his company were built on the Model T, which continued in production through 1927. It was well designed for its time, with high clearance and ruggedness for the rutted roads then common, high reliability and durability, and a simplicity of design that enabled farmers and others with mechanical skills to make their own repairs. It offered high quality for the price—approximately one-tenth to one-twentieth that of the early cars already mentioned—making the Model T widely affordable. Ford's focus on the manufacturing process enabled the company to continue to drive down the sale price, from $950 in 1908 to $290 in 1924. Of the twenty million cars on American roads by the late 1920s, more than half were Model Ts. By the time the Model T was taken out of production, more than fifteen million had been built.

When it was time to sell the new Model T, Ford mounted a national advertising campaign in October 1908 in the *The Saturday Evening Post* and other magazines. For publicity reasons, two Model Ts were entered against three other cars in a 4,000-mile race from New York City to Seattle that began in June 1909. Although the conventional wisdom was that heavier cars could better endure the bad roads, and all three of the competing cars were heavier than the Model T, the Ford entries placed first and third in the race (the winner finishing the course in twenty-two days). The race was national news for several weeks.

The annual automobile show played an important role in advertising cars to the public. Henry Ford had success at the 1906 New York Automobile Show, where he introduced the stripped-down Model N to the public and gained widespread popular interest in his company, which had not yet distinguished itself from the competition. When Ford introduced the more substantial Model T two years later, auto shows were central to the publicity campaign. In anticipation of the shows, most of which occurred in January, the company sent numerous photos and publicity releases to the newspapers. In fact the publicity was too successful, developing many more sales orders than the company could fulfill in its early years.

During the last two months of 1927, the biggest news story in American papers—generating more coverage than the coming U.S. presidential election—was the introduction of Ford's first new car (the Model A) in almost twenty years. Henry Ford wrote an open letter in November 1927 published in two thousand newspapers. The advertising budget for the week prior to the car's unveiling was $1.7 million. The company also ran special "salons" in major cities around the country to allow the public to see the new car. On the first day, December 2, 1927, crowds were so large that riot police were called out to restore order in four cities. The company's (probably inflated) figures claim that over ten million people—a full 10 percent of the U.S. population—came to the salons or a Ford dealership on the day of launch. In January 1928 the company ran a special Ford Industrial Exposition in connection with the New York Auto Show. 223,000 people were recorded as coming through the Ford exhibit the first day, and more than a million people attended during its six-day run. The American public was clearly eager to see Ford's new offering.

The pace of innovation in the auto industry through 1945 was perhaps greater than in the postwar years. Early cars were noisy and dirty, offered a bone-jarring ride, and were hard to shift, liable to break down due to failure of internal components and tire blowouts, and physically challenging to start. A series of innovations vastly improved the car-driving experience for Americans: manufacturing accuracy and standardization in new and replacement parts (1908), electric starter (1911), all-metal body (1914), electric-powered windshield wipers (1919), hydraulic brakes (1920), balloon tires (1922), leaded gasoline (1923), metal disk wheels (1920s), automatic transmission (1937), and air conditioning (1940).

Some innovation was driven by safety concerns, although the auto industry did not make safety a foremost consideration in this era. There was an expectation in the early auto industry that the buyers would assume the financial and safety risks for their cars. This began to change in the 1920s, when the manufacturers implemented better inspection processes for both components and finished vehicles, in response to liability claims made by car owners in the courts. The change continued in the 1930s when the auto manufacturers, especially General Motors, established research laboratories, which devoted at least part of their effort to safety issues. Numerous government and nongovernment organizations, such as the Society of Automotive Engineers and various state motor vehicle administrations and legislatures, pushed for better safety features in lights, windshields, and bumpers.[21]

One thing that made the Model T vulnerable to competition was Henry Ford's opposition to people borrowing money to buy a car; in fact his company did not offer company financing to buyers until 1928, when it created the Universal Credit Corporation in order to finance sales of the new Model A. By this time the majority of Americans were already buying their cars on installment loans. The first organizations to offer car loans were the Morris Plan banks in 1910 and on a much wider basis Guarantee Securities, a company formed in 1915 specifically to finance cars. But more significant competition for the Ford Motor Company was created by the formation of General Motors Acceptance Corporation (GMAC) in 1916, which enabled GM to wrest away a significant portion of Ford's market share. By 1924, a third of the cars sold by GM were financed by GMAC because of the convenience to buyers of having the dealer arrange on site for the loan. GM made new car sales even easier by allowing people to trade in their used car as a down payment on their new car. These used car trade-ins gave GM an advantage over Ford because many of the cars traded in at GM dealerships were Model Ts. The used Model Ts could be advantageously priced by the GM dealers against new Model Ts, and had comparable features since the Model T did not change much from year to year. In this way, the GM dealers were able to undermine new car sales by Ford. Unlike Ford, new Chevrolets (GM's most directly competitive product with Ford) were somewhat protected from competition from used late-model Chevrolets because GM had instituted annual model changes in their cars.

The following story may be an extreme, but is nevertheless illustrative of early car purchases. One family reported having purchased seventeen cars—all on credit—between 1917 and 1934:

Never in all these years have I actually purchased a motor car for hard cash, and I have always wished for the day to come when I would be able to lay the money on the line, take my motor car home and evade interest. My interest payments, called carrying charges, have pushed me back about $3000 during the seventeen years, or four fair cars. . . . There has scarcely been a month during all these years when I have not been paying for a machine. The lowest monthly payment was fifty dollars, which I suppose is rock bottom, and the highest was $225 every thirty days. That $225 hit me every month for a year and caused much home moaning and remorse." (Preston 1934)

Another competitive vulnerability of the Ford Model T was the fact that rival manufacturers, especially General Motors, were willing to give customers choices of interior fabrics and exterior colors whereas Henry Ford focused on offering a utilitarian product at a rock-bottom price, enabled by offering few customer choices. Black paint dried faster than other colors, making it more suitable for the Ford manufacturing line. Alfred Sloan, head of General Motors, introduced annual model years in 1927 as a way to make previous car offerings obsolete and encourage buyers to trade for a new car. These developments, not coincidentally, happened at the same time as the rapid increase in the use of advertising to sell cars.

It is not generally recognized today how extensive competition in the auto industry was in the period between 1900 and 1930. Eighty-seven companies displayed cars at the National Automotive Show of 1921. The capital barrier to industry entry was low because cash-flow problems could be avoided by buying components from suppliers on credit and selling cars for cash to dealers and distributors before they were manufactured. This low barrier resulted in hundreds of American firms manufacturing automobiles in the first decade of the twentieth century. European competition was kept out by U.S. trade barriers, with a 45 percent tariff on imported cars prior to 1913 and a 30 percent tariff for the remainder of this era.[22] By the late 1920s, the market for new cars was saturated, and the capital barrier for manufacturing became high enough that the number of new entrants greatly diminished. By 1929, only forty-four carmakers remained in the United States and 80 percent of new car sales were made by the Big Three: Ford, General Motors, and Chrysler.

Ford was the dominant auto company in America in the 1910s and early 1920s. It entered the 1920s with a thirteen-to-one sales advantage over second-place General Motors but lost that lead by 1927. GM grew to be a powerful competitor not only through business practice innovations such as model years, planned obsolescence, enhanced customer choice in interior and exterior colors, and company-offered financing, but also through acquisitions of other companies. The plan of Alfred Sloan, as president and later chairman of GM, was to offer through its different brands "a car for every purse and purpose."[23] If Ford showed how to use the system of mass production to good effect in the automobile industry, GM perfected the system by replacing Ford's single-purpose machine tools with general-purpose tools that enabled more flexible manufacturing. This resulted in less costly changeovers to new models and less down time in the manufacturing plant. GM did not relinquish the market lead after it wrested it away from Ford in 1927.

The next largest competitor in the U.S. auto industry was Chrysler, created by Walter Chrysler, who had once been the head of General Motor's Buick Division. In 1923 he took over the Maxwell automobile firm and introduced a technically advanced car, the Chrysler Six, which sold well in the high end of the market. He rounded out the Chrysler product line in 1928 by acquiring Dodge to sell as a mid-range product and creating the low-end Plymouth division. The Plymouth model became a big seller, attracting many former Model T owners. Chrysler was particularly effective at using advertising in the late 1920s and early 1930s to gain market share. The company kept costs low by relying on its suppliers to do much of the innovation that went into its automobiles rather than relying, as General Motors did, on operating an expensive research laboratory.

There were many other competitors in the 1920s. At the high end, competing with the General Motors Cadillac and Buick divisions and with Chrysler, were Duesenberg, Franklin, Marmon, Packard, Pierce-Arrow, and Stutz. In the more mainstream market, competing with the General Motors Oldsmobile and Chevrolet divisions, Chrysler's Dodge and Plymouth divisions, and Ford were Auburn, Essex, Hudson, Hupmobile, Kissel, Nash, Reo, Studebaker, and Willys-Overland. A shakeout in the 1930s created by limited sales during the Depression, the high cost of mass-market advertising, and large capital outlays for equipment to manufacture solid metal bodies caused Auburn, Duesenberg, Franklin, Marmon, Peerless, Pierce-Arrow, Reo, and Stutz to exit the market.

Whereas only 1 percent of American households owned cars in 1910, 60 percent owned them by 1930. The car industry could not have grown as rapidly as it did during the first three decades of the twentieth century without advances in other industries that provided an infrastructure for the car industry. One was the wide availability of fuel. Major oil discoveries in Texas, Oklahoma, and California in the late nineteenth and early twentieth centuries made the United States the world's leading producer of crude oil. A series of innovations, such as thermal cracking in 1913 and catalytic cracking in the mid-1930s, greatly increased the yield of gasoline from oil refining.

A second important part of the infrastructure was the highway system. Because of the lessons learned about the weaknesses in the highway system from the transportation of materiel during World War I, the government built a network of national highways between the two world wars. The Federal Highway Act of 1921 matched with federal funds any state funds made available for the road system. Gasoline taxes, first introduced in 1919 (by Colorado, New Mexico, and Oregon), provided a politically palatable way to continue road building in the 1930s during hard economic times. These highway construction projects were popular because they employed large numbers of workers at a time when unemployment rates were near 25 percent.

A third element of the infrastructure was the emergence in the 1920s and 1930s of shopping malls, self-service food markets, drive-in fast food stands, and auto camps and motels. These innovations provided facilities on the outskirts of towns, with ample parking and reasonable costs. They were a welcome alternative to the high prices of the downtown shops, formal restaurants, and hotels, and made it more affordable for American families to participate in the popular vacation activity of touring the country in their cars.

Although automobile sales grew rapidly in the years after the Model T was introduced, the gas-powered automobile was not without competition. At first the competition was from horse-drawn vehicles, which were still used widely on farms and in the cities. Even a decade after the introduction of the Model T, only about 30 percent of farms had any kind of automobile. There was also competition, especially before World War I, from both electrical and steam-powered vehicles, such as the Stanley Steamer. The electrical cars had many advantages over gas-powered cars; they were not so noisy and smelly, and they did not require hand cranking to start or hand shifting. However, the electric vehicles could not travel at fast rates of speed and were regarded widely as women's automobiles, which made marketing them to young adult males difficult. Even so, Detroit Electric, which was building cars before World War I, continued in business selling electric vehicles until 1938.

Steam-powered cars had a long tradition, going back to steam-powered vehicles built for the French military in the 1760s, followed by a series of steam-powered tractors, carts, and bicycles in the nineteenth century. They were quiet and vibrated less than gasoline-powered automobiles. They also provided the greatest speeds, as witnessed by a Stanley Steamer clocking 127 mph on a Florida beach in 1906. In the year 1900, more electric and more steam automobiles were produced in America than gasoline automobiles. However, "the gasoline-fueled internal combustion engine offered the best compromise. It combined high-performance potential with reasonable practicality, and the rapid strides made in improving efficiency, quelling vibration, and simplifying operation, made it even more acceptable for average motorists" (Volti 2004, 9).

The high point of car sales in the pre-1945 era was 1929, when more than four million cars were sold. After the stock market crash of 1929 and throughout the Depression of the 1930s, sales were deeply suppressed, in several years reaching only one-quarter what they had been in 1929. Families worried about employment in hard times were unlikely to buy a new car. Nevertheless, the car was important to their lives and livelihoods, and older cars were nursed along. Three days after the United States entered the war following the bombing at Pearl Harbor in December 1941, the federal government suspended the sales of new tires, followed three months later by a suspension of automobile production so that the auto plants could be repurposed to build tanks, jeeps, bombs, airplanes, and other war equipment. Driving was severely limited during the war because of rationing of both gasoline and tires. By the end of the war in 1945, there was a strong pent-up demand for automobiles.

A popular magazine article described the main features of a typical car ownership experience from 1917 to 1934 (Preston 1934):

• the high-pressure salesman who sold the buyers a "snap" (the common name for a used car the dealership almost "gave away" for $350 in order to get it off the lot; it fell apart before the new owners made it home from the dealership),

• the unwillingness of the dealership to stand behind the cars it sold,
• the fact that none of the cars lasted more than 74,000 miles,
• the total cost of ownership approaching $40,000 for seventeen years of driving,
• the need to be a "gentleman mechanic" in order to keep the cars running despite the owners having limited mechanical abilities,
• the need to replace tires two hundred times during that seventeen-year period,
• the frustrations of having to pay for fire, theft, and liability insurance,
• the role of buyer buddy to less experienced car buyers ("Acquaintances asked for advice, and I gave it freely, and probably with a touch of hateur. I actually told simple-minded friends which cars to purchase and which to scorn, how much to pay, and whether the engine had years of service or was worn out. They believed me."), and
• the bad roads ("We nearly lost our automobile that day—a new one, the apple of our eye. The road was so rutty that we took turns driving in low gear, and the one not driving climbed out and walked beside the struggling fenders. The engine smoked, smelled, steamed and threatened to catch fire and explode; the body groaned.").

Information Issues for Car Buyers before 1945

Although the general histories of the automobile prior to 1945 discussed in the previous section do not focus primarily on the automobile buyer, they identify a number of issues that buyers faced:

• Which make and model to buy
• The cost of operation
• The ability of their vehicle to drive on bad roads (e.g., clearance and pulling power)
• Durability of the automobile
• Reliability of the automobile
• Ride comfort
• Drivability (e.g., physical ability to crank start the car and shift gears)
• New technologies
• Availability of financing
• Monthly cost of car payments
• Choice of exterior color and interior fabrics
• Top speed and sportiness
• Range on a tank of fuel
• Gasoline, steam, or electric engine

In order to identify additional issues faced by American car buyers prior to 1945, the author examined every article about cars published in the *The Saturday Evening Post* between 1908 and 1945.[24] In addition to the issues previously listed, readers of this popular magazine would have read about the following buyer concerns:

- The need for mechanical skills in order to handle frequent breakdowns
- Availability, cost, and coverage of insurance
- High taxes on automobiles (gasoline taxes, auto registration fees, etc.)
- Alternative fuels
- Safety issues
- Dubious sales practices for new and used cars
- Traffic issues
- The common practice of overly fast and reckless driving
- Downtown parking for work and shopping
- Keeping an old car in running order at times when new cars were hard to come by (especially during the Depression and the war years)

Prior to 1945, sources of information about cars were limited. The first auto trade journal published in America, which also doubled to some degree as an enthusiast magazine, was *Horseless Age*. It appeared monthly from 1895 until 1918 and carried numerous car ads as well as information on such topics as the different kinds of propulsion systems.[25] The magazine had a following among wealthy car buffs, but its circulation was limited (only eight hundred copies printed of its early issues). After its demise, no car enthusiast magazine for a general readership was published in the United States until after World War II although some car enthusiasts read the British magazine *The Automobile* during these years or the publications of the car clubs many enthusiasts joined.[26]

Popular magazines such as *The Saturday Evening Post* and *Colliers* published occasional articles about cars. Often their articles offered stories about colorful figures in the industry or gave retrospective looks at collectible cars. These stories did not provide information that helped the average prospective buyer. Beginning in 1936, *Consumer Reports* provided reviews and practical advice, but its focus until after World War II was more on keeping one's car and tires in running order. One of the first how-to books for car owners was a pamphlet distributed by a parts supplier in 1903, entitled *Diseases of a Gasolene [sic] Automobile—and—How to Cure Them.*[27]

If you lived near a city, you might join an auto club, where you could talk with other enthusiasts, read the club's magazine, go on driving tours it sponsored, and obtain guidebooks listing garages and hotels. The American Automobile Club was formed in New York City in 1899, and five years later there were almost a hundred clubs in the United States. You were likely to engage in these activities, however, after you bought a car—not as part of your information gathering during the buying process.

If you lived near a large city, you might also attend an auto show, usually held each January. As one might expect, Detroit had its own auto show, the Detroit Auto Show (now called the North American International Auto Show), held annually from 1907, with a hiatus from 1942 to 1952. The most important auto show was the one held each January, beginning in 1903, at Madison Square Garden in New York City (with a hiatus from 1941 to 1955). Other major cities later introduced their own annual auto shows. These shows gave auto executives a chance to find out what interested the public as well as to talk with one another; and they profiled the current and (in the postwar years) concept cars for owners and prospective owners.

From the early years of the industry, newspapers reported on manufacturers' introductions of their new car models. The first such coverage appeared in 1897. However, most newspaper reports—at least until the 1960s—simply covered the event (e.g., the auto show) or provided a gloss of information provided by the manufacturer. There was essentially no independent coverage of automobiles by the newspapers during this era. This author has found no evidence of radio shows that discussed car buying in the years prior to 1945. In sum, there was essentially no place between the world wars where the American car buyer could read objective reports of test drives and the comparative strengths and weaknesses of competing automobiles.

Another source of popular information came from the car races, auto tours, and publicity stunts by the manufacturers, and their coverage by the press. The first organized car race in America was held in 1895. Some of these races featured ordinary passenger cars that one could buy and drive for normal use. But many of the races, including the most important American race of all, the Indianapolis 500, involved specialty cars. The Indianapolis 500 garnered enormous popular interest, with eighty thousand spectators at the first race on Memorial Day 1911 (with the winning car introducing a new car accessory, the rearview mirror). Today, the Indianapolis 500 attracts an in-person audience of more than four hundred thousand and twenty million television viewers. While these races attract general interest in cars and may fuel interest in speed and handling ability, they have done little to inform buyers about which cars to buy.

Cars after 1945

Personal automobile production resumed in May 1945, shortly after Germany surrendered to the Allied Forces. Most cars on the road then were holdovers from the 1920s and 1930s that had been nursed along during the war. Of the cars on the road in 1942 when the prohibition on building new passenger cars went into effect, almost twelve million had been built prior to 1937 and another thirteen million had been built between 1937 and 1942. After the war, there was a pent-up demand for new cars; by the end of 1947, the waiting list for a new car was thirteen million long. But the manufacturers had trouble delivering new cars at first because of shortages of supplies, the need to retool or refurbish manufacturing equipment that had been idle during the war or used for war production, logistical problems of getting each of the 15,000 parts that went into the typical automobile (over one billion parts per month) to the right plant at the right time, and difficulties in hiring labor fast enough (especially as many women were removed from the factories once the war ended). Most of the first postwar cars were copies of prewar models; for example, Chevrolet's first three years of offerings after the war were versions of its 1942 model. It was not until around 1950 that the auto industry began to catch up with demand. But once the manufacturers caught up, it was a golden period for the industry, with rapid growth in sales. In 1951 there were about seven million cars sold in America, three times the number sold twenty years later. By 1970, America was approaching a hundred million cars on the road.

The cars sold in the 1950s are significantly more recognizable to today's car buyers than those sold before the war. The postwar cars typically were built in a modern sedan style, with a trunk in the back, a large V8 engine, and running boards and headlights integrated into the body.

The manufacturers also would seem mostly familiar today. The war ended with General Motors in the lead position in new car sales, followed by Chrysler, then Ford. All of the other carmakers sold many fewer cars than each of these three firms. There were a few new entrants after the war, notably Tucker and Kaiser-Fraser, but neither lasted beyond the 1950s. Several of the smaller firms merged in an (unsuccessful) effort to have adequate resources to compete. Nash and Hudson merged in 1954, creating American Motors, which held a distant fourth position in the industry until being purchased by Chrysler in 1987. Studebaker and Packard merged in 1954, but they stumbled along until going out of business in 1966. Although there were fewer car makes, each maker offered many more models than before the war. For example, Chevrolet increased the number of models from one before the war to eight in a typical year of the 1950s.

The most notable feature of cars of the 1950s was the focus on style, influenced primarily by General Motor's chief stylist, Harley Earl, who introduced tail fins and other style features modeled after the Lockheed P-38 fighter plane. Cars were large and gas-hungry throughout most of the 1950s, but perhaps because of the recession late in the decade, the Big Three automakers began to lose sales to smaller cars such as the American Motors Rambler, the Studebaker Lark, and even the German imported Volkswagen Beetle. The Big Three were reluctant to build small cars because the profit margins were much larger on large cars, and many customers preferred large cars so long as gas was plentiful and inexpensive. But in response to lost sales to these smaller cars, the Big Three introduced their own compact models in 1960: the Chevrolet Corvair, Ford Falcon, and Plymouth Valiant. Given the popularity of the Volkswagen Beetle in America and the lack of a market for large, gas-hungry American cars in Europe, America began to import more cars than it exported for the first time in 1958.

Foreign competition began in earnest in 1955, when Volkswagen opened an American subsidiary. The company sold four million Beetles and vans in the United States by 1970.[28] Between 1955 and 1975 the U.S. auto industry's principal foreign competition was from Europe. Foreign competition continued to mount in the 1960s, with imports numbering over a half million for the first time in 1965. Japanese manufacturers, which did not have the advantage of a strong home market that the European makers did, increased U.S. imports in the 1970s to 400,000 cars in 1975 and three million in 1980. By the late 1970s it was clear that the Japanese companies were the Big Three's major competition, not the Europeans.

Although the compact cars offered initially by the Big Three did not generally sell well, the automakers had great success when they offered small, inexpensive sporty cars such as Chevrolet's Camaro, or Ford's Mustang, introduced in 1964. Ford marketed the Mustang heavily, selling 400,000 units in its first year. The evening before the Mustang's official launch, the company ran multiple ad spots throughout the principal evening viewing hours on all three major television networks—reaching an audience of 29 million people. The following morning, ads appeared in both the news and "women's" sections of 2,600 newspapers. Millions saw the new car in the Ford Wonder Rotunda at the 1964 New York World's Fair. Mustangs were also placed on display in fifteen major airports and in the lobbies of 200 Holiday Inns.

The rise of compact cars changed the market structure for the dealers and the buying habits of American families. In the past, the only new cars available were large and somewhat expensive. Many low-income and lower-middle class American families could not afford these new cars, especially before credit laws were relaxed in 1952 (ending periods of federal regulation to limit consumer borrowing during World War II and the Korean War as a means to control inflation) to allow lower down payments and longer-term loans.[29] This meant that poorer American families had a choice of the new, compact cars as well as their traditional market of larger used cars. These compact cars posed a challenge for the dealerships, which had to be able to sell the used cars they took in as trade-ins on new cars.

One of the major auto stories of the 1960s was the emerging public concern over safety. The signal event was the publication by Ralph Nader, a young Washington, D.C. lawyer, of his book *Unsafe at Any Speed*, which indicted Chevrolet for safety problems with the Corvair—steering problems and a propensity to roll over.[30] As a result of Nader's attack, Corvair sales dropped 93 percent. Ford had a similar problem in the 1970s, when it was slow to make an inexpensive repair to its Pinto subcompact, which had a propensity for catching fire when hit from the rear. The press fueled these concerns with numerous articles in the 1960s and 1970s not only about the safety of vehicles produced by the Big Three manufacturers, but also about their new- and used-car selling practices such as high-pressure sales, bait and switch, expensive but unnecessary add-ons, and packaging undesirable features with ones buyers wanted.[31] This consumer movement may be one of the most important but neglected factors in changing the information calculus for car buyers because the press was for the first time willing to tell all, thus reducing the information asymmetry that car buyers previously had been subject to.

State and local governments—and later the federal government—began to address car safety issues. The National Traffic and Safety Act in 1966 mandated safety standards for cars and gave the National Highway Traffic Safety Administration the power to force manufacturer recalls when there was a safety issue. One of the important changes mandated by the federal government was the requirement for seat belts on all new cars sold beginning in 1968. Safety changes led to a steady decline in car fatalities. In the 1970s the National Highway Traffic Safety Administration recalled millions of cars, to the displeasure of the auto industry. Recalls have continued over time, with one of the largest recalls occurring in 2000, when millions of Firestone tires mounted on Ford Explorer SUVs were recalled. In 2009 and

2010, there were extensive recalls of most of the Toyota line of vehicles for accidental acceleration.

Playing out at the same time were concerns about automobiles harming the environment. The problem first surfaced in Los Angeles, where smog appeared in the 1940s. In the early 1950s a Cal Tech professor, A. J. Haagen-Smit, showed that the smog—a polluting combination of smoke and fog—was mainly due to automobile emissions. The word *smog* itself entered popular parlance in the 1950s. California was the state most affected by auto emissions. Changes to California law required all new cars sold in the state from 1963 on to have PVC valves to reduce emissions; and in 1966 the state set maximum emission standards for carbon monoxide and other pollutants on all new cars sold there. The state continued to be an environmental leader when, in 1975, it required new cars to have catalytic converters to reduce tailpipe emissions.

The federal government also became involved in regulating automobile emissions. The 1965 Motor Vehicle and Air Pollution Act limited new car emissions (although it was less strict than the California law). The Clean Air Act, passed in 1970, limited emissions of carbon monoxide and other pollutants from new cars sold, beginning with the 1976 models, and gave enforcement powers to the newly created Environmental Protection Agency. The 1975 Energy Policy and Conservation Act set minimum-gas-mileage fleet standards on the automobile manufacturers. In 1978, the federal government imposed a "gas guzzler" tax on cars with high fuel consumption.

The technological challenges of building environmentally friendly cars were significant, and the first offerings from Detroit were underpowered, consumed a lot of gas (although fewer pollutants appeared in the emissions), and frequently stalled. These problems eventually were overcome through the use of fuel injection, computerized engine-management systems, and cleaner gasoline. Environmentally friendly cars helped to lessen, if not eliminate smog in Los Angeles.

Closely related to these environmental issues in public policy debates was the issue of fuel availability. At the end of World War II, the United States was a net exporter of gasoline. As late as 1970, the United States was still able to ramp up domestic production to meet domestic consumption each year. Through that period of time, gasoline was inexpensive in the United States, unlike the situation in Europe. By 1973, however, the United States was importing 27 percent of its oil. From October 1973 to March 1974, The Organization of Petroleum Exporting Countries (OPEC) temporarily cut off the supply of oil to the United States to protest U.S. support of Israel in the Yom Kippur War, creating widespread gas shortages across the United States and rapid fuel price increases. Another fuel shortage occurred in the United States in 1979, when Islamic revolutionaries toppled the U.S.-assisted government of the Shah of Iran.

Faced with higher fuel prices, consumers began to seek out smaller cars with better fuel consumption. The Detroit automakers resisted building smaller cars, and in the 1970s they lost market share first to Volkswagen, then to Toyota and Nissan, and finally to Honda. In the face of stiff foreign competition and regulated fleet fuel standards, the American manufacturers eventually introduced smaller, front-wheel drive cars such as the Chevrolet Citation and the Plymouth Reliant; and these companies were able to double their fleet mileage over time. However, they waited almost too long. The federal government had to bail out Chrysler with $1.5 billion in loan guarantees in 1979, while the company retooled to build smaller, more fuel-efficient cars.[32] To protect the American manufacturers from growing Japanese competition, the Reagan administration persuaded the Japanese automobile companies to impose voluntary import limits to the United States. This strategy backfired because the Japanese companies, starting with Honda in 1982, opened manufacturing plants in the United States, leading to even larger U.S. sales for the Japanese firms.

Partly as a result of the need to reduce pollution and increase fuel economy, the American manufacturers moved to new body styles and configurations in their vehicles. But their offerings were also intended to meet changing demographics and attract new audiences. In addition to new, smaller, front-wheel drive, V6 engine vehicles, "practicality" body styles were introduced, including the hatchback, minivan, and sports utility vehicle (SUV). There was also an increase in the number of trucks sold as passenger vehicles.

Although technological innovations were made in automobiles after 1945, some were slow to find their way into regular production because of economics or the Detroit carmakers' lack of interest. In 1940 Packard introduced the first air-conditioning unit in cars, but it was expensive and hard to implement. By the mid-1960s only 10 percent of cars were air conditioned, although over the next thirty years this became a standard feature in most cars. A powerful engine was a big selling point of the American manufacturers. They introduced V8 engines in the 1950s and gas turbine and turbocharger technologies in the 1960s. However, the American manufacturers were slow to build safety features into their cars. Disc brakes were introduced by Chrysler in 1949 but were not widely available until the 1970s, first on sports cars, only later on family sedans. Radial tires were introduced on European cars in 1947 and provided improved road-gripping ability, but American manufacturers resisted them until the 1970s because radial tires undermined the soft ride that these manufacturers believed American consumers wanted.

Several American companies attempted to use safety features as a major selling point in the 1950s but quickly abandoned the strategy. The 1948 Tucker sedan was the first postwar car sold on the basis of its safety features, but it was a commercial failure and the company soon went out of business. Nash introduced seat belts in 1952, but many other makes did not have seat belts until they were federally mandated in 1968. Various companies, notably Ford in the 1956 model year, sold safety packages as an option. These packages included breakaway rearview mirrors, padded dashboards, and crash-proof door locks—all of which became standard features on cars thirty years later. Although the safety padding was popular with purchasers, Ford decided that prospective buyers should not be hearing anything that implied the cars might not be safe (although the company did introduce these features into its high-end Lincolns). In recent years, additional safety features have been introduced into most cars, such as four-wheel disc brakes, automatic anti-lock braking systems, and smart airbags.

The infrastructure for American driving continued to grow in the postwar years, in many cases building on the infrastructure that had been built in the interwar years but neglected during World War II. There was growth in roads. The Federal Aid Highway Act of 1944 authorized $1.5 billion in matching federal funds for highway construction during the first three years after the war. The Federal Highway Act of 1956 authorized $25 billion over a twelve-year period to build the interstate highway system, paid for mainly by taxes on fuel, tires, and other automotive consumables. Massive building of roads to meet suburban growth occurred throughout the 1950s and 1960s. Some new roads, although administered by states, were paid for by tolls—beginning with the Pennsylvania Turnpike opened in 1940, followed by the Ohio, Indiana, and New Jersey turnpikes all opened in 1956. By 1956, one could use these toll roads to drive from New York City to Chicago without having to stop for a traffic light.

Higher-octane gasoline that led to more powerful combustion and cleaner-burning fuels also improved the driving infrastructure. Motels had existed before the war, but motel chains emerged later, beginning with the first Holiday Inn (in Memphis) in 1952. The chains standardized cleanliness levels, amenities, and pricing schemes. The first budget motel chain was Motel 6, started in 1962 in Santa Barbara, California. Emerging motel chains were accompanied by the rise of fast-food restaurant chains. The first was Howard Johnson's, established on Cape Cod in Massachusetts with a single restaurant in 1935, but expanded rapidly after the owner won the food concession contract for the Pennsylvania Turnpike in 1941. The quintessential fast food restaurant, McDonald's, opened its first store in Des Plaines, Illinois, in 1955. In the 1943s, Duncan Hines came out with his guide to eating on the road, and in the 1950s, the American Automobile Association (AAA) introduced travel guides that listed restaurants, lodging, and places of sightseeing interest.

The annual auto shows continued after World War II to be an important way to advertise new cars to the general public, even in the age of television. A main themes of these shows was the automobile of tomorrow—so-called "concept cars"—following on the success of the auto show at the 1939 New York World's Fair. These cars were prototypes, not intended for immediate, regular production but instead to generate interest in the brand and give the manufacturers a venue for judging the public's interest in new engines or styling features. Of particular note were the General Motors Motoramas of the 1950s. Modeled after the extravagant industrial luncheon shows mounted by GM chairman Alfred Sloan in the 1930s in connection with the New York auto show, the company's chief designer Harley Earl produced the first Motorama in the hotel ballroom of the Waldorf Astoria at the time of the National Auto Week Show in 1949. This Motorama, entitled "Transportation Unlimited," was a thirty-five-minute-long musical show in the style of Busby Berkeley, with showgirls framing autos rotating on turntables. Between 1949 and 1961, GM took these shows on national tours eight times, using either the name Autorama or Motorama. It was an effective marketing tool. More than two million people saw the 1956 Motorama, and more than ten million people attended these eight shows altogether. This tradition of showing concept cars endured, in examples such as the all-stainless steel Pontiac Club de Mer sports car introduced in 1956, plans for a nuclear-powered Ford Nucleon shown in 1958, and the Volvo YCC designed by an all-women team in 2004.

Although the late 1970s and early 1980s were hard times for the American automobile industry, the Big Three were markedly more successful as the 1980s unfolded. Chrysler introduced the highly popular minivan in 1983, while Ford introduced the bestselling Taurus family sedan in 1985. That same year General Motors opened its Saturn division to positive public reaction; Saturn sold small cars with no-haggle pricing, built in nonunion factories. The U.S. makers did relatively well until the Japanese companies began to eat into their market share with cars having fewer defects. Always happy to reenter the big-car business, the American companies did well with SUVs and trucks throughout the 1980s and 1990s, until gas prices climbed early in the new century. The last few years have been dominated by the auto industry's race to develop hybrid technologies (starting with the Honda Insight in 2000) and other alternatives that do not pollute as much or consume as much fuel as the gasoline internal combustion engine.

Information Issues for Car Buyers after 1945

While it does not focus primarily on car buyers, the summary of automobile history in the preceding section helps us identify issues Americans faced in the postwar period when considering a car purchase. Among the issues:

- Availability of cars after the war (before production had fully resumed)
- How to acquire a car in a reasonable time (with or without a bribe)
- New technologies introduced into cars
- Selecting the right make and model
- Offerings from foreign manufacturers
- Safety needs and features
- Avoiding the worst practices of dealers such as bait and switch and high-pressure sales
- Environmentally friendly cars
- Availability of fuel (during oil crises)
- Fuel economy of vehicles
- Concerns over the viability of American manufacturers in light of foreign competition
- Car type (sedan, minivan, truck, SUV)
- Where and how to buy a car, e.g., new distribution methods (no-haggle pricing) and venues (auto malls)

We can supplement this list of issues by examining those raised in the *The Saturday Evening Post*.[33] Why look at this magazine in particular? In the 1930s, General Motors began to collect statistical information in order to sell its cars on a "scientific" basis. A study of owners of cars built by Chevrolet, Buick, and Cadillac showed that more than half the people who owned each of these three brands read the *The Saturday Evening Post* on a regular basis.[34] This magazine persisted as a major source of information about cars for many years, through both its articles and its advertisements.

In 1938, for example, automobile manufacturers spent more money to advertise in this magazine than in the next five combined.[35] In addition to those issues just listed that were of concern to post-1945 car buyers, readers of this magazine would have found out about the following issues:

- Conditions of the highways
- Where to get one's car serviced
- Dubious practices of used-car salesmen
- Dubious practices of finance companies
- Station wagons, vans, convertibles (functionality of new car types)
- How the new compact models compare to traditional offerings
- Cars of the future
- Alternative motors (electric, gas-turbine, Wankel engine)
- Styling and planned obsolescence
- Customization
- Product quality
- Warranties, existence and coverage
- The importance of buying American

The introduction of *Car & Driver*, *Motor Trend*, and *Road & Track* magazines in the period between 1948 and 1953 signaled a new interest in cars as a hobby. It also indicated a period in American auto history when the various makers sold a wide selection of competitive cars, so there was more need for buyers to have information that enabled them to differentiate among the new cars.

Perhaps the most important source to car buyers in the postwar era has been *Consumer Reports* and in particular its annual April car issue (first published in 1953). Other consumer magazines also published car reviews, most notably *Consumers Digest*, which has been published continuously since 1961.

Americans showed strong interest in general-purpose magazines after World War II, although that interest has waned in the past several decades. Many of the general-interest magazines (including *The Saturday Evening Post*, *Time*, *Newsweek*, and *The Nation*) frequently contained stories about automobiles and society, which educated readers about cars in general but only occasionally gave them specific information that would inform their purchase. More common topics were stories about the automotive industry (not individual new cars), foreign competition, the oil crisis, cars and the environment, important or colorful personalities in the auto industry, and general trends. *Business Week* and *U.S. News and World Report* published these general kinds of articles too, but they would also occasionally review

a specific car or class of cars. Some other popular general-interest magazines, such as *New Republic* and *National Review*, seldom contained stories
about cars or the automobile industry.

As described previously, *Money* magazine regularly carried articles giving
practical advice to car buyers, often about financial aspects of car buying
but sometimes about technologies or other practical matters. *Kiplinger's*,
published since 1947, contains a mix of practical and more general articles
about cars, similar to what is found in *Money* magazine.[36] *Forbes* has often
published practical buying-advice articles as well as a number of more
general articles, especially about the auto industry. *Fortune* publishes much
less about cars than any of the other three personal finance magazines
mentioned here.

There were other sources as well. The auto shows, which had been
suspended during the war, were resumed in the early 1950s, after car
production caught up with pent-up consumer demand. The postwar auto
shows displayed not only the new car offerings but also the concept cars
of the future. These concept cars received considerable attention in the
press. Television gradually supplanted radio as an important advertising
medium. Popular magazines have remained an important advertising
venue—although magazine and newspaper readership has declined in the
past twenty years partly as a result of the Internet. The World Wide Web
has become a major source of news for car buyers, especially sites such
as www.edmunds.com. Programs about cars have also become popular on
radio and television, such as *Motor Week* and *Car Talk*.

Newspaper classifieds have long been a common way to find used cars
and car parts. They were a staple revenue for the newspapers. Even today,
with newspaper classifieds greatly reduced compared to in the past, the
typical Sunday newspaper devotes an entire section to ads from dealers
and private parties wanting to sell new and used cars. In the Internet era,
newspaper classifieds increasingly are being replaced by Internet services.
Craig's List offers an online service that is a direct replacement for newspaper classifieds and in fact is better because it is free, has fewer restrictions
on length, and photos can be included. There are separate advertising sections on Craig's List for more than three hundred different metropolitan
areas in the United States. A recent section for Denver, for example,
included approximately two hundred new listings of cars and car parts for
one particular day (December 31, 2008), plus similar numbers listed for
prior days.

eBay provides an inexpensive auction service for selling cars. On December 31, 2008, for example, there were almost 40,000 cars listed on eBay—

23,000 for sale by dealers, 12,000 for sale by private parties, and the remainder that didn't specify the seller. Within fifty miles of the author's home, there were more than 3,000 cars for sale on eBay. One service that goes well beyond what was ever available with newspaper classifieds is CarFax, which provides a national database that a prospective used-car buyer can access online to learn a vehicle's history. This is a way to avoid "lemons"—defective cars that have suffered flood damage, had their airbags deployed, have odometer or title problems, or were used heavily as a rental or fleet car. This helps to even out the information asymmetry for used-car buyers.

Methodology and Literatures

This chapter was inspired in part by the literature on the Internet and everyday life that has appeared since 2002.[37] The earlier social science literature on the Internet often focused on what happened online (e.g., in MUDs and MOOs); one important difference in the more recent literature on the Internet and everyday life was to situate online behavior in the context of the home and investigate the interaction between online and offline behavior. This chapter is similar to that recent literature in that it explores how the family and home situation impact the questions that are raised and the sources that are examined when an American family considers the purchase of an automobile. The Internet and every-day-life literature focuses on the user rather than the producer of the technology and considers the user as someone who shapes the technology, not merely accepts it as it is produced. Similarly, this chapter focuses on the user and on the user as an active agent, not on the manufacturer of cars or even the producers of car magazines and Web sites that provide fixed technologies and static information.

This chapter differs in one key respect from prior Internet and everyday-life literature. Most of that literature is based on either ethnographic examination through the use of surveys or site visits, or through the use of critical or cultural theory.[38] Both of these approaches offer a snapshot of what is happening at a given period of time. They are less useful at understanding change over time. Unlike that literature, this chapter employs a historical method that enables one to study changes over time in both the information issues and the information sources. The challenge is to find a uniform set of historical sources, so that our representations of what is happening at two different times are comparable and so that the sources do not introduce a bias into the comparisons we are making from different times. We handle that challenge in two ways, by using sources that are uniform over long periods of time (such as long runs of particular magazines such as the *The Saturday Evening Post*), and by examining directly the change agents (both internal and external to the auto industry) that affect the nature of the car-buying experience.

This chapter is meant to introduce the subject of information issues in car buying, not to definitively study it; and as a result, the chapter samples various pertinent literatures rather than examine them exhaustively. The published and archival literatures are enormous; see as illustration the numerous and wide-ranging sources cited in Clarke's thoroughly researched history, *Trust and Power* (2007). This author believes that he has examined enough literature to reach some general conclusions (as stated in the next section), but another goal of this sampling is to demonstrate the value of different kinds of sources for a study of this sort. Some of the reading for this chapter has been in the historical literature about automobiles (see the first section of the bibliography), even though little of this literature is directly about users and essentially none of it is about car buying specifically. However, reading this literature has helped to identify the exogenous and endogenous forces that have shaped the auto industry and the practice of car buying. This literature has been less helpful in identifying the information sources available to car buyers or their information-seeking behavior at different points in time.

This chapter has also sampled the economic, psychology, marketing, and information science literatures on consumer behavior, especially as they apply to the purchase of cars and other large-ticket items.[39] (See the second section of the bibliography.) The reading in this literature did not pay dividends in proportion to the time expended—although this literature might be productively used if one wanted to conduct a more in-depth analysis of contemporary car-buying behavior. Some of this literature focuses on economic issues such as the impacts of credit restraints on the car-buying search process; the economic efficiency of a search measured by how much better a price might be rewarded by extra time spent searching; or building an abstract economic model of car-buying behavior. Some of this literature focuses on psychological issues, such as the role of self-confidence and persuasion in the car-buying process, or how families manage conflict when family members differ in their car preferences. Other social issues were also addressed, such as cross-cultural values and how they affected the car-buying process. The principal utility of this literature for the writing of this chapter has been primarily to broaden the list of the information issues and the information sources consulted by car buyers today.

There is a growing information science literature on information-seeking behavior in everyday life. Case (2006) and Fisher and Julien (2009) provide good reviews of this literature. On the role of the Internet in changing information sources in everyday life, see Case et al. (2004) and Kaye and Johnson (2003). For comparison, there are studies on information seeking in everyday life in Sweden (Hektor 2003) and Finland (Savolainen 2001a, 2001b; Savolainen and Kari 2004a, 2004b). Scholars are producing a growing literature on information-seeking behavior among hobbyists (e.g., Hartel 2003; Yakel 2004; Duff and Johnson 2003). There is also a body of literature attempting to theorize information-seeking behavior (Pettigrew, Fidel, and Bruce 2001; Pettigrew and McKechnie 2001; and Fisher, Erdelez, and McKechnie 2005).

Finally, the author has reviewed a sample of the massive primary source materials, including mass-market periodicals, consumer guides, automobile enthusiast magazines, personal finance magazines, radio and television car shows, Internet Web sites, literature from automobile shows, print advertising, archival material from an automobile manufacturer, and stories reported about the buyer experience in the showroom from both buyers and car salespeople and their managers. This chapter has hardly scratched the surface of this enormous literature.

Conclusion

The purchase of a car has been an important issue in everyday American life since 1908, when the introduction of the Ford Model T made the car a purchase item for the masses rather than only for the wealthy. Cars were much less complex before World War II than they were afterward, but from their inception they were sufficiently complex that most potential buyers did not fully comprehend them in a technical sense. As the automobile became increasingly complex in the postwar decades—incorporating, for example, new computerized engine management, pollution controls, anti-lock braking, stability control, and many other technological changes—it became even harder for consumers to understand the technology and to determine which new features would matter to them. However, in one sense, the car-buying decision today is easier to make than it was in the past; thanks to the Japanese manufacturers' innovations in safety and technology, and competitive responses from American and European manufacturers, most cars today are highly reliable, so that the buyer does not have to worry nearly so much about the issue of technical quality; cars can be chosen according to other criteria.

After World War II, the number of models, colors, fabrics, and accessories grew tremendously, so that cars could be greatly customized to an individual consumer's desires. This change made the buying issue more complex. New concerns about safety, fuel consumption, and environmental friendliness of cars, emerging in the 1960s and 1970s, also increased the complexity of the car-buying task.

Although the specific issues may have changed (e.g., one no longer need wonder if one has sufficient physical strength to turn the starting crank), it is remarkable how consistent questions about the driving experience have been over the past hundred years. Then and now, potential buyers have been concerned about ride comfort, handling, speed, sportiness, and driving range on a tank of gasoline. Concern about obsolescence has been a part of the purchase decision since General Motors introduced the practices of the model year and the annual styling change in the 1920s.

There has always been segmentation in the car market: inexpensive small cars, middle-class family cars, and luxury vehicles. Today, those segments persist but there is much greater choice and fragmentation. Lifestyle choices provide an option not only of sedans and coupes, but also of passenger trucks, SUVs, station wagons, convertibles, minivans, and various crossover models. Thus the information issues concerning which type of vehicle to purchase are much more significant than they were in the past.

The time of the Model T introduction in 1908 witnessed the largest number of providers of automobiles at any time in the last century, approximately three hundred, but a shakeout dropped the number to about twenty serious suppliers by the end of the 1920s. The number kept dropping throughout the Depression of the 1930s and the war years of the 1940s, until the mid-1950s when pent-up demand was overcome and buyers could become more selective—to a point where there were only four major players in the U.S. auto market. The number of contestants grew once again from the 1960s through the 1990s, through the establishment of European and Asian competition, so that today there are about as many makes to choose from as there were at the end of the 1920s.

Until the 1920s, most car buyers paid cash up front to buy a car. In the 1920s, financing through the preferred credit organizations of the carmakers, such as General Motors Acceptance Corporation, became a viable option. After World War II, the number of suppliers of car loans multiplied, terms of credit eased (making loans available to a wider segment of the population), and leasing became another option to buying with cash or on installment, so that the financial aspects of car buying became more complicated.

Once dealership networks became the normative business practice in the American automobile industry during the 1920s, almost every new car buyer bought through a dealership contractually bound to a manufacturer. The distribution network has changed somewhat over time, through dealerships handling multiple brands, auto malls (where dealers for different makes concentrate in a geographic region), competition among dealers handling the same brand, and multiple locations for a single dealership. But the age-old problem persists of determining which dealer to work with based on pricing, availability of cars, business integrity, quality of service, and other factors. With the option for buyers to go to multiple dealerships of the same carmaker after geographical restrictions to dealer competition were eliminated, the buyer has more choice but also more information to consider.

With the demise in 1918 of the magazine *Horseless Age*, there was no American publication that reported on test drives of automobiles until 1948. From that time on, the number of car enthusiast, personal finance, and general-purpose news magazines reporting on cars grew rapidly. These magazines were followed by radio, television, and Internet programming that provided similar information. Today, the number of media sources for car buyers to consult on cars is in the hundreds, if not the thousands. Perhaps equally important is the quality of the information provided. Not only is much of it independently arrived at (for example, by independent testing of vehicles), but also since the consumer movement began in the 1960s as a product of the activism of Ralph Nader, Rachel Carson, and reactions to the Vietnam war, the media is more willing to reveal information that was unavailable to buyers in the past, such as hidden defects, recalls, and rebates. Today, there is less information asymmetry between car buyer and seller than at any time in the past.

Between the two world wars, automobile advertising and auto shows played a major role in informing potential car buyers about their options. Ford claims, for example, that over 10 percent of the American population saw its Model A on the first day of its introduction in 1928 by visiting dealerships or auto shows.

A number of changes within the automobile industry have shaped the issues that potential car buyers address. These include the structure of the car manufacturing industry (especially the shakeout to only a few providers in the late 1950s), foreign competition, interactions between dealers and manufacturers, the relationship between the automobile industry and interlocking industries (oil, highways, fast food, motels), changes in places to advertise autos (auto malls, airports, shopping malls), and changes in negotiating practices (no-haggle pricing, CarMax individual vehicle histories).

If anything, there are an even greater number of exogenous forces that have shaped the American car-buying experience. These include the two world wars, the Depression, competition from mass transit, the 1950s baby boom, suburbanization, the consumer movement, the environmental and safety movements and government regulation of them, oil embargoes, women entering the workforce in greater numbers, the growing wealth of Americans, changes in the effectiveness and pervasiveness of advertising, and the rise of the Internet.

This chapter is merely suggestive of a much more expansive research and writing project that would be required in order to do full justice to the changes in the information needs, information sources, information-seeking behaviors, and endogenous and exogenous forces that caused changes. It would require looking in more detail at automotive history and its immersion in American culture, and its intersection with the endogenous and exogenous forces mentioned in the preceding paragraphs (and no doubt other forces as well). In consideration of length, there are some topics that have not received here the attention they deserve. These include the story of buying, selling, and trading used cars; issues related to gender; the role and process for selecting buyer buddies; and the differences caused by demographic differences in buyers (wealth, rural vs. urban, minority, etc.). To do full justice to this topic would require writing multiple books.

The literature is focused on the technology and the industry, and to a lesser extent on the culture of cars. There is not much scholarship on car buying or the buyer experience that one can build upon. This means that any complete study will involve substantial archival work. It is also possible, and perhaps valuable, to conduct more empirical research on car buying and car information-seeking practices, as is suggested by the psychology, economics, information studies, and marketing literature mentioned previously.

Acknowledgments

Thanks to Jon Bill, director of education and archives, at the Auburn-Cord-Duesenberg Museum, for his help with archival materials. For their suggestions, thanks to Harrison Archer, Barbara M. Hayes, and many of the students in my Fall 2008 graduate seminar on Information in Everyday American Life at the University of Texas at Austin. Thanks to Gesse Stark-Smith and Ellie Kemple for their research assistance.

Notes

1. Background information (including this statistic) on American car culture in this section and the sections surveying the automobile before and after 1945 is not individually cited but taken from the general histories of the automobile in America listed in the bibliography.

2. Ford built a tractor body prototype in 1907 and a production model introduced as the Fordson F in 1917. An enterprising clergyman decided to take his religion on the road in 1922 and placed a wooden room on the back of his Model T chassis, with stained-glass windows, a folding hinged steeple, and an organ inside. There are many other examples of personal customization of Model T cars, and of various body styles offered commercially by Ford and the aftermarket suppliers.

3. Inasmuch as the focus is on car buying in everyday life, the chapter mainly focuses on cars and trucks sold to the general buying public. It does not consider the interesting issues of ultraluxury cars such as Maserati or specialty-performance cars such as drag racers, stock-car racers, monster trucks, or swamp buggies; nor does this chapter consider motorcycles, motor scooters, or motorized bicycles.

4. According to a Zillow.com survey, people spend more time on choosing their car loan than they do on choosing their home mortgage. See Extra Realty, a real-estate blog, April 3, 2008, http://extrarealty.blogspot.com/2008/04/americans-spend-more-time-researching.html (accessed September 29, 2008).

5. This information is taken from the *Consumer Reports* Web site. See http://www.consumerreports.org/cro/aboutus/history/printable/index.htm (accessed August 16, 2008).

6. For example, see a list of several dozen auto and cycle enthusiast magazines at http://www.magazinediscountcenter.com/category/Auto_and_Cycles.html (accessed September 19, 2008).

7. See http://www.ischool.utexas.edu/%7Ebill/magazine_descriptions_reviews.html (accessed September 17, 2008) for more information on what the magazine trade and fans say about these car enthusiast magazines. There are also many other automobile magazines and trade journals, e.g., *American Rodding, The Automobile, Automobile Quarterly, Automobile Review, Auto Topics, Automotive Industries, Automotive News, Car Life, Cycle and Automotive Trade Journal, Hot Rod, The Marque and Shelby American Magazine, Mustang Monthly, Special Interest Automobile,* and *U.S. Auto Scene.*

8. For a more detailed analysis of these issues of *Money,* see http://www.ischool.utexas.edu/%7Ebill/content_analysis_money_mag.html (accessed September 19, 2008). You will also find there a listing of the articles from *Money* reviewed for this purpose.

9. For example, see a list of sixty-six television shows on automobiles at http://
en.wikipedia.org/wiki/Category:Automotive_television_series (accessed September
19, 2008).

10. See http://www.ischool.utexas.edu/%7Ebill/sample_internet_sites.html (accessed
April 24, 2010) for a sample of Internet sites about car buying and what information
and advice each offers.

11. Major auto shows in the United States in 2008 were held in Washington, DC,
Houston, Philadelphia, Chicago, Detroit, Cleveland, Minneapolis, Nashville, New
York City, Westchester County (NY), Scarsdale (NY), Pebble Beach (CA), Seattle, Las
Vegas, Tampa Bay, Miami, Birmingham (AL), Los Angeles, San Francisco, and San
Diego. A list of auto shows by continent is available at http://en.wikipedia.org/
wiki/List_of_auto_shows_and_motor_shows_by_continent (accessed September 19,
2008).

12. This section relies heavily on Stevenson 2008 although other sources including Franklin 2002, Ikita 2000, Laird 1996, Roberts 1976, Stern and Stern 1978,
and Williams 1997 are also taken into consideration.

13. See Clarke 2007, 162–163, for a discussion of advertising that highlights
safety.

14. For a harsh critique of the American practice of selling cars on style, see Keats
1958.

15. GM pushed its dealers to use GMAC, which insisted on monthly payments
rather than a payment schedule that might be more suitable to farmers or school-
teachers based on when they had the most cash available. Another issue was
whether the dealer or the manufacturer would get the major portion of the profit
from the car loan.

16. "Sales-Promotion Outline for Introduction of the Cord Front Drive," T. M. S.
Gibson, General Sales Manager, Cord Front Drive, Auburn Automobile Co., Auburn,
IN, 1930. Courtesy of the Auburn-Cord-Duesenberg Museum Archives, Auburn,
Indiana.

17. Given that we do not have the original letter requesting testimonials and we
do not have any information about whether testimonials were omitted from the
book or edited, we have to be careful in the use of this evidence.

18. For a history of the Cord 810, see http://www.automaven.com/index.html. Also
see Clarke 2007, 197–202.

19. "Notes on Demonstrating the Cord," Auburn Automobile Co. Courtesy of the
Auburn-Cord-Duesenberg Museum Archives, Auburn, Indiana.

20. "Twenty Minutes with the Cord Prospect," Auburn Automobile Co. Courtesy of
the Auburn-Cord-Duesenberg Museum Archives, Auburn, Indiana.

21. For more detail, see Clarke 2007, chapter 4.

22. U.S. tariffs on imported automobiles continued to drop—to 10 percent in 1950 and 3 percent in 1973.

23. As quoted in Flink 1988, 234.

24. For a bibliography of *The Saturday Evening Post* articles surveyed for this purpose, see http://www.ischool.utexas.edu/%7Ebill/kiplingers_money_SEP.html (accessed April 24, 2010).

25. There were some other early auto magazines and journals in the United States, including *Cycle and Automotive Trade Journal* (introduced in 1897), *Automobile* (1899), *Motor Age* (1899), *Motor World* (1900), and *Motor* (1903). None of these was as successful as *Horseless Age*, most of them did not last long, and some were intended primarily for the trade rather than for enthusiasts. For example, *Motor World*, which did persist into the 1920s, had the subtitle "for Dealers, Jobbers, and Garagemen," not for car enthusiasts. See also, for example, the specialty trade journals *Motor Body* or *Dealer and Repairman*.

26. Historian Sally Clarke (2007, 42) reports on regional clubs appearing as early as 1910 and having their own journals for members only: *Automobile Club of Hartford Bulletin*, *Automobile Club of Philadelphia Monthly Bulletin*, the Automobile Club of America's (based in New York) *Club Journal*, *Auto News of Washington*, *Automobile Journal* (later *New England Automobile Journal*), and *Empire State Motorist* (later *Motordom*). There were also local Ford clubs in the 1910s, which sponsored their own journal, known as the *Fordowner*.

27. Andrew Lee Dyke, *Diseases of a Gasolene [sic] Automobile—and—How to Cure Them*. A.L. Dyke Automobile Supply Company, 1903, Baker Library, Harvard Business School, as cited in Clarke 2007, 50.

28. In 1972, the Beetle became the first car to surpass the Model T in sales, and eventually more than twenty million were sold worldwide.

29. However, it was not until 1974 that the federal government enacted the Equal Credit Opportunity Act, which made it easier for women and minorities to obtain credit for auto loans.

30. Eventually the Corvair received a clean bill of health from the National Highway Traffic Safety Administration, with the proviso that the car be properly maintained, including ensuring correct tire pressure.

31. High-pressure sales go back far in time, as this caricature of the salesman attests: "Q: . . . How much do you want for that wabbly wheels over there?
    A: Say, that ain't no wabbly wheels. That's my shiner. It's a car with a history—a schoolteacher used to own it. She never drove it over fifteen miles an hour, and she put it to bed with a hot-water bottle every night. It's a clean piece and I'll let you have it for a short down and three bills to go. There's plenty of soup left in her and

a lota miles left in them doughnuts [tires]. Here you are; sign right here." (Elliott Curtis, Jr., *The Saturday Evening Post*, September 7, 1937).

32. History repeated itself in 2008 with the much larger federal bailout to General Motors and Chrysler, which had been slow to build fuel-efficient vehicles in a time of rising fuel prices.

33. We have examined every car article that appeared in *The Saturday Evening Post* since 1945. See http://www.ischool.utexas.edu/%7Ebill/content_analysis_SEP.html for a content summary and http://www.ischool.utexas.edu/%7Ebill/kiplingers _money_SEP.html for a list of the articles reviewed.

34. See the discussion in Clarke 2007, 132.

35. Commercial Research Division, Curtis Publishing Company, *Who Buys Automobiles and Automotive Products* (Philadelphia: Curtis Publishing Company, 1939), 24, as discussed in Clarke 2007, 136.

36. For a list of the *Kiplinger's* articles reviewed for this purpose, see http://www .ischool.utexas.edu/%7Ebill/kiplingers_money_SEP.html.

37. See, for example, Bakardjieva 2005; Lally 2002; and Wellman and Hay-thornthwaite 2002. The chapter was also inspired in part by my teaching and research on health information-seeking behavior. See, for example, Barbara M. Hayes and William Aspray, eds., *Health Informatics: A Patient-Centered Approach to Diabetes* (MIT Press, 2010). The first few chapters of Bakardjieva's book do a good job of identifying various approaches to these Internet studies of everyday life, such as social construction of technology, philosophy of technology, phenomenological sociology, reader reaction of media, and other critical theories of the everyday.

38. For a discussion of the methods used in the Internet in everyday life literature, see the first three chapters of Bakardjieva 2005.

39. There is considerable overlap in methods in this study of large-ticket-item consumer behavior, so we resist breaking these literatures in the subcategories of economics, information science, marketing, and psychology. See the listing in the bibliography of the items consulted during this study; all four types of scholarship are listed in the same subset of the bibliography.

Bibliography

Histories of the Automobile and Other General Literature

Alvord, Katie. 2000. *Divorce Your Car!* Gabriola Island, BC, Canada: New Society Publishers.

Brinkley, Douglas. 2004. *Wheels for the World*. London: Penguin.

Calder, Lendol. 1999. *Financing the American Dream: A Cultural History of Consumer Credit.* Princeton, NJ: Princeton University Press.

Clarke, Sally H. 2007. *Trust and Power.* Cambridge, UK: Cambridge University Press.

Eisbrenner, Kenneth N. 2005. *The Complete U.S. Automobile Sales Literature Checklist 1946–2000.* Hudson, WI: Iconografix.

Flink, James. 1988. *The Automobile Age.* Cambridge, MA: MIT Press.

Foster, Mark S. 2003. *A Nation on Wheels.* Belmont, CA: Wadsworth.

Franklin, M. J. 2002. *Classic Muscle Car Advertising.* Iola, WI: Krause Publications.

Genat, Robert. 2004. *The American Car Dealership.* St. Paul, MN: MBI Publishing Co.

Halberstat, David. 1986. *The Reckoning.* New York: Avon.

Holder, William. 1957. *The Automobile Dealer: Yesterday, Today, and Tomorrow.* Costa Mesa, CA: N.A.D.A.

Ikita, Yasutoshi. 2000. *Cruise-O-Matic.* San Francisco: Chronicle Books.

Keats, John. 1958. *The Insolent Chariots.* Philadelphia: Lippincott.

Laird, Pamela Walker. 1996. The Car without a Single Weakness: Early Automobile Advertising. *Technology and Culture* 37 (October): 796–812.

Lewis, David L., and Lawrence Goldstein, eds. 1980. *The Automobile and American Culture.* Ann Arbor: University of Michigan Press.

Marling, Karal Ann. 2002. America's Love Affair with the Automobile in the Television Age. In *Autopia: Cars and Culture,* ed. P. Wollen and J. Kerr, 354–362. London: Reaktion Books.

McIntyre, Stephen L. 2000. The Failure of Fordism: Reform of the Automobile Repair Industry, 1913–1940. *Technology and Culture* 41 (April): 269–299.

Nader, Ralph. 1965. *Unsafe at Any Speed.* New York: Grossman.

Nelson, Walter Henry. 1970. *Small Wonder: The Amazing Story of the Volkswagen.* 2nd ed. London: Hutchinson.

Preston, James M. 1934. Seventeen Years behind the Windshield. *Kiplinger's Personal Finance* 113 (9): 27–32.

Rae, John B. 1965. *The American Automobile.* Chicago: University of Chicago Press.

Rae, John B. 1984. *The American Automobile Industry.* Boston: G. K. Hall.

Roberts, Peter. 1976. *Any Color So Long as It's Black.* New York: William Morrow.

Robinson, Jon G. 2003. *Classic Chevrolet Dealerships: Selling the Bowtie.* Saint Paul, MN: MBI Publishing.

Sandler, Martin. 2003. *Driving around the USA*. New York: Oxford University Press.

Stern, Jane, and Michael Stern. 1978. *Auto Ads*. New York: Random House.

Stevenson, Heon. 2008. *American Automobile Advertising, 1930–1980*. Jefferson, NC: McFarland.

Stevenson, Matthew. 1982. Why Buy an American Car? *The Saturday Evening Post* (July/August): 66.

Volti, Rudi. 2004. *Cars & Culture*. Baltimore: Johns Hopkins University Press.

Williams, Jim. 1997. *Boulevard Photographic: The Art of Automobile Advertising*. Osceola, WI: Motorbooks International.

Wollen, Peter, and Joe Kerr, eds. 2002. *Autopia: Cars and Culture*. London: Reaktion Books.

Economics, Information Science, Marketing, and Psychology Literatures

Attanasio, Orazio P., Pinelopi Koujianou Goldberg, and Ekanterini Kyriazidou. 2008. Credit Constraints in the Market for Consumer Durables: Evidence from Micro Data on Car Loans. *International Economic Review* 49 (2): 401–436.

Bakardjieva, Maria. 2005. *Internet Society: The Internet in Everyday Life*. London: Sage Publications.

Bell, Gerald D. 1967. Self-Confidence and Persuasion in Car Buying. *Journal of Marketing Research* 4 (February): 46–52.

Bennett, Peter D., and Robert D. Mandell. 1969. Prepurchase Information Seeking Behavior of New Car Purchasers—The Learning Hypothesis. *Journal of Marketing Research* 6 (November): 430–433.

Berry, Steven, James Levinsohn, and Ariel Pakes. 2004. Differentiated Products Demand Systems from a Combination of Micro and Macro Data: The New Car Market. *Journal of Political Economy* 112 (1): 68–105.

Burns, Alvin C., and Donald H. Granbois. 1977. Factors Moderating the Resolution of Preference Conflict in Family Automobile Purchasing. *Journal of Marketing Research* 8 (February): 77–86.

Case, Donald O. 2006. Information Behavior. In *Annual Review of Information Science and Technology* 40, ed. Blaise Cronin, 293–327. Medford, NJ: Information Today Inc.

Case, D. O., J. D. Johnson, J. E. Andrews, S. Allard, and K. M. Kelly. 2004. From Two-Step Flow to the Internet: The Changing Array of Sources for Genetics Information Seeking. *Journal of the American Society for Information Science and Technology* 55: 660–669.

Cosse, Thomas J., and Terry M. Weinberger. 1997. Saturn Buyers: Are They Different? *Journal of Marketing Theory and Practice* 36 (Fall): 77–86.

Duff, W. M., and C. A. Johnson. 2003. Where Is the List with all the Names? Information-seeking Behavior of Genealogists. *American Archivist* 66 (1): 79–95.

Duncan, Calvin P., and Richard W. Olshavsky. 1982. External Search: The Role of Consumer Beliefs. *Journal of Marketing Research* 19 (February): 32–43.

Farley, John U., John A. Howard, and Donald R. Lehmann. 1976. A "Working" System Model of Car Buyer Behavior. *Management Science* 23 (3): 235–247.

Fisher, Karen E., and Heidi Julien. 2009. Information Behavior. In *Annual Review of Information Science and Technology* 43, ed. Blaise Cronin, 317–358. Medford, NJ: Information Today Inc.

Fisher, K. E., S. Erdelez, and E. F. McKechnie, eds. 2005. *Theories of Information Behavior*. Medford, NJ: Information Today.

Furse, David H., Girish N. Punj, and David W. Stewart. 1984. A Typology of Individual Search Strategies among Purchasers of New Automobiles. *Journal of Consumer Research* 10 (March): 417–431.

Giese, Joan L., and Joseph A. Cote. 2000. Defining Consumer Satisfaction. *Academy of Marketing Science Review* no. 1, http://www.amsreview.org/amsrev/theory/giese01-00.html.

Gupta, Pola, and Brian T. Ratchford. 1992. Estimating the Efficiency of Consumer Choices of New Automobiles. *Journal of Economic Psychology* 13: 375–396.

Hartel, J. 2003. The Serious Leisure Frontier in Library and Information Studies: Hobby Domains. *Knowledge Organization* 3: 228–238.

Hauser, John R., Glen L. Urban, and Bruce D. Weinberg. 1993. How Consumers Allocate Their Time When Searching for Information. *Journal of Marketing Research* 30 (4) (November): 452–466.

Hektor, A. 2003. Information Activities on the Internet in Everyday Life. *New Review of Information Behaviour Research* 4: 127–138.

Hendel, Igal, and Alessandro Lizzeri. 2002. The Role of Leasing under Adverse Selection. *Journal of Political Economy* 110 (1): 113–143.

Hoffer, George E., Stephen W. Pruitt, and Robert J. Reilly. 1992. Market Responses to Publicly-provided Information: The Case of Automotive Safety. *Applied Economics* 24: 661–667.

Hubbard, Thomas N. 2002. How Do Consumers Motivate Experts? Reputational Incentives in an Auto Repair Market. *Journal of Law & Economics* 45 (2): 437–468.

Johnson, Eric. J., and J. Edward Russo. 1984. Product Familiarity and Learning New Information. *Journal of Consumer Research* 11 (June): 542–550.

Kaye, B. K., and T. J. Johnson. 2003. From Here to Obscurity? Media Substitution Theory and Traditional Media in an On-line World. *Journal of the American Society for Information Science and Technology* 54: 260–273.

Kim, J.-O., S. Forsythe, Q. Gu, and S. J. Moon. 2002. Cross-Cultural Values, Needs and Purchase Behavior. *Journal of Consumer Marketing* 19 (6): 481–502.

Klein, Lisa R., and Gary T. Ford. 2003. Consumer Search for Information in the Digital Age. *Journal of Interactive Marketing* 17 (3): 29–49.

Lally, Elaine. 2002. *At Home with Computers*. New York: Berg.

Lehmann, Donald L. 1977. Responses to Advertising a New Car. *Journal of Advertising Research* 17 (4): 23–27.

Maheswaran, Durairaj, Brian Sternthal, and Zeynap Gurhan. 1996. Acquisition and Impact of Consumer Expertise. *Journal of Consumer Psychology* 5 (2): 115–133.

Mannering, Fred, Clifford Winston, and William Starkey. 2002. An Exploratory Analysis of Automobile Leasing by US Households. *Journal of Urban Economics* 52 (1): 154–176.

Mizerski, Richard W. 1982. An Attribution Explanation of the Disproportionate Influence of Unfavorable Information. *Journal of Consumer Research* 9 (December): 301–310.

Moorthy, Sridhar, Brian T. Ratchford, and Debabrata Talukdar. 1997. Consumer Information Search Revisited: Theory and Empirical Analysis. *Journal of Consumer Research* 23 (March): 263–277.

Newman, Joseph W., and Richard Staelin. 1972. Prepurchase Information Seeking for New Cars and Major Household Appliances. *Journal of Marketing Research* 9 (August): 249–257.

Peterson, Robert A., and Maria C. Merino. 2003. Consumer Information Search Behavior and the Internet. *Psychology and Marketing* 20 (2): 99–121.

Pettigrew, K. E., and L. McKechnie. 2001. The Use of Theory in Information Science Research. *Journal of the American Society for Information Science and Technology* 52: 62–73.

Pettigrew, K. E., R. Fidel, and H. Bruce. 2001. Conceptual Frameworks in Information Behavior. *Annual Review of Information Science & Technology* 35: 43–78.

Punj, Girish N., and Richard Staelin. 1983. A Model of Consumer Information Search Behavior for New Automobiles. *Journal of Consumer Research* 9 (March): 366–380.

Ratchford, Brian T., and Narasimhan Srinivasan. 1993. An Empirical Investigation of Returns to Search. *Marketing Science* 12 (1): 73–87.

Ratchford, Brian T., Debabrata Talukdar, and Myung-Soo Lee. 2001. A Model of Consumer Choice of the Internet as an Information Source. *International Journal of Electronic Commerce* 5 (3): 7–21.

Reilly, Robert J., and George E. Hoffer. 1983. Will Retarding the Information Flow on Automobile Recalls Affect Consumer Demand? *Economic Inquiry* 21 (July): 444–447.

Savolainen, R. 2001a. "Living Encyclopedia" or Idle Talk? Seeking and Providing Consumer Information in an Internet Newsgroup. *Library & Information Science Research* 23: 67–90.

Savolainen, R. 2001b. Network Competence and Information Seeking on the Internet: From Definitions towards a Social Cognitive Model. *Journal of Documentation* 58: 211–226.

Savolainen, R., and J. Kari. 2004a. Placing the Internet in Information Source Horizons: A Study of Information Seeking by Internet Users in the Context of Self-Development. *Library & Information Science Research* 26: 415–433.

Savolainen, R., and J. Kari. 2004b. Conceptions of the Internet in Everyday Life Information Seeking. *Journal of Information Science* 30: 219–226.

Vanden Bergh, Bruce G., and Leonard N. Reid. 1980. Puffery and Magazine Ad Readership. *Journal of Marketing* 44 (2): 78–81.

Wellman, Barry, and Caroline Haythornthwaite, eds. 2002. *The Internet in Everyday Life*. Oxford: Blackwell.

Wetzel, James, and George Hoffer. 1982. Consumer Demand for Automobiles: A Disaggregated Market Approach. *Journal of Consumer Research* 9 (September): 195–199.

Xia, Lan, and Kent B. Monroe. 2004. Consumer Information Acquisition: A Review and an Extension. *Review of Marketing Research* 1: 101–152.

Yakel, E. 2004. Seeking Information, Seeking Connections, Seeking Meaning: Genealogists and Family Historians. *Information Research* 10, http://informationr.net/10-1/paper205.html.

Zimmerman, Linda, and Loren V. Geistfeld. 1984. Economic Factors Which Influence Consumer Search for Price Information. *Journal of Consumer Affairs* 18 (1): 119–130.

3 Informed Giving: Information on Philanthropy in Everyday Life

Barbara M. Hayes

Introduction

Every day, Americans are asked to engage in philanthropy. We open the mail and receive an invitation to a fundraising dinner for the local hospital. We answer the door for Girl Scouts selling cookies and teenagers selling books of coupons for carwashes for the soccer team. We see appeals on television to help earthquake victims in Haiti or fund famine relief in Somalia. We make online donations to teams of coworkers running fundraising marathons for cancer research. We bring donations to the church bazaar, which redistributes used items from the well-off to those in need.

Philanthropic activity is embedded in American culture and is so important to the U.S. economy that it is often referred to as the "third sector." An important economic engine in addition to the public and private sectors, it is known variously as the charitable, voluntary, tax-exempt, or independent sector. *The Nonprofit Almanac 2008* defines the third sector as "all 501(c) (3) organizations, including all religious organizations and congregations, as well as all 501(c) (4) organizations. The 501(c) (4) organizations are termed 'social welfare' organizations in the IRS code and include a disparate collection of organizations, including health maintenance and medical plans, civic leagues, many advocacy organizations and a range of others" (Wing, Pollak, and Blackwood 2008, 3). *The Nonprofit Almanac 2008* estimates that nonprofit institutions were responsible for 5 percent of the U.S. gross domestic product (GDP) in 2006. The United States relies on philanthropy to fund many of its social services and even some aspects of its foreign aid. It supports our religious institutions, our educational systems, our humanitarian aims, and our cultural lives.

Philanthropy has become part of everyday American life. Some acts of charity are impulsive and require very little information. Others are complex transactions that require a great deal of information and months to complete. We sometimes seek information about charitable giving, but more often, it seeks us. Information about philanthropy is shaped by a complex mix of social, economic, and political forces. Over time, those forces have produced new ways to engage in philanthropy.

This chapter will examine information about philanthropy in everyday American life. It will explore the following questions: What motivates us to engage in philanthropy and how does information affect our motivations? How have Americans, from colonial times to present day, obtained information about charitable giving? What information and communication technologies have been used to persuade them to give? What exogenous and endogenous forces have shaped information about charitable giving, and consequently, shaped American opinions and behavior around these topics?

Distinctions are sometimes made between charity and philanthropy. Over time and within the realm of organized fundraising, charity has come to be more narrowly defined as tangible assistance to the poor, while philanthropy has been defined as a broader collective effort to transform society through social change. When European colonists arrived in North America, the vast majority of giving could easily be categorized as charity: material relief for the needy. As the country grew and matured, we expanded our definition of philanthropy, accepting it as a means of defining and addressing broad societal issues. This kind of philanthropy may include politicized movements that redefine for us who and what are worthy of assistance. While these distinctions are important to formal study of the field, they are less important to this chapter because common (nonexpert usage) does not emphasize the distinctions. They are, therefore, less important to the everyday American making a decision to give.

Philanthropic Activity and Its Presence in Everyday Life

Philanthropy is tied to our social, cultural, and political identity, as a nation and as individual Americans. Many of us, rich or poor, espouse a cause at some time in our lives. We like to think of ourselves as generous, individually and collectively. Relative to many countries in the world, the United States is wealthy and the world looks to us to distribute some of that wealth in times of trouble. We expect the wealthiest among us to contribute to a degree commensurate with their financial success.

Pleas for charitable giving prompt many questions. Some are philosophical: Why should we engage in philanthropy? Should the needy accept help or "pull themselves up by their bootstraps?" Do we create or foster indolence when we assist others? Who deserves our philanthropy? Are we striving for an equitable and just society or do we believe "the poor you will always have with you"? Is it more important to save an endangered species or help flood victims rebuild their homes? Should we seek recognition or remain anonymous when we give?

Other questions are more practical: How much should we give? Should we give cash donations, in-kind donations, or volunteer? Should our philanthropic efforts go to public or private institutions? How do we avoid wasting money in poorly managed philanthropic efforts? How do we avoid outright scams? Are there advantages to the giving, such as tax relief? Should gifts be planned or can some giving be spontaneous? Should we leave our money to our children, create some kind of philanthropic legacy, or both?

Philanthropic activity is not well studied by social scientists, although the amount of scholarly literature is increasing.[1] The interests and behavior of wealthy donors and large foundations have received the most attention. There is scant literature on the giving behavior of less wealthy individuals in everyday American life and even less on information seeking as a part of that behavior.

One exception to the lack of information is the survey activity of The Independent Sector, a coalition of charities, foundations, and corporate giving programs. In 2001, the group published *Giving and Volunteering in the United States 2001* (Toppe, Kirsch, and Michel 2002). The study focused on the giving and volunteering habits of 4,000 Americans. It found that fully 89 percent of American households gave in 2000 and that those households gave an average of 3.1 percent of their income. The study also found that 44 percent of Americans volunteer. An example of ongoing research about the "average" American's giving behavior is the Indiana University Center on Philanthropy Panel Study (COPPS) that is following families' philanthropic behavior over the course of each family's life cycle.

Philanthropic giving tends to increase with wealth and therefore, to some extent, with age: "The amount that people give is more closely related to their income than to other demographic characteristics such as age, marital status, or race/ethnicity. Within each income range, the percentage that gives increases with age, as does the average amount donated. Unlike age, marital status and race/ethnicity do not appear systematically related to charitable giving once income differences are taken into account" (Wing, Pollak, and Blackwood. 2008, 76).

The evidence indicates that giving is an extremely common activity in everyday American life. Americans gave a record $314.07 billion in 2007[2]

to charitable organizations or 2.3 percent of the GDP (Giving USA Foundation 2008). Most of that amount—a full 75 percent—was the result of
individual giving. Over the course of 2008, the United States slid into
recession. American giving dropped to $307.65 billion, a 2 percent drop
in current dollars over 2007 and the first real drop in giving since 1987
(Giving USA Foundation 2009, 1). Although philanthropy represents a
relatively small segment of the GDP, these billions are extremely important
to our quality of life.

Motives for Philanthropy

Perhaps the most basic discussion of philanthropic motivations starts with
whether there is an endogenous drive to be altruistic. Altruism may be
defined as an unselfish regard for the welfare of others. When acting altru-
istically, we assume there is no reward or recognition for actions taken to
safeguard or improve the welfare of others. For the first three hundred years
of American history, we assumed that genuine altruism exists and that
people make choices to be altruistic of their own accord. Is there true altruism or is altruistic behavior genetic, existing to ensure the viability of a
community or the continuation of the species? Today, there is a growing
literature in social psychology, social biology, evolutionary biology, and
neuroeconomics that investigates whether there is a biological basis for
altruism (Okasha 2008).

If there is brain-based altruism, we would then have a biological imperative to seek information about philanthropy and to engage in giving, at
least on an intermittent basis, to ensure our own survival and the survival
of our communities.[3] If we accept that there is an impulse to give, there
are still a multitude of individual beliefs, social cues, and social connections that mediate and complicate the decision to give. A "hardwired"
impulse to give is just the beginning of analyzing an individual charitable
act—say, for example, a shopper's decision to toss a few coins into a
Salvation Army kettle at Christmas.

In this instance, the shopper did not seek information about giving.
Rather, her act was situational. It may have been prompted by the sight
of the familiar kettle, by a sense of holiday generosity, or even by concern
about how she might be judged by other shoppers if she walked by without
giving. She had no real sense of saving the species at that moment. She
may have acted to avoid embarrassment or to have a moment's worth of
"warm glow." "People don't just give money to save the whales—they give
money to feel the glow that comes with being the kind of person who's

helping to save the whales" (Leonhardt 2008, 47). Researchers in economics call this motive "impure altruism" (also known as the warm glow motive). It benefits the donor by allowing him to think of himself as a good person, to show off his wealth, or to reduce his guilt (Andreoni 1989, 1990).

Some of our decisions to give, even if they are not emotional or impulsive ones, do not appear to be particularly rational. For example, we are more likely to give money to assist a single needy individual than we are a group of people, even if that money would have far more impact when applied to the group (Small, Loewenstein, and Slovic 2007). In fact, being presented with two children in need (rather than one) actually decreases the likelihood of giving (Slovic 2007). Writing about why good people have repeatedly ignored mass murder and genocide, Slovic hypothesizes that we have cognitive limitations that make it difficult to conceptualize the suffering of large numbers of people. Those who create information campaigns to generate interest in philanthropy are well aware of the emotional response that individual stories can generate.

At other times, our giving behavior involves rational self-interest. We give to improve the communities we live in by supporting libraries, building hospitals, creating public parks, or funding drug addiction treatment programs to improve public safety. We contribute to charitable causes to make sure that we too will have adequate social services should we need them. This kind of "reciprocal altruism" provides a social safety net.

On occasion, we participate in giving because we feel disenfranchised by the larger society and want to effect social change. In this sense, philanthropy may be seen as a form of civic engagement. "'Giving is not about a calculation of what you are buying . . . It is about participating in a fight.' It is about you as much as it (is) about the effect of your gift" (Karlan, quoted in Leonhardt 2008, 47).

American tax law also encourages charitable contributions as a way to save money. Although very wealthy individuals do not endorse tax savings as a motivation (Rooney and Frederick 2009), tax advisors often urge last-minute donations as a way to ease tax burden. When tax rates are higher, deductions are more valuable. Money affects giving in other ways. Researchers have studied the effects of providing matching funds for charitable gifts (Karlan and List 2007); the effects of providing seed money and refunds (List and Lucking-Reiley 2002); the psychology of lump sum matching versus one-to-one matching (Baker, Walker, and Williams 2009); the effect of anonymous giving on the amount given (Soetevent 2005); and matching funds versus rebates (Eckel and Grossman 2003).

Motives for philanthropy include group affiliation and inclusion. They are linked to our relationships to God or morality; to family, community, and country; and finally, to our own identities. Our religious institutions instruct us in our duty to give. We see our parents volunteering at school and donating to the United Way at work. Our civic leaders encourage us to make our communities better. We absorb this information, add personal experiences as we grow, and over the course of our lives, fashion highly individualized philanthropic "identities." We make decisions about the importance that philanthropy will have in our lives, from writing a few checks each year to careers in not-for-profit organizations. Philanthropy often brings people together and can even provide opportunities for recreation through charity balls, fundraising auctions, and walkathons.

The various motives we have for giving direct us to different sources of information. For example, we may seek the obituary of a friend so that we can make a contribution to his or her favorite charity. We may seek information from the Alzheimer's Association after a parent's diagnosis and subsequently make a donation to medical research into the disease. Sometimes we respond to information we encounter passively: the disturbing images and stories broadcast after the 9/11 disaster prompted millions of Americans to search the Internet for information on ways to help those affected by the attack. Motivations shape where information is sought and determine its relevance to the task at hand.

Younger individuals may be more motivated to change society for themselves and their children. Older individuals may be more interested in leaving a legacy than changing the world; they may seek relationships in their giving or wish to impart the value of giving to their children. "Young or old, when people perceive time as finite, they attach greater importance to finding emotional meaning and satisfaction from life and invest fewer resources into gathering information and expanding horizons" (Carstensen 2006, 1915). A professional fundraiser might well develop different information approaches when appealing to different age groups.

Philanthropic organizations today study potential donors carefully and explore every avenue of communication that might engage them. In effect, they create elaborate information systems. These professionals have access to psychological research, marketing research, the growing literature in philanthropy, and, more recently, to research in information-seeking behavior. They design campaigns that take into account such behavior and diverse motivations of different demographic groups. They pursue the question: "Is there a "psychology of giving?"

In sum, a decision to engage in philanthropic behavior may be motivated by dozens of considerations. It is informed by a variety of sources
and is embedded in social relationships and social context. Our motives
and our social connections shape where we seek and encounter information about giving. As individual wealth increases and our social, economic,
and political environments become more complex, the factors in any decision to give multiply. More information and more information sources
may be required: perhaps more information than the average American
can reasonably sort and act on.

Historical Forces and Their Influence on Information about Philanthropy

The next several sections of this chapter focus on historical events and
forces, both exogenous and endogenous, shaping Americans' access to
information about philanthropy and ultimately, their view of appropriate
philanthropic objects and goals. Information about philanthropy almost
always is presented in a social context. In information science, this context
is well described by research into "small worlds" and "lifeworlds." Small
worlds describe the interconnections we feel with the small social groups
of which we are a part. We have much in common, for example, when we
are interacting with other scholars, or parents, or workers, or athletes, or
volunteers. We have more in common if we live in close geographic proximity, grew up observing these activities, and assumed these roles over
time. These small worlds may have dense information connections or quite
limited ones.

The concept of "lifeworlds" implies a broader context. "A lifeworld . . .
is the collective information and communication environment—the social
tapestry—of a society, as information and communication continue to tie
everything more closely together in the modern technology-driven environment" (Burnett and Yaeger 2008, 4). Information that arises inside our
small worlds may be more important than information that comes from
the larger society. If information from the larger society threatens relationships inside our small worlds, we may be less likely to act on it (Spink and
Cole 2006).

Exogenous and endogenous historical forces have changed the way
charitable giving is viewed over time. They shaped the small worlds and
lifeworlds of American colonists just as they shape the small worlds and
lifeworlds of modern Americans, whose reach is dramatically extended
by technology. Philanthropy in everyday life and in historical context

provides a rich opportunity to explore these concepts. We will begin that exploration with religion, important in colonial times and important today.

Religion

Religion is an exogenous force that has shaped giving behavior throughout American history. Religious institutions both support and derive much of their funding from the activity of philanthropy. Among the various categories of giving, such as education, healthcare, and humanitarian aid, in 2007 religious organizations received the largest share, $106.89 billion,[4] 35 percent of the total given (Giving USA Foundation 2008).

Religious training is so closely tied to philanthropy that it could reasonably be considered an endogenous factor affecting information dissemination as well. Most of the world's religions and spiritual philosophies maintain that we have a moral obligation to assist the poor and sick. A number of rationales for individual philanthropy flow from religious beliefs. Many Americans accept the idea that acts of philanthropy fulfill God's commandments. Even if they are not particularly religious, others give out of sense of moral duty.

European colonists brought ideas about charity and philanthropy with them to their new homes in North America. Many also brought idealized views of a communal society and a strong ethic of reciprocity. Indeed, the Native Americans groups they encountered practiced gift exchange, a form of reciprocity. Gift exchanges with Native Americans ensured the survival of some early colonists.[5]

Religious institutions and religious leaders have been information brokers throughout American history. They bring American worshippers, rich and poor, information and exhortations about the importance of giving in everyday life. Americans who are actively affiliated with a religious institution may be exposed to information about philanthropy on a weekly or even daily basis. Presumably, Americans who attend religious services have established some level of trust with their place of worship. Individuals may be more likely to act on information about philanthropy when it is obtained through this trusted connection.

The sermon has been the most common medium chosen to disseminate religious teachings and their philanthropic corollaries. The sermon is useful for a number of reasons. When delivered from a pulpit, sermons can be understood by both literate and illiterate populations. Sermons may be delivered spontaneously or on a regular, recurring basis. They lend themselves to emotional appeals (impulsive giving) and to rational appeals (improving the community over time and sustaining the congregation as an important community resource). Sermons can be recorded in writing and circulated to further the spread of ideas. Today, we capture them in audio and video formats and post them on church Web sites.

The early colonists typically obtained information about the need to give and how to give through close association with their neighbors and through sermons, pamphlets, and the occasional book. Colonial ministers, who were often literate, were respected community leaders. They influenced their congregations and shaped attitudes toward giving. One example is Cotton Mather, the well-educated, influential minister whose life bridged the seventeenth and eighteenth centuries. His book, *Bonifacius: An Essay Upon the Good by Cotton Mather* (Mather 1966), published in 1710, taught that philanthropy is not only a duty to God, but also a reward in and of itself. Mather made distinctions between the deserving poor and the undeserving poor. He believed the undeserving poor (whose poverty he judged to be a result of their own actions) should receive indirect assistance: instruction, encouragement, carefully designed incentives to work, and sometimes stern correction. Mather's life and interests illustrate the fact that religious leaders, then and now, act as filters of information about philanthropy.

As the country expanded, the nature and character of individual American engagement in religion began to change. There was tension between the secular thinking in the young Republic and its religious traditions. The Great Awakening of the 1730s and 1740s and the Second Great Awakening that lasted until the end of the Civil War "rank as the largest, strongest, most sustained religious movement in U.S. history" (Butler, Wacker, and Balmer 2008, 172). The revival emerged from this era as a method of connecting Americans with God on a more direct, intimate level than might otherwise occur through ordinary church attendance. Revivals were great theater and became another important way to encounter information and absorb attitudes about philanthropy.

Orator-ministers such as Jonathan Edwards and George Whitefield helped spark the Great Awakening. They inspired large numbers of "unchurched" Americans through emotionally charged conversion experiences. Itinerant preachers traveled the countryside, delivering powerfully affecting appeals. An American of that time who experienced a personal religious conversion was more likely to accept charity as an individual responsibility rather than a communal one. American families did not go to revivals seeking information about philanthropy, but they received it anyway. Whitefield raised money for a wide array of causes, including an orphanage, disaster relief, higher education, and other good works. Revivals subsequently became an important component of reform movements.

The Benevolent Empire, a phrase used to describe the activity of large numbers of voluntary associations that sprang up between 1790 to 1840, attracted significant numbers of Americans interested in societal reform. Women and the clergy were among them, as described by Friedman and MacGarvie (2003, 206): "During the 1830s, two groups of Americans that were disenfranchised in varying degrees in the early republic combined to assert their political wills in private actions. Women, unable to vote, and the clergy, stripped of its political authority during the process of disestablishing state religions, advocated . . . an alternative societal ideal in which slavery, alcoholism, disease, poverty, illiteracy, penal suffering, and crime would be eliminated through public attention to social problems." These two groups found that they could wield considerable influence when they joined together in voluntary associations. Women, in particular, were believed to be more sensitive to suffering and therefore suited to charitable work. The first female charitable society, the Female Society for the Relief of the Distressed, was formed in Philadelphia in 1795. Association work gave middle- and upper-class women the opportunity to socialize and spend time out of their homes. They began to encounter information about philanthropy in their own small social groups. Gradually, women's charity work became an outlet for women that was sanctioned by society. Social norms hold small worlds together (Burnett, Besant, and Chatman 2001). It became normal and even expected that middle- and upper-class women engage in some kind of charitable work.

Women's most important donation was their time. If they had little money of their own, they could donate their work. Women gained a degree of respect in their communities through their volunteerism. A few even found careers in the reform movements. Depending on the size of the association, they learned to manage major bureaucracies. Even though women did not have the vote, they could still petition the government for a redress of grievances. "Groups of women used their volunteer time to create 'parallel power structures' to the political and for-profit sectors controlled by men, and to carve out a significant place for themselves in the public arena" (McCarthy 2003, 179).

By the mid-1800s, reform organizations had become sophisticated "media machines," employing many forms of media to extend their vision of civilized society to the West and, to some extent, the South. They created information systems for their causes that were designed to inform people whether they wished to be informed or not. They brokered information designed to transform society.

Tracts were the principal mode of disseminating information encouraging temperance and discouraging prostitution. In a sense, it was information warfare. Evangelizing tract societies first organized locally, then combined into what was called "the nationwide American Tract Society" in 1825. "The sheer number of pages that its partisans distributed without charge was startling by any standards. Between 1829 and 1831, for example, the American Tract Society turned out sixty-five million pages of tracts, five pages for every person in America" (Butler, Wacker, and Balmer 2008, 188). Just as the colonists had felt justified in converting Native Americans, the reform organizations had no qualms about dictating their view of civilized society.

One such group, the American Temperance Society (ATS) grew from local roots in Boston into a nationwide organization. Traveling agents of the ATS distributed written tracts and gathered donations from congregations, granting honorary membership to those who contributed (Blocker and Tyrrell 2003, 42). Although the temperance movement is a historical footnote today, "in the antebellum years at least, it attracted the largest, most diverse collection of supporters of any reform. They ranged from shy, pious churchwomen to militant feminists, from free thinkers to fundamentalists, from the high and mighty to the low and degraded" (Walters 1978, 123–124). Reform organizations used many forms of media for moral suasion, including newspapers, pledges, plays, songs, and novels about the ravages of alcohol.

The philanthropic efforts of American religious groups, whether liturgical or evangelical, have continued to grow over time. They played an important role in relief efforts during the Civil War. During World War I and World War II, they provided information about and resources to devastated populations overseas. The social programs of the New Deal and the Great Society have made it less necessary for religious groups to provide subsistence relief in the United States. However, because they can act more quickly than government bureaucracies, they continue to provide important "stop-gap" funds for individuals suffering sudden disruption (e.g., disaster aid, assistance for the suddenly homeless, or aid for battered women).

Religious organizations have expanded their role in subsistence relief in other countries. When President Truman outlined a campaign against world poverty in 1949, there were already hundreds of religious organizations operating various types of medical, agricultural, and educational missions in multiple countries. Missions and missionaries, building blocks of many religions, are powerful agents for philanthropy. While some still travel overseas to proselytize, others are content with providing tangible relief and alleviating human suffering through education and culturally sensitive programs. They can offer firsthand information and a sense of emotional connectedness when communicating with donors interested in international relief (using such personal language as "our mission in Kenya"; "our youth trip to Guatemala"). By the 1970s, religious organizations such as the Church World Service, Catholic Relief, and the American Jewish Joint Distribution Committee were the means through which the U.S. government distributed many of its donations for foreign aid (Bremner 1988, 197).

In addition to information about philanthropy derived from sermons, revivals, tracts, and missions, American religious organizations now employ modern media to disseminate information about giving: by radio, beginning in the 1920s; by television, beginning in the 1930s but not widespread until after World War II; and by the Internet, beginning in the 1990s and progressing to support streaming video, podcasts, and other interactive media. One of the pioneers of radio ministry was S. Parkes Cadman. The first of the radio pastors, "he reached the ears of millions" (New York Times 1936).

Fulton John Sheen was a Catholic bishop who made a successful transition from radio ministry to television in the 1950s. By 1956, his *Life is Worth Living* television show reached 30 million people each week (Catholic Post 1979). By the end of his active career, the American evangelist Billy Graham had enormous influence, extended in part by his active use of radio and television.

Sophisticated, modern-day televangelism is closely linked to the rise of the "megachurches." These churches, such as the Crystal Cathedral in Garden Grove, California, and the Lakewood Church in Houston, Texas, use media to expand their influence far beyond their local communities. "These churches' use of the Internet was nearly 100 percent throughout this eight-year period (2000–2008), but our experience suggests that more megachurches are adopting web based streaming media to broadcast their message than the more costly radio and TV approaches" (Thumma and Bird 2008, 9).

The Crystal Cathedral broadcasts the popular television show *Hour of Power*. The church's Web site allows users to subscribe to video broadcasts and to receive email alerts about upcoming guests. Joel Osteen, pastor of the Lakewood Church, provides weekly podcasts and streaming video, and allows users to sign up for daily devotional emails. Religious institutions receive more donations than any other type of nonprofit endeavor, but the share going to religion peaked in 1985 and has been declining steadily since (Wing, Pollak, and Blackwood 2008, 72). All of the megachurches employ sophisticated multimedia techniques to reach their congregations and conduct fundraising.

Involvement in giving through a religious institution does appear to have an impact on giving overall: "Americans who give to or volunteer with religious congregations give more time and money than those only involved in secular charitable activities" (Toppe et al. 2002, 8).

In sum, Americans today continue to rely on their religious institutions as important brokers and filters of information about philanthropy. We still listen to sermons, but some information tools have dropped by the wayside: the tracts, pledges, plays, and novels of the revival movement and the revival itself have certainly receded in importance. Young Americans are contributing their knowledge of new and emerging communication tools—audio, video, and the Internet—to transform both their religious services and charitable activities.

Americans who practice a religion no longer rely on their place of worship as their only source of information about philanthropy. Individuals and families also seek and receive information from many secular organizations that wish to promote good works and social reform. In fact, a large number of Americans now look to religious institutions for philanthropic activity that supports traditional values more than social reform. Organized religion and organized religious societies are perhaps the most widespread and consistent information mediators in existence today. They inform members and constituents about their duty to give and also help to match those in need with those who want to give.

War and Depression

Like religion, war and economic crises are exogenous forces that change the nature and quality of information about philanthropy and our engagement in it. The previous section suggests that religion and religious institutions act as brokers of information about philanthropy. War and economic crises lessen the need for such brokers. When they occur, we are more likely to encounter firsthand and even face-to-face information about needs for assistance. Indeed, Americans may find themselves personally acquainted with the effects of war or serious economic downturn.

These events, which sometimes last for several years, tend to alter our patterns of giving in predictable ways. We have touched on the notion of the deserving and undeserving poor. In times of plenty, that information filter is important to decisions about philanthropy. Americans who want to give in "normal" times look for information and information brokers who can direct their giving to worthy recipients and causes. War and economic depression lessen these concerns. We know people like us who are affected. If one is caught up in war or jobless through no fault of one's own, distinctions between the deserving poor and the undeserving poor become less meaningful.

The chaotic conditions inherent in war and economic crises create a number of information problems. Normal patterns of communication are disrupted. Reliable information is difficult to get, particularly early on. It may take several months to comprehend the depth of the crisis and the types of people affected. The severity of need overwhelms private giving and additional help from government is required. Government information systems must grow or be created in response to the crisis. Private charitable groups may be eager to provide assistance, but need information on how to do so effectively.

During the major wars in American history, the war effort has taken precedence over other philanthropic concerns. During the misery of major depressions, dollars have been redirected to the subsistence needs of those most affected. These far-reaching events change the information culture around charitable giving, disrupting routine habits of giving and redirecting them. They upend the prevailing social order and alter the norms of both our small worlds and lifeworlds. They blur the lines between upper- and under-classes. They change philanthropy's focus from the poor and the sick to the suffering of entire societies.

For example, the early colonists found that the New World could be harsh. There was little real wealth and little stratification of social class in the seventeenth century. Some people owned land and other people were servants, but every colonist lived on a thin edge. A settlement could be wiped out in a season or an instant. Neighbors helped neighbors so that all could survive. Settlers pulled together to help the displaced and injured who arrived on their doorsteps during the four French and Indian wars spanning 1689 to 1763. Refugees were taken in. Fatherless families were given assistance. There was little separation between the needy and the rest of the community. There were no institutions; no writing of checks. Assistance was tangible, in the form of gifts of food, clothing, shelter, and nursing care.

The need for war relief was similarly obvious during the Revolutionary War. Although reliable information was difficult to obtain (the usual mode of communication was the horse and rider who sometimes had to cross enemy lines), colonists often learned about need when the war touched them personally. The war could appear suddenly in the next town. Individuals, churches, and local governments joined forces to do what they could. Many authors have noted this public-private partnership (and consequent mixing of sources of information) as an important factor in the evolution of American philanthropy.

The resources of private philanthropy are rarely adequate to deal with war and major economic crises. These events require large-scale fundraising efforts as well as the rapid organization of new information and material infrastructures to manage massive relief operations. For example, relief agencies proliferated during the Civil War, but were unable to coordinate effectively. President Lincoln addressed some of the disorder by creating the United States Sanitary Commission. The commission coordinated information and organized thousands of independent efforts to provide relief. Its work foreshadowed the work of the American Red Cross (Bremner 1988, 78).

African Americans had had few resources and few opportunities to participate in philanthropy or any kind of social activism before the Civil War.[6] After the war, private philanthropic groups proved unequal to the task of providing food, clothing, and medical aid to thousands of former slaves and poor whites who had lost their homes during the war. At the request of a number of freedmen's aid societies, President Lincoln established the Bureau of Refugees, Freedman and Abandoned Lands to coordinate information and efforts.

When federal support diminished after the initial years of the Reconstruction, African Americans took up the cause of black education in the South: "The African-American churches . . . established and maintained hundreds of secondary schools and colleges in the decades after the war" (Finkenbine 2003, 166). African American churches have been and continue to be an important mediator informing African American philanthropy. African Americans had vastly different small worlds and lifeworlds before and after the Civil War. Over time, African American participation in philanthropy evolved to include the social activism of the 1960s.

Post war, as the country began to change rapidly from an agricultural economy to an industrial economy, painful economic dislocation occurred. The depressions of 1873 to 1878 and the 1890s completely overwhelmed the infrastructure of private philanthropy, but the government response was not as coordinated as when the country was clearly focused on a war effort. Suffering was widespread and severe. The depression of 1893 to 1897 was a jolting setback (Hoffmann 1970, 272). Railroads failed, banks failed, and millions were unemployed.

World War I was not fought on American soil, but the scale of the war prompted the contribution of astonishing sums of money. The Red Cross stepped to the forefront, organizing and distributing fundraising information. The fundraising campaign was orchestrated to convince every American that giving to the Red Cross was a patriotic duty. The Red Cross tapped Charles S. Ward, the YMCA's best fundraiser, and seventy-five YMCA secretaries to organize thousands of volunteers who prepared posters, advertising, and publicity releases (Bremner 1988, 124).

Assisted by the government's war council, the Red Cross surpassed its goal of raising $100 million in one week. Membership in the organization soared. Ward and his fellow fundraiser Lyman L. Pierce are considered the originators of the modern fundraising campaign, the elements of which were evident in theirs: "careful organization, picked volunteers spurred on by a team competition, prestige leaders, powerful publicity, a large gift to be matched by the public's donations, careful records, report meetings, and a definite time limit" (Cutlip 1990, 44).

To raise war funds, the government deliberately undertook the promotion of an information culture that made giving a social norm. It used posters advertising Liberty Bonds, another important fundraising vehicle during World War I. It tapped silent-movie stars to host bond rallies; Mary Pickford, Douglas Fairbanks, and Charlie Chaplin helped to popularize the idea of bonds. In 1918, Chaplin made a film, *The Bond*, which used a series of short sketches to sell war bonds (Chaplin 1918).

When the Great Depression began in the United States, President Herbert Hoover relied heavily on private efforts by the Red Cross and the Community Chests, federated fundraising organizations that developed toward the end of World War I. Community Chests were often organized by city or region. They combined fundraising activities for multiple organizations into single overarching campaigns. Hoover's Organization for Unemployment Relief mounted a carefully planned fundraising campaign in the fall of 1931. It employed a number of communications media to prompt the average American to give: "Movie theaters and college football teams gave benefit performances. Radio broadcasts carried messages of inspiration and hope to millions of homes. Advertising agencies contributed a series of high powered advertisements which appeared, free of charge, on billboards and in newspapers and magazines" (Bremner 1988, 139).

Hoover resisted tax-supported relief, but the enormity of the Great Depression overwhelmed him. President Franklin Roosevelt legitimized large-scale governmental response to the emergency by focusing not on its morality, but on its practicality. He told the American people that it was necessary for business recovery; therefore, giving generously would be a wise investment. Although there was concern about "big government" and paternalism, Roosevelt pushed ahead, implementing new tax systems requiring that the employed pay more to support the needs of the unemployed or dependent. Later in the century, the Nixon and Ford administrations tried to reverse the trend by attempting to control the growth of social programs and putting additional emphasis on private philanthropy.

Once again, during World War II, the pendulum swung from private philanthropy to federal organization. At the start of the war, organizations fell over each other trying to help. Utilizing professional fundraising expertise, President Roosevelt's War Relief Control Board took charge of both foreign and domestic appeals. The board scheduled national appeals, preventing competition among the agencies and raising huge sums of money. The World War II United War Fund is an example of a successful public and private fundraising effort that combined appeals for the usual concerns of the Community Chests and the war effort. It helped to secure monies for Community Chest projects that did not have the glamour of the war effort.

There have been a number of economic crises in the years after World War II, but none have lasted as long as the Great Depression. As new social programs are put in place, Americans have had to worry less about their subsistence needs. The smaller conflicts of Korea, Vietnam, Afghanistan, and Iraq have not engaged American philanthropic "muscle" in the same way that the world wars did, and there has been a clear rise in skepticism about when government provides information about conflict or war. While there was an outpouring of donations to support the victims of the 9/11 attacks, in general, private philanthropy efforts have not been directed toward fighting Al Qaeda, an enemy that transcends national boundaries.

In 2008 and 2009, the United States began suffering a recession of significant magnitude. At this time of this writing, private philanthropic groups are concerned about dwindling contributions. Once again, the federal government has stepped in to stimulate the economy, this time with the American Recovery and Reinvestment Act. Interestingly, President Barack Obama's 2009 inaugural address to the American people echoed Roosevelt's justification for government intervention: "The success of our economy has always depended not just on the size of our gross domestic product, but on the reach of our prosperity; on the ability to extend opportunity to every willing heart—not out of charity, but because it is the surest route to our common good" (Obama 2009).

In sum, war and economic crisis are powerful "levelers." The average American has less need for information brokers because he or she encounters information directly from family, neighbors, and coworkers who are affected. The information resources utilized in these situations have changed over the course of our history. Rather than a slow trickle of information gathered at church or in the town square, we are able to turn to television and the Internet to get immediate information on philanthropic efforts. Another change is that we tend to receive and evaluate that information alone rather than in groups. Most recently, Americans, particularly younger Americans, have used the Internet to respond generously in a crisis: donors large and small contributed millions to assist those affected during the 9/11 attacks.

When war takes place on American soil, our information lifeworlds change dramatically. Although we looked to our local communities historically, we now look to the federal government for information on how we can help ourselves and our neighbors. After televised accounts of the Vietnam War, Americans became somewhat more skeptical of government information about war, particularly wars fought overseas. We no longer see the type of patriotic information campaigns promulgated during World Wars I and II. Movies and serials used to popularize war efforts have disappeared. In addition, we now have nearly unfettered access to information from foreign regimes via video sites like YouTube, blogs, and other Web-based applications. Many Americans now struggle to sort out the reliability of information about our conflicts, trying to discern reliable information from propaganda.

Information may be more difficult to obtain in an economic crisis than it is during a war. We are not all affected equally in an economic downturn. Some of us are deeply affected; some of us are unscathed. The colonists were largely dependent on their own small communities, so they did not have to go far to get help in hard times. Today, we must negotiate the various social programs that have been put in place as economic safety nets if we need help. If we do not qualify for federal or state assistance, we may need to ask "what charity should I go to?" Even though the government and private philanthropic groups have gained a great deal of expertise in managing economic crises, they still create chaos and make reliable information difficult to obtain, at least in the first few months. They still overwhelm private philanthropy, necessitating intervention by government and new governmental information systems.

The Professionalization of Philanthropy

Having examined forces exogenous to philanthropy, we now consider an endogenous force that has shaped information about American giving: the professionalization of philanthropy. Today, students can take courses in philanthropic studies. Professional fundraisers run sophisticated campaigns. Executives trained in nonprofit management run large foundations. Social workers, skilled in helping the disadvantaged, are licensed in the states in which they practice. One can be paid for efforts to identify important societal causes, raise money to ameliorate them, and work to alter the conditions that spawn social problems. While we are accustomed to such occupations today, these remunerated professions are relatively new in the United States, coming into existence at the turn of the twentieth century.

Why are these occupations important to individual American giving? The individuals in these occupations devote their working lives to creating the information-rich "culture" of philanthropy that surrounds us and permeates our everyday lives. They seek to educate and influence us. These are the people who create the appeals, develop the Web sites, make the phone calls, and systematically study our every impulse to give. They do everything they can, on a daily basis, to reinforce expectations that American should give and give often. As is the case with religious institutions, professional philanthropic workers and their organizations now act as powerful brokers of information about giving.

After the deprivations of the Civil War and the Reconstruction, in the late nineteenth century there was a gradual return to the view that charity had the potential to weaken its recipients. As more communities funded institutions for orphans, the mentally ill, blind, and deaf with public monies, the demand for wise financial management increased. Boards of volunteers were chosen to oversee the management of institutions. A new information "filter" emerged. Philanthropy now needed to become more "scientific," or carefully managed. The attitude grew that organizations that were not scientifically managed should not receive funding.

Calls for scientific philanthropy resulted in the formation of Charity Organization Societies (COS). These societies were characterized by exacting business methods. They screened applicants for assistance and attempted what they regarded as the moral uplifting of the poor. Initially, the screening of candidates for relief (casework) was carried out by unpaid workers. "Friendly visitors," usually middle- and upper-class women, were recruited to call upon the disadvantaged. However, paid charity workers began to appear during the depression of 1873 to 1878 as private organizations became overwhelmed by the numbers of people who needed help. By the 1890s, salaried workers began overseeing the volunteers. Soon, there were enough of these individuals to suggest a new profession.

At about the same time, the U.S. population began a major shift from rural and small towns to urban areas, which also accepted a large influx of immigrants. "Vast urban slums appeared, accompanied by high population density, horrendous health conditions, rampant crime, and a loss of the sense of community that characterized the small town of the past" (Ehrenreich 1993, 2231). Conditions in the cities were so bad that Edward Devine, general secretary of the New York Charity Organization Society, was prompted to say: "We may quite safely throw overboard, once and for all, the idea that the dependent poor are our moral inferiors, that there is any necessary connection between wealth and virtue, or between poverty and guilt" (Trattner 1998, 102).

Early social workers helped the urban immigrant populations who suffered during the depression of the 1890s. Paid workers gained face-to-face knowledge of the poor. They learned that some people were in trouble or destitute through no fault of their own. Inevitably, many of these workers became interested in issues of social justice and began pursuing social change. In 1897, Mary Richmond, a leader in the charity organization societies movement and an early founder of social work practice, called for the formation of a "school of applied philanthropy" to train such workers (Ehrenreich 1993, 2237). It is generally agreed that American social work traces its educational roots to 1898, when the first formal courses in "social philanthropy" were offered under the auspices of the Charity Organization Society of New York City (Feldman and Kamerman 2001, 1).

Rather than continuing to separate themselves from the poor, early social workers responded by starting the settlement house movement, which spread throughout the country in urban areas. Settlement workers began to examine the causes of poverty. These workers began to promulgate a broader definition of "scientific" philanthropy (broader than good fiscal management). They conducted surveys, gathered data, and hypothesized about society's failings, rather than moralizing about individual deficiencies. They challenged the idea that "the poor will always be with us" and suggested that perhaps the causes of poverty could be eliminated. They fought for better education, labor laws, and healthcare. Americans now had the option of engaging in "palliative" charity, transformative philanthropy, or both.

At nearly the same time as the arrival of the professional caseworker, professional fundraisers and fundraising consultancies began to appear: "just before the turn of the century the paid solicitor for charitable funds appears, one on whom great reliance was to be placed by philanthropic and eleemosynary agencies for the next two decades" (Cutlip 1990, 16). In 1919, two veterans of World War I fundraising, Harvey J. Hill and Charles S. Ward, opened Ward & Hill Associated, a fundraising firm to serve multiple clients such as colleges and hospitals on a paid basis. "By the end of the 1920s, there were twenty firms of this sort in New York City alone" (Bremner 1988, 133).

"There were few organized drives, in the modern sense, before 1900. World War I and the decade that followed provided the seedbed for the growth of today's fund raising and today's people's philanthropy" (Cutlip 1990, 3). The success of the war drives during World War I led to the establishment of Community Chests, which provide an excellent example of the impact of professional fundraising. Community Chest fundraising allows several philanthropic organizations to join together for appeals, thereby reducing the number of appeals and splitting the proceeds among organizations that have been prescreened for merit and good management.

The Community Chest approach provides both a source of information about charitable organizations and a way to filter that information. This approach is advantageous to the average American for a number of reasons: it decreases the number of appeals, reducing the amount of information that must be gathered; it prescreens charities to ensure they are well run; and it provides a single, easy way to give. The Community Chests became the United Fund in the 1950s, and the United Way in the 1970s.

A culture of giving at work has grown up around the United Way solicitation. Today, many business and industry leaders sponsor United Way fundraising in the workplace. Executives set an example of giving for their employees. Many companies provide matching dollars for employee donations. As much as possible, United Way fundraisers try to make giving at work an annual expectation. They promote such giving as evidence of civic commitment. Similarly, the federal government permits one charitable solicitation among government employees and military personnel each year, for the Combined Federal Campaign (CFC).

One drawback to this approach to fundraising is that newer charities or causes that are not yet fully accepted by the United Way do not get the same opportunity to raise funds. The United Way tends to support the status quo. This kind of combined fundraising has both a positive and negative impact on information about philanthropy when it limits choice and access to information about new groups.

As philanthropy has become a profession, it has found a voice. There are educational programs available in social work and philanthropic studies. The numbers of fundraising professionals has grown enough in the ensuing decades to support a nationwide Association of Fundraising Professionals and an Association of Donor Relations Professionals. There are now two organizations that undertake the study of philanthropy: the Association for Research on Nonprofit Organizations and Voluntary Action (ARNOVA) and the International Society for Third Sector Research (ISTR). Both organizations publish journals. This cadre of educated professionals in our midst strongly influences Americans' access to information about philanthropy and their willingness to engage in it.

In sum, the professionalization of philanthropy has dramatically increased the richness and reach of information about charitable giving. The average American now receives multiple, sophisticated fundraising appeals every year through every conceivable communication channel. No matter what our demographic group is, we are studied, profiled, and targeted with appeals designed to increase our likelihood of giving. Bridges are built to our small information worlds and efforts are made to redefine our lifeworlds so that frequent, repeated giving becomes the social norm.

Repeated appeals increase the risk that we will become immune or overwhelmed by societal needs. They may even reduce our own spontaneous attempts to seek information about giving. If we throw away an appeal that arrives in the mail, we know another one will follow shortly. This is information that seeks us.

Professional fundraising both expands and limits our information resources. Over time, demands for professional management and screening of charities have prompted the building of information brokers like the United Way. Many Americans now rely on such agencies to filter information and aggregate appeals. These umbrella organizations pre-screen charities for busy Americans and offer a "seal of approval" for some and not for others. As a result, they limit our access to new or unproven charities as well as poorly managed ones. Once, we heard appeals primarily from our religious institutions. The professionalization of philanthropy has now helped to define both workplace and home as additional venues for soliciting philanthropic dollars. We are asked to commit to lifelong relationships with favorite charities and even include them in our wills.

Reformers and Citizen Philanthropists

It is difficult to characterize the effect of the opinions and actions of society's influential thinkers and wealthy leaders as either endogenous or exogenous to philanthropy. At some point in their lives, many of the intellectual and economic elite of the country turn their attentions to "doing good." Since the beginning of the country, a few men and women have been so moved by social problems that they have stepped forward to solve them as part of their life work. Their influence and their money have had profound effects on the way the average American engages in philanthropy.

In the early days of the country, ministers and authors were most likely to shape the average person's view of his or her philanthropic duties. Benjamin Franklin (1706–1790) is emblematic of another type of influence. Franklin, who was a statesman, printer, inventor, and prolific author, had a decidedly secular approach to charity. He believed in hard work and thrift, self-governing institutions, and community spirit. His writings allowed him to influence American attitudes about charitable works. Like many reformers, he had a vision of a good society and he used philanthropy to try to bring that vision about.

Writing in the *London Chronicle* in 1766, Franklin said:

I am for doing good to the poor, but I differ in opinion of the means. I think the best way of doing good to the poor, is not making them easy in poverty, but leading or driving them out of it. In my youth I travelled much, and I observed in different countries, that the more public provisions were made for the poor, the less they provided for themselves, and of course became poorer. And, on the contrary, the less was done for them, the more they did for themselves, and became richer. (Franklin 1766, 430)

Franklin's *Poor Richard's Almanack*, a popular series of pamphlets published from 1732 to 1758, provided its readers with practical instructions for living, serial stories, and aphorisms. Franklin did not discourage charity: "Great almsgiving, lessens no man's living" (Franklin 1999). However, his almanacs gave many, many more instructions on frugality and hard work.

Franklin's writings contributed to the kind of independent work ethic and general skepticism about charity many Americans still feel today. In 1727 he formed Junto, a club that had as its goals the improvement of its members and of the city of Philadelphia. Franklin and his fellow club members worked to establish a library, a fire brigade, a hospital, and a militia. They demonstrated the usefulness of voluntary associations to accomplish community aims. Voluntary associations continue to be an important means of Americans learning about opportunities to contribute to various causes.

Another early "citizen philanthropist" was Stephen Girard. Girard was a millionaire businessman in a time when there were few real fortunes. He had an unsentimental view of charity. Still, he assisted in a yellow fever epidemic that afflicted the city of Philadelphia in 1793 by assuming responsibility for managing a makeshift hospital. Using his business acumen, he made the hospital a model refuge for the sick and dying. Girard's charitable work continued after his death, for "never before had an American bequeathed such a large estate to charitable and public purposes" (Bremner 1988, 39).

While many great men and women have helped to define American philanthropy, the reformers who preceded the Civil War deserve special attention. Between 1820 and 1860, several dedicated, consciousness-raising crusaders stepped forward to redefine society's responsibility to a number of disadvantaged groups. Examples include Thomas Gallaudet, who championed the cause of education for the deaf; Dorothea Dix, who worked to improve the condition of the mentally ill; Samuel Gridley Howe, who campaigned for education for the blind; Charles Loring Brace, who advocated that children be placed in foster homes rather than sent to live in orphanages; and William Lloyd Garrison, who used his newspaper to crusade against slavery and eventually, to advocate for women's suffrage.

Working with the government, and on occasion in opposition to it, these men and women played transformational roles in American philanthropy. They used speeches and writings to put their causes before the American people. Through compelling stories and tireless campaigning, they offered firsthand knowledge of the dreadful conditions in which orphans, the deaf, the blind, the mentally ill, and slaves lived. They identified problems and then insisted that addressing and solving these problems was everyone's responsibility.

These philanthropic "entrepreneurs" promoted social safety nets of programs and services paid for by local and federal governments. They believed that American society and ultimately the U.S. government should guarantee services for certain categories of people, particularly children, the elderly, and the sick. Ultimately, all Americans have taken on these additional burdens through the national tax system. Through taxes, the reformers were able to secure a certain level of care and permanence of care. Thus, caring for these disadvantaged groups has become the social norm.

The industrialists who prospered in the second half of the nineteenth century are another important group that influenced American thinking and behavior about philanthropy. Andrew Carnegie made his money in steel and railroads. John D. Rockefeller made his money in oil. Cornelius Vanderbilt and Leland Stanford made their fortunes in railroads. Others were financiers who amassed their fortunes through the banking industry. All of these men began to practice philanthropy on a scale that had never been seen before. They endowed universities; they built museums and public libraries; they supported symphony orchestras; they donated art collections.

Society tends to accord a degree of respect to those who are able to make large sums of money. Businessmen and women, then and today, emulate them. Carnegie became one of the most prominent philanthropists in American history. In 1889, he wrote an essay entitled "Wealth" that later became a book, *The Gospel of Wealth*. In it, Carnegie suggested that self-made men have a responsibility to distribute significant amounts of the wealth they acquire. Without apology, he assumed that those who had acquired wealth were best suited to distribute that wealth in order to shape society to better ends.

This, then, is held to be the duty of the man of Wealth . . . to consider all surplus revenues which come to him simply as trust funds, which he is called upon to administer, and strictly bound as a matter of duty to administer in the manner which, in his judgment, is best calculated to produce the most beneficial result for the community—the man of wealth thus becoming the sole agent and trustee for his poorer brethren, bringing to their service his superior wisdom, experience, and ability to administer—doing for them better than they would or could do for themselves. (Carnegie 1889, 13)

Carnegie was a Scottish immigrant with a rag-to-riches life story. When he died in 1919, he had given away over $350 million to philanthropic causes. It is interesting to note that entrepreneurs who have made their own wealth tend to give away more of their resources than those who inherit them (Rooney and Frederick 2009, 4).

The fortunes of men like Carnegie and John D. Rockefeller were so extraordinary and so visible that they changed the average American's opportunities for giving. The presence of Carnegie-built libraries, concert halls, museums, parks, and schools gave ordinary people new places where they might donate their time and monies. The philanthropic foundations these men left behind had ambitious societal agendas.

For example, the John D. Rockefeller Foundation targeted the treatment and eradication of a number of important diseases that affected both individuals in the United States and in other countries. "Between 1913 and 1929, the International Health Commission of the Rockefeller foundation

. . . lent its assistance to campaigns against hookworm, yellow fever, pellagra, malaria and tuberculosis" (Bremner 1988, 129). Foundations such as Rockefeller's set aside the traditional goals of charity—relief for the destitute and response to disaster—and began extended social experimentation. Over time, although the great foundations continued to provide relief during war and economic crises, they concerned themselves more with education, medical research, and issues of social welfare. Much like the early social workers, the foundations and their professional managers became interested in the root causes of social problems: for example, their aim became to eradicate a disease rather than ameliorate it. They funded research and gave Americans new ways in which to give.

Americans almost immediately became suspicious of the industrialists who engaged in large-scale philanthropy. There was (and is) speculation that very wealthy individuals give away significant amounts of their fortunes to atone for the way in which they acquired this wealth. Others see large gifts as a self-serving way to avoid income, inheritance, or corporate taxes.

Still other critics saw and continue to see large-scale philanthropy as a means for wealthy individuals to exert undue influence on society. For example, the Slater Fund, established after the Civil War, distributed monies for African American education in the South, but its managers would only fund programs that trained African Americans in agricultural, domestic, and industrial skills. This brand of "scientific" philanthropy assumed that African Americans would not need a liberal arts education but rather should be trained for a subservient role in American life. Time did not improve matters. Eduard C. Lindeman, a professor of social philosophy at the New York School of Social Work, studied the operation of 100 foundations between 1921 and 1930. One of the major conclusions of his book *Wealth and Culture* (Lindeman 1936/1987) was that foundation money is used primarily to reinforce the status quo.

Relatively few individuals have commanded enough wealth to attempt to influence the direction of society with their donations. The extraordinary wealth concentrated in today's Silicon Valley entrepreneurs echoes that of men like Carnegie and Rockefeller. Like those early philanthrocapitalists, these entrepreneurs are applying their business acumen and their new digital tools to their favorite causes.

Tech Billionaires (Solomon 2009) discusses several new approaches the information elite are taking to philanthropy. The first is illustrated by Bill Gates and the Melinda & Bill Gates Foundation. The Gates Foundation takes a traditional approach, focusing on long-term cures in public health and improving public education. However, Gates has called for a revision of capitalism that he terms *creative capitalism*. Gates argues, "It is mainly corporations that have the skills to make technological innovations work for the poor. To make the most of those skills, we need a more creative capitalism: an attempt to stretch the reach of market forces so that more companies can benefit from doing work that makes more people better off" (Gates 2008, 40).

Others who have made their fortune in computing applications support Gates's approach. Jeffrey Skoll of eBay provides longer-term funding for "social entrepreneurs" who apply business techniques to solving social problems. Steve Case of AOL is also interested in social entrepreneurship. He funds mission-driven for-profit businesses. Sergey Brin and Larry Page of Google have similar goals of facilitating business-driven solutions for social problems for their philanthropic arm, Google.org: "Building on Google Inc.'s information management and engineering prowess, Brin and Page are pioneering in creating a brand-enhancing, profit-making, taxpaying type of philanthropy designed to tackle systemic problems" (Solomon 2009, 6).

Pierre Omidyar, founder of eBay, has taken another approach. The Omidyar Network blurs the distinction between business and philanthropy with a hybrid infrastructure. The network has two arms, one that funds tax-exempt organizations and one that invests in for-profit businesses with positive social aims. The Omidyar-Tufts Microfinance Fund grants poor people in other countries small loans for starting or expanding businesses.

Over the course of American history, visionary reformers and highly successful business people have applied their considerable powers of persuasion on behalf of philanthropic causes. Using their skills, they have tried to eradicate some societal ills. Those from the pre–Civil War reform years redefined the federal government's responsibility (and ours today) to assist certain disadvantaged and dependent groups. Today as taxpayers we have less direct choice about which groups we will support because our tax contributions are distributed to specific initiatives through government channels. Wealthy individuals, then and now, have built cultural and service facilities in our communities to which we now donate both time and money. Today, as we have noted, a new generation of entrepreneurs, steeped in information technology, has begun experimenting with business models that marry social activism to profit. That may result in new businesses and jobs that have both profit and the improvement of society as their goals. These IT entrepreneurs have also provided the average American with the means to select and promote a cause of his or her own. For example, the social networking site Facebook sponsors an application named "Causes" that allows any individual, nonprofit group, or company to build a network around a cause. "Causes was founded on the belief that in a healthy society, anyone can participate in change by informing and inspiring others. . . . We strive for equal opportunity activism, which means that causes created by nonprofits and causes created by concerned individuals share the same level-playing field" (Project Agape 2009).

Although wealthy donors will continue to be important, the hope is that digital tools will make it less necessary to depend upon wealthy donors. These tools may allow the exploitation of the "long tail of philanthropy,"[7] enabling fundraisers to solicit many small donations instead of pursuing a few large ones. We now have inexpensive digital tools that allow us to collect information about a cause, communicate with others about it, and dream of transforming the world ourselves.

The Arrival of Modern Communications and Media

Communication was difficult in the early years of the country. "As late as the 1760s, probably no more than 5 percent of the population had more than the most incidental contact with either institution (the postal system and the press). . . . Only in the nineteenth century would the postal system and the press come to exert a palpable influence on the pattern of everyday life" (John 1993, 2352–2353). In the half century following the ratification of the U.S. Constitution, there was huge growth in the book trade and religious-works publishing. In 1790, there were 92 newspapers in the United States; in 1800, there were 242; in 1860, there were 3,725. Newspapers assisted in building a sense of community (John 1993, 2355). The telegraph knit the country together by 1861. The telephone was invented in 1876, with the first coast-to-coast telephone call taking place in 1915.

At the turn of the twentieth century, the population of the United States population grew numerically and expanded geographically and the rise of the American corporation began to have a significant impact on the ways in which Americans could receive information. Transportation improved dramatically. Communication provided "improved techniques of printing, the increased literacy of a people achieving more education, and heightened interest in public affairs combined both to reflect and stimulate the growth of mass media. The mass circulation daily newspaper and the growing numbers of popularly priced, widely read magazines were coming to be important factors in accelerating these profound changes" (Cutlip 1990, 30–31).

These innovations allowed information to be disseminated throughout populations in new ways. In fact, everyday information exchanges were utterly transformed by these sorts of large-scale developments.

Today, radio, television, and the Internet have made coast-to-coast, grassroots fundraising campaigns a practical reality. Not only can these modes of communication be used to present compelling stories of suffering and hope that emphasize the need for charitable giving, they also transmit these stories instantaneously. Americans can now have direct, immediate knowledge of a single child who needs help in Somalia, or a Peruvian family that has no access to clean water.

One historical example of the interaction between American philanthropic behavior and media of the day is found in the fight against polio. The nationwide polio epidemic of 1916 raised American awareness of the disease. Franklin Delano Roosevelt contracted polio in 1921. In 1932, he became president of the United States. In 1937, Roosevelt announced the establishment of the National Foundation for Infantile Paralysis, which later became the March of Dimes. The successful eradication of polio in the United States, which took place over the next forty years, was accomplished by many, many small donations from ordinary Americans across the country. The campaign utilized nearly every new mode of communication available in its era to overcome substantial economic obstacles.

It was difficult to generate large philanthropic gifts during the Great Depression. Between the establishment of the National Foundation for Infantile Paralysis in 1938 and the suppression of the disease in 1979, grassroots-level fundraising was essential to finding a cure. Every donation counted. Stories appeared in *The Saturday Evening Post* and *Good Housekeeping* about individuals who had learned to live successfully with the disease. Those stories implied that the disease could be temporary in nature, creating hope (Orshinsky 2005, 45–46). Fundraising campaigns featured intentionally inspiring accounts of "poster children" who had recovered from the disease and learned to walk again.

Eddie Cantor, a well-known entertainer, coined the phrase "the March of Dimes" (a play on the newsreel, *The March of Time*). In a radio broadcast he asked the public to send their dimes to the White House to fight polio. The campaign was hugely successful: millions of dimes were sent.

Other fundraising efforts included "Birthday Balls," thrown on President Roosevelt's birthday. They were attended by numerous celebrities and employed the slogan "We Dance So That Others Might Walk." The first ball, held in 1934, consisted of more than six thousand separate parties held around the country and raised over $1 million (Orshinsky 2005, 50). Fundraisers had fewer than six weeks to organize the balls, since there was no central organization at the time, and relied heavily on contacts with newspaper editors to publicize the events. In the 1950s, Mothers Marches were organized to respond to an outbreak of the disease. The message was that every dime counted and every American could make a difference. "Tens of millions of Americans (as many as 80 million a year)" gave and "more than four billion dimes later after the foundation program started, the Salk vaccine became available" (Carter 1992, 91–92).

No one questioned the value of raising money to fight polio, a disease that primarily affects children. Engaging sympathy and prompting donations for a more recent epidemic, HIV/AIDS, has presented different challenges because of the stigma attached to the disease. Advocates for HIV/ AIDS have embraced numerous formats for publicizing their cause. In the United States, the campaign has been inextricably linked with the social agenda of destigmatizing homosexuality.

Theater productions, movies, television programs, and awareness concerts have helped to humanize victims of the disease. The Broadway hit musical *Rent* depicted, among other things, friends affected by AIDS. The movie *Philadelphia*, starring A-list actors Tom Hanks and Denzel Washington, told a harrowing but compassionate story about a lawyer afflicted with the virus. Television has taken on the crisis in specials and through characters in weekly shows. Like polio, the cause of HIV/AIDS has been embraced by a number of celebrities, who have used their reputations to publicize the effects of the virus and the need for research. A number of famous musicians lend their names and talents to HIV/AIDS philanthropy.

HIV/AIDS has also spawned cause-related marketing of commercial products. Product Red, an initiative launched in 2006 by singer Bono and Debt, AIDS, Trade, Africa (or DATA), allows partner companies to use the "Product Red" brand name to increase their sales of goods as long as part of the proceeds are designated to the Global Fund to Fight AIDS, Tuberculosis and Malaria. Companies such as Starbucks, Microsoft, Gap, Hallmark, Dell, Motorola, and Converse have created products or services to sell under the Product Red brand name. The Product Red initiative has been deliberately tied to popular culture. Americans who buy the products are encouraged to feel both "cool" and socially responsible. These product/ charity partnerships are seen as good business practice, increasing companies' reputation, for social accountability.

The advent of the Internet allows dedicated professional and amateur fundraisers alike to try a number of strategies to provide information that

will engage the average American in the fight against HIV/AIDS. Fundraisers post moving public service announcements on YouTube and other
video Web sites. They blog about advocacy, research, and activism. They
increase interactivity on their Web sites.[8] They create participatory communities via social networking sites like Facebook and MySpace. In January
2009, there were more than forty-five thousand charity efforts under way
on the Facebook site (Chronicle for Philanthropy 2009). The hope is that
information about favorite causes will become "viral"—embed itself in
preexisting social networks and therefore be easily found and easily
adopted. All of these efforts are designed to create a sense of participation
and personal engagement that was rarely possible before the advent of
interactive media.

In sum, these new communication channels extend rather than replace
older ones. Americans will still receive appeals in church, through direct
mail, and by phone solicitation, all important fundraising tools. There is
now, however, a new generation of Americans that has grown up with
information technology. In order to deliver information to them, philanthropic organizations are adopting Internet tools at a rapid clip. Ninety-five
percent of the top four hundred charitable fundraising organizations
in the United Way currently accept online donations (Waters 2007). A
younger American who wants to donate to a cause is quite likely to turn
to its Web site. An intermediary is no longer necessary; the potential donor
can avoid having to speak to someone directly to gather information.

Modern media have profoundly changed Americans' everyday interactions with information about philanthropy. We have more direct access
to information about philanthropy at the same time that we are overwhelmed with information in dozens of new formats. We can purchase
products that support our favorite causes. We can give online. We are able
to hear and see stories about needs thousands of miles away. We can participate in a contemporary culture of philanthropy that defines giving as
both trendy and responsible.

Information Challenges in Philanthropy Today

As communication media multiplied, so did philanthropic organizations.
Between 1900 and World War I, a number of important philanthropic
organizations formed: "These years saw the start of the Boy Scouts, Girl
Scouts, Campfire Girls, National Tuberculosis Organization, American
Cancer Society, the National Association for the Advancement of Colored
People, National Urban League, the Lighthouse, the National Child Labor

Committee, and scores of other leagues, associations, and committees" (Cutlip 1990, 38).

John D. Rockefeller referred to all this as the "business of benevolence." Today, Americans have a dizzying array of options for giving. In 1996, the nation had 1.1 million nonprofit organizations; by 2006, that number had grown to 1.4 million (Wing, Pollak, and Blackwood 2008, 1). Tax-exempt organizations are the fastest-growing sector in the U.S. economy. Thousands of new charities are created each year. The competition for funds has become intense.

Complexity and Overload

Since there are so many philanthropic entities, each charity must compete to secure the attention and ongoing engagement of donors. The current information environment, particularly on the Internet, is incredibly rich and dense. There has been a steady increase in interactive features and functionality as philanthropic organizations invest in their Web sites over time. Table 3.1 catalogues the growth of interactivity on one important philanthropic Web site over the last decade.

Table 3.1
American Red Cross Web site: Growth in Web site interactivity*

2000	1-800 numbers; news reports; secure online donations; inspirational stories posted with an invitation to email the user's own inspirational story
2003	Ability to find a local Red Cross chapter by Zip code; ability to sign up for emails with monthly news, tips, and disaster updates; ability to search the site; a gift calculator for people planning bequests; an online gift store with Red Cross-branded merchandise
2006	Multiple online videos posted of tsunami recovery efforts; interactive map of the United States allowing access to state chapter Web sites; the ability to send a virtual postcard
2009	Ability to sign up for courses at a local chapter; to find a blood drive, register a personal profile, select a location to give blood, make an appointment to do so, and view a calculation of the number of people who have needed blood since the user began scanning the page; functionality that allows the user to designate himself or herself "safe and well" if he or she has been in a disaster; use a volunteer matching service that sorts by Zip code, distance, keyword, skills, city, state and age range; bookmark and share the site in a number of ways, including email, printing, Digg, MySpace, Facebook, Twitter, Delicious, Google, Live, Stumbleon and fifty-three other sharing choices, from instant messaging to blogging sites (American Red Cross 2009).

*All pages except 2009 sampled through the Wayback Machine of the Internet Archive (http://www.archive.org/index.php).

In essence, the American Red Cross has created a sophisticated information system which can, when donors indicate an interest, push information out as well as "pull" (provide) it when they seek it. Interactive tools play the role of matchmaker, particularly when an individual has already decided to expand his participation in philanthropy.

Table 3.1 shows the increase in interactivity, as well as the increase in complexity with which donors must cope. The "helpfulness" of philanthropic Web sites can be overwhelming. An interested donor visiting the site must now decide how involved he or she wants to be. (Do I sign up for emails? Do I want to give blood? Should I share the date of that life-saving course with my friend?) Donors now have to decide how much time they can invest, and in what ways they want to engage with an organization.

Further, there are completely novel ways to engage in philanthropy via the Internet. Table 3.2 illustrates some of the great variety that can be found.

All these new modes of giving require a substantial investment of time: time to review and evaluate the information they present and time to learn how to interact with them: "In an information-rich world, the wealth of information means a dearth of something else: a scarcity of whatever it is that information consumes. What information consumes is rather obvious: it consumes the attention of its recipients. Hence a wealth of information creates a poverty of attention and a need to allocate that attention efficiently among the overabundance of information sources that might consume it" (Simon 1971, 40–41).

This "information overload," caused both by the number of charitable organizations and the increasing sophistication of their digital tools, can be problematic. There are limits to the number of causes the average American wants to or has time to espouse. Researchers are attempting to estimate the total number of meaningful relationships any one individual can maintain in his or her social networks.

Relationship Building and the Decline in Volunteerism

Charitable organizations want to create both short-term relationships with donors that prompt them to make impulsive gifts and lasting relationships that result in repeated gifts over time. "Fundamental to most fundraising activity is its relational nature. People do not give money either to strangers or to unknown causes" (Ragsdale 1995, 2). This chapter has described the importance to philanthropy of voluntary associations and social activism throughout American history, but must also note that, since 1970,

Table 3.2
Web-enabled approaches to philanthropy

Web site	Approach
Charity Click Donation	"This site is dedicated to sites that allow you to give money to them by simply clicking on a button and then viewing advertisers. The advertisers pay a small amount to the charity when you view their ad so you are able to donate to a cause of your choice without spending a cent of your own money." Charity Click Donation. 2009. http://www.charityclickdonation.com (accessed June 30, 2009).
Charity Guide	"We help you Volunteer On Demand.™ Charity Guide presents service projects that are entirely flexible and immediately actionable. You can volunteer whenever you want, from wherever you want, in as few as fifteen minutes." Charity Guide. 2009. http://Charity guide.org/ (accessed June 30, 2009).
Free Rice (An online word definition game site)	"For each answer you get right, we donate ten grains of rice through the UN World Food Program to help end hunger." Free Rice. 2009. http://freerice.com/ (accessed June 30, 2009).
Global Giving	"An online marketplace for philanthropy that connects you to grassroots projects all over the world." Donors select different levels of giving and "project leaders" can engage in competitive fundraising; includes ability to give gift cards. Global Giving. 2009. http://www.globalgiving.org/americanopen/ (accessed June 8, 2009).
Kiva	"Kiva lets you lend to a specific entrepreneur, empowering them to lift themselves out of poverty." Kiva Dream Builders. 2009. http://kivabcdreambuilders.com /mission-statement.php/ (accessed July 29, 2009).
Network for Good	"This is a website where you can give to your favorite charity/charities and have all your donation records stored and accessible at any time." Network for Good. 2009. http://support.networkforgood.org/default .asp?a=4&q=279 (accessed July 29, 2009).
Charity Mall	"CharityMall.com is an online shopping system that showcases 100s of merchants (i.e. Target, Home Depot, Sony, Gap, REI, and more) that have agreed to donate to cancer research every time an individual purchases something from their online stores." Charity Mall. 2009. http://www.aboutus.org/CharityMall.com (accessed July 29, 2009).
Change. org	"Change.org aims to [serve] as the central platform informing and empowering movements for social change around the most important issues of our time." http://www.facebook. com/change.org?v=info (accessed July 28, 2009).
Charity Folks	"Managing hundreds of successful charity auctions since its inception, Charity Folks has raised millions for nonprofit organizations around the world, all while offering consumers highly coveted auction lots donated by today's hottest celebrities, musicians, sports stars, high-profile personalities, and top-tier brands." The Charity Folks Blog. 2009. http://www.blogcatalog.com/blogs /the-charity-folks-blog.html (accessed July 28, 2009).

there has been a significant drop in the number of Americans who join organizations of any sort, including philanthropic ones.

"Bowling Alone: The Collapse and Revival of American Community" (Putnam 2000) describes how American participation in civic, religious, and other community networks began to decline precipitously after the 1960s. Volunteerism likewise declined. Today philanthropic organizations have to work much harder to form relationships with potential donors and develop a sense of trust and affiliation. They can no longer count on membership alone to drive donations.

Some fundraising campaigns have adopted customer relationship management systems. These software applications ensure that potential donors are treated respectfully and contacted through multiple means, but not so often that they become annoyed at repeated appeals. These systems can be Web-based:

For nonprofit organisations the Internet represents an unprecedented and highly cost-effective opportunity to build and enhance relationships with supporters, volunteers, clients and the community they serve. As ePhilanthropy has emerged, organisations have discovered that consistent and deliberate email communication that drives traffic to the organisation's well organised and informative website has become the key to success. Charities should approach the Internet as a communication and stewardship tool first and a fundraising tool second. Success will come not from an emphasis on the technology, but on cultivating and enhancing relationships. (Hart 2002, 353)

In lieu of membership, some charitable organizations have attempted to create long-term relationships with donors through the sale of branded merchandise. One early example was the Christmas Seals campaign of the National Tuberculosis Association. This remarkably simple device, borrowed from a Danish crusader who had lost six brothers to tuberculosis, became part of everyday American life. The first American seal was sold in 1907. The purchase of Christmas Seals became a highly effective, widely recognized, inexpensive new method of giving during the holidays for many American families. "The mere mention of Christmas seals is enough to make fundraisers for other voluntary health agencies sob with envy" (Carter 1992, 75).

Another early example of relationship building that has become a permanent part of American culture is the Girl Scout Cookie sale. Starting with informal sales in the 1920s, "Girl Scout cookie time" has become part of American life. It would be difficult to find an American who does not associate good feelings with Girl Scout cookies, even though membership in the organization has dwindled. The information and positive feeling the cookie sales convey about the organization is embedded in our culture.

Relationship building, branding, and cultural acceptance have become more difficult to achieve over time as the numbers of philanthropic organizations increase. An example of this crowding is found in the today's popular wristband sales. Yellow "awareness bracelets"—silicone wristbands—imprinted with the phrase *Live Strong*—were developed for Lance Armstrong's LIVESTRONG Foundation, which raises money for cancer research. The bracelets were immediately so popular that other charitable organizations adopted them in dozens of different colors and styles. It is now difficult to remember which color is associated with which charity.

Donor Fatigue

Another challenge to contemporary philanthropy is donor fatigue. There is evidence that too many disasters in a row or too many appeals in a row cause Americans to feel overwhelmed and give less. The international disasters that occurred in 2004 and 2005 illustrate the problem:

The December 2004 tsunami alone generated over $7 billion in international donations from a full range of public and private sources. In August 2005, when Hurricane Katrina caused widespread destruction across the southeastern United States, international donors responded again, providing over $1 billion in financial and other assistance. But less than two months later, a 7.6 magnitude earthquake hit South Asia, leaving almost 80,000 people dead and 3 million homeless. Donors did not respond with such generosity. . . donors are having trouble concentrating on more than one humanitarian effort at a time. (Coppola 2006, 530)

During the 1980s, charitable organizations tried to stand out from the general philanthropic "noise" by enlisting celebrities in a cause. "Star power" has increasingly been used to create a sense of glamour and connection. In 1971, ex-Beatle George Harrison and sitarist Ravi Shankar staged the Concert for Bangladesh in Madison Square Garden. By the mid-1980s, several well-known, philanthropically active musicians were staging concerts to raise money for international relief work. Concerts and other large group events create camaraderie. They are compelling ways to gain attention and build a sense of relationship, but the number of concerts began to dwindle as the public became inured to the everyday, ever-present nature of war, famine, and disaster.

Fundraisers also attempt to create smaller-scale, more intimate emotional connections with donors. It is particularly challenging to create empathy in donors for causes in far-flung places. One excellent example of small-scale engagement is sponsoring a child. Some agencies allow donors to select an individual child (or allow them to feel they are sponsoring an individual child while funds actually support programs that sustain multiple children). For example, the Web site of World Vision provides a searchable database of "hope children." Potential donors can select a child based on several attributes: gender, age, birthday, country, and whether or not that child is affected by HIV/AIDS. Photographs of children matching those attributes are displayed and a message appears: "[Child's name] has been reserved for you to sponsor within the next five minutes. Just click the 'Sponsor Now' button above to become his sponsor right now!" (World Vision 2009)

Fundraisers have always probed the social networks of wealthy donors to find ways to approach and engage them. Theoretically, Web-based tools can be used to reach any potential donors, large or small, who use social networking software. "The probability of success increases by sharing researched names with people and finding out about their personal contacts and their six degrees of friends, family, and colleagues" (Hart et al. 2006, 14). Most philanthropic organizations are well aware of the importance of relationships. Of those wealthy donors who stopped supporting a charity in 2008, 57.7 percent surveyed cited no longer feeling connected to the cause as their reason for stopping (Rooney and Frederick 2009, 56).

Differences in Online Giving

As philanthropy on the Internet increases, researchers are beginning to study the characteristics of online donors. Such donors tend to be younger and have higher incomes than direct-mail donors. They have slightly lower retention rates than non-Internet donors. They may be more difficult to cultivate. They may begin a relationship online, but migrate away from online giving to other channels over time (Flannery, Harris, and Rhine 2009).

In a study called "The Young and the Generous," Network for Good, partnering with GuideStar, reviewed the characteristics of donors of $100 million distributed through the site over approximately five years. The study found that online is "becoming donors' charitable avenue of choice" at times of disaster. With each successive disaster—9/11, the Asian tsunami, and Hurricane Katrina—a larger percentage of total relief dollars came from online donations (Network for Good 2007, 3). The Internet may become an important avenue for donations made on impulse. Remarking on fundraising for the Obama campaign, one Internet fundraising expert said, "Raising money online draws on the capacity of the Internet to convert people's momentary enthusiasm or sense of whimsy or outrage into money" (Mosk 2008).[9]

Potential for Fraud

Americans sorting out these new modes of giving need information about whether they are legitimate and wise ways to give. The Internet has increased the possibility of encountering fraudulent pleas for philanthropy. Unfortunately, with minimal investment, an unethical individual or organization can put up a professional-looking Web site. It is much more difficult for the average citizen to evaluate the worth of an organization by relying on its Web site alone. These activities fall into the familiar theoretical frameworks on information-seeking behavior and the literature on consumer behavior.

Some donors protect themselves from fraud by giving within their own communities, both so they can see the effect of their dollars and so that they have some idea of how those dollars are spent. Presumably, individual donors are better able to evaluate the merit of local charities than of national or international charities. They can also rely on their local United Way fund to prescreen charities for merit and good management.

As soon as potential donors are separated from direct knowledge of the ways in which charity work is conducted, they are at an information disadvantage. There is "information asymmetry": they have no way to confirm firsthand that their dollars are being spent wisely, and the charities (or scammers) can do much more to control what donors know about them.[10] It is even more difficult to discern when money is spent wisely if it is sent overseas for disaster relief or international social programs, because standards of oversight differ from country to country.

Most charities make a serious effort to police themselves. One organization, the Center for Effective Philanthropy, provides comparative tools for charitable organizations to help them "define, assess, and improve their effectiveness and impact" (Center for Effective Philanthropy 2009). Many philanthropic organizations participate in efforts to make their management transparent to donors. GuideStar is a nonprofit that has collected, organized, and analyzed information contributed by nearly two million tax-exempt charitable organizations and made much of that information directly accessible online. Nonprofits voluntarily contribute their information to the database. The organization reports, "A high percentage (93 percent) of nonprofits are embracing the Internet to disclose information about their programs and services" (GuideStar 2009).

There are also nonprofit organizations that serve as watchdogs on charitable organizations. Charity Navigator, CharityWatch, and the Better Business Bureau provide information for potential donors through extensive Web sites. Charity Navigator rates charitable organizations via a four-star system, assessing them on financial acumen, long-term sustainability, and privacy policies governing donor information. The American Institute of Philanthropy provides CharityWatch, which assigns a letter grade to charitable organizations based on a number of financial measures. CharityWatch measures such things as how much money an organization spends in order to raise $100. The Better Business Bureau provides services to help donors evaluate organizations that have requested contributions and to report unethical solicitation practices.

For potential donors attempting to avoid fraud and abuse, we begin to see some divergence in the information resources available to wealthy donors and those available to the average American. Truly wealthy individuals, who are likely to receive personal attention from fundraisers, are able to employ financial planners, lawyers, and planned giving experts to investigate any organizations with which they may become involved. Middle- and lower-income Americans are more likely to receive mass mailings and other broad-spectrum appeals for donations. They must search for more generic information on how to avoid fraud and abuse.

The Federal Trade Commission (FTC), the nation's consumer protection agency, provides a checklist and suggests, for example, that donors should avoid any charity or fundraiser that refuses to provide written information about its activities or will not provide proof that a contribution is tax deductible (Federal Trade Commission 2009). In addition to the FTC, the Better Business Bureau, the National Crime Prevention Council, Charity Navigator, GuideStar, and the American Institute for Philanthropy all maintain Web pages with advice on avoiding scams.

Tax Issues

Individual donors need information about the tax laws related to charitable giving. Some seek information about how to give in ways that create tax advantage. Again, in this area, there is some divergence of information sources between upper and lower classes. The wealthy can pay specialists who have such information. The less wealthy must sort through a number of issues with less assistance.

The first decision that must be made is whether to take a standard deduction or itemize deductions on taxes. Americans can only deduct charitable contributions if they itemize. Then, if a decision to itemize is made, the donor must learn the many rules associated with deducting donations promulgated by the Internal Revenue Service (IRS) (Internal Revenue Service 2009). Within each income range, a higher percentage of itemizers than non-itemizers gives to charity; itemizers give a higher percentage of their income (Independent Sector 2003).

Most individual donors are interested primarily in whether their donations will be tax deductible. The IRS lists over thirty different types of charitable organizations that are exempt from corporate income taxes. Religious congregations and small voluntary nonprofits such as parent-teacher associations, community theaters, and neighborhood associations that have less than $5,000 in annual revenue do not have to register with the IRS.

There are rules for cash donations, rules for non-cash donations, rules on documentation, rules on the limits of deduction, and rules on eligibility of different charitable organizations before the deduction can be taken. The information burden is considerable. Some Americans turn to popular financial "gurus" such as Suze Orman. Orman makes frequent appearances on public television and nationally syndicated shows including *Oprah*. She provides a Will and Trust kit on her Web site and wrote a foreward to Gary and Kohner's *Inspired Philanthropy: Your Step-by-Step Guide to Creating a Giving Plan and Leaving a Legacy* (Gary and Kohner 2002).

Educating Children

Finally, Americans need information on ways to impart the value of philanthropy to their children. Wealthy donors teach children about philanthropy by discussing it with them, allowing them to participate, establishing criteria for their participation, and giving them money to donate (Rooney and Frederick 2009, 50). They can employ specialists to guide their children when they first take on the philanthropic responsibilities associated with family- or company-owned foundations.

Many Americans, including those of moderate means, want their children to perceive giving as an important moral and ethical duty. Religion has been a traditional mode of imparting philanthropic values to children; there are secular ones as well. Most schools attempt to inculcate values about giving. Many American children first encounter charitable giving through school projects. The organization Learning to Give provides a Web-based wealth of resources: lesson plans for teachers; resources and books for parents; games for students, including resources for students who want to learn how to include philanthropy in a spending plan; activities for youth workers; a searchable database of quotes on philanthropy and related topics; and essays on the importance of philanthropy in different religions (The League 2009).

Conclusions

While Americans do, on occasion, seek information about the particulars of giving, for the most part, they do not have to look very hard for information about philanthropy. It seeks them out. Many people and institutions take it upon themselves to provide us with information about the thousands of philanthropic activities occurring in America today. If we were to imagine a single American family and its descendents engaging in philanthropic activity over the last four hundred years, we would see some continuity in information sources, some replacement of information sources, but also a vast increase in the number and complexity of information sources. We would also see the number and complexity of their questions increase.

In colonial times, individual Americans and their families were most likely to encounter information on philanthropy in church. If literate, the members of that family might also encounter the secular writings of some of the country's influential statesmen who committed their thoughts to pamphlets and newspapers produced on colonial printing presses. After the Revolutionary War, whether the descendents of that family attended church regularly or participated in the revivals of the Great Awakening, they continued to learn about worthy causes from committed religious and secular reform leaders who were expert at engaging their emotions.

In the nineteenth century American women took on a more formal role in philanthropy, beginning with the great surge of voluntary associations in the early 1800s and continuing through the Civil War. They extended the work of many groups through donations of their time and became important brokers of information about charitable giving. They shared information about their causes with other women in their communities. Without benefit of the vote, American women were able to influence the social agenda, to some extent, in their favor and in favor of other disadvantaged groups. After the Civil War, African American philanthropic activity became part of everyday life. African American churches and schools became increasingly important hubs of philanthropy.

Americans became wealthier as the country moved from an agricultural to an industrial economy. During the Gilded Age, a few fabulously wealthy and civic-minded men, by their example, made philanthropy a visible symbol of business success that was to be emulated by less successful men.

By the 1930s, the Community Chest movement had further institutional-
ized that notion that businessmen have a civic duty to contribute to their
communities.

Today American businessmen and women may obtain information on
charitable causes that has been prescreened by federated fundraising organizations such as the United Way. They can also control, somewhat, the
number of appeals and the amount of information they receive if they
limit their giving to the United Way-endorsed charities.

In addition to our religious institutions, Americans now have professional fundraisers as a second powerful information broker attempting to
offer us the right story at the right time, in order to move us to give but
not overwhelm us so that we fail to act. We are the targets of powerful,
sophisticated "customer relationship management systems," designed to
help individual donors when they actively seek information about giving,
but also to push information to them when they are not. The average
American may experience donor fatigue when appeals occur too often or
too aggressively.

Social workers, professional fundraisers, and large foundations expanded
the American definition of philanthropy beyond subsistence relief to
include social justice. They pursued a vision of social change that went far
beyond traditional charity. Most Americans have accepted this broader
definition of philanthropy and are willing to donate money for societal
change as well as subsistence relief. After the implementation of the New
Deal and the Great Society social programs, American families were less
likely to encounter severe poverty or deprivation in their neighborhoods
or communities. The information challenge then became connecting them
to national and international needs that were not directly present in their
communities.

Unlike the early colonists, Americans today encounter an overabundance of information about local, national, and international giving, delivered via digital media such as radio, television, cell phones, and the
Internet. Today's wealthy philanthropists, particularly those who have
made their fortunes in information technology, are creating new ways to
give and hybrid structures that combine business and philanthropy. Individual Americans now have tools at their disposal that make it possible for
any one of us, rich or poor, to raise money for a cause.

The need to protect individual donors from fraud and abuse has
increased dramatically over time. Americans must also seek information
about tax law as it affects philanthropy; information about taking charitable deductions must be gathered and understood. Many families now

seek information and opportunities to educate children in the value of philanthropy, although there is little research available on how best to accomplish that aim.

Throughout our history, individual motivations and social connections have affected where Americans obtain information about philanthropy. American attitudes and views of philanthropy have been shaped by many societal forces, both exogenous and endogenous. At any given time, in different economic and political circumstances, there are patterns in the flow of information around philanthropy. Religion remains an important and trusted information mediator for charitable giving. War and economic crises decrease the information distance between donors and recipients. Attempts to provide assistance increase in times of crisis, but are often poorly coordinated. Information about giving then requires centralized coordination by the federal government. We derive new insights, new targets, and new technologies with which to give from committed reformers and wealthy philanthropists.

The Internet hasn't changed everything, but it has accelerated the pace of change. Nearly every philanthropic organization is using it to push information to Americans who are computer literate. Consequently, there is a significant information burden in participating in philanthropy today. The process has become much more complex and it takes time to assess the large numbers of philanthropic organizations and their highly interactive Web sites.

Everyday information about philanthropy and charitable giving is embedded in each day's political, social, and economic context. This is a dynamic, information-rich field, where amateur and professional fundraisers are experimenting with technology to push information, build relationships, and create connections to people in need in far-flung places so that they seem like next-door neighbors we'd like to help. We find information that motivates us to give; we use information to decide where and how to give; we interact with information that creates lasting relationships with causes that become part of our identities. It is information that can change our lives, and the lives of others, for the better.

Notes

1. Friedman writes that the reformer John Gardner created the Independent Sector in the early 1980s "to forge a self-consciousness among grant-making and voluntary organizations—a sense that they occupied a distinct third space between the government and the private market economy" (Friedman and MacGarvie 2003, 1). In 1983,

the Independent Sector Research Committee recommended that philanthropy become an interdisciplinary research field in American higher education.

2. This is a revised estimate; the original report estimated $306.39 billion.

3. In fact, there is a related discussion taking place in information science about a biologically based drive to forage for information. Researchers are hypothesizing that our adaptive success depends on "increasingly sophisticated information-gathering, sense-making, decision-making, and problem solving strategies" (Pirolli and Card 1999, 643).

4. This is a revised estimate; the original report estimated $102.32 billion.

5. Despite those exchanges and the help they received, some colonists considered it their religious or charitable duty to convert Native Americans to Christianity and to otherwise "civilize" their new neighbors. Most settlers had no compunction about imposing their religious and political views on native populations. Throughout American history, there have been instances of compassionate, caring philanthropy and instances in which philanthropy was used as a thinly veiled instrument of social control. There were many instances in the next three hundred years in which Native Americans were "helped" out of their lands and cultural identity.

6. A small group had supported preabolitionist efforts to "repatriate" African Americans, endorsing a colonization movement that resulted in a settlement of freed African Americans in Liberia in 1822. Others African Americans were suspicious of colonization, viewing it as a way to remove blacks altogether from the American landscape. Later, African Americans such as Frederick Douglass and David Walker became prominent in the drive to abolish slavery.

7. The "Long Tail" is a theory popularized by Chris Anderson of *Wired* magazine. In this context, it refers to the potential of inexpensive digital tools to capture and aggregate the contributions of many small donors into a powerful and transformative lump sum.

8. All these efforts endeavor to transform the individual who encounters information from passive observer to active, committed donor who defines HIV/AIDS as one of his or her own causes. An inventory of several HIV/AIDS organizations in June 2009 showed a variety of interactive features including the ability to make a donation; sign up for email messages; join Facebook and MySpace pages; view photos on Flickr and videos on YouTube; receive instructions on creating a video testimonial; download a banner to use on the visitor's own site; sign up for an RSS feed; volunteer for a clinical trial; create a memorial to a loved one; and sign up to use Microsoft Live Messenger for instant messaging so that a portion of its advertising revenue is diverted to AIDS research.

9. He also pointed out that it is hard to budget around impulses.

10. George Akerlof, Michael Spence, and Joseph Stiglitz received the Nobel Prize in 2001 for work begun in the 1970s analyzing various markets affected by conditions of asymmetric information.

References

American Red Cross. 2009. http://www.redcross.org/ (accessed June 27, 2009).

Andreoni, J. 1989. Giving with Impure Altruism: Applications to Charity and Ricardian Equivalence. *Journal of Political Economy* 97: 1447–1458.

Andreoni, J. 1990. Impure Altruism and Donations to Public Goods: A Theory of Warm-Glow Giving. *Economic Journal* 100 (401): 464–477.

Baker, R. J., J. M. Walker, and A. Williams. 2009. Matching Contributions and the Voluntary Provision of a Pure Public Good: Experimental Evidence. *Journal of Economic Behavior & Organization* 70 (1–2): 122–134.

Blocker, J. S., D. Fahey, and I. R. Tyrrell. 2003. *Alcohol and Temperance in Modern History: An International Encyclopedia*. Santa Barbara, CA: ABC-CLIO.

Bremner, R. H. 1988. *American Philanthropy*. Chicago: University of Chicago Press.

Burnett, G., and P. R. Yaeger. 2008. Small Worlds, Lifeworlds, and Information: The Ramifications of the Information Behaviour of Social Groups in Public Policy and the Public Sphere. *Information Research* 13 (2): paper 346.

Burnett, G., M. Besant, and E. Chatman. 2001. Small Worlds: Normative Behavior in Virtual Communities and Feminist Bookselling. *Journal of the American Society for Information Science and Technology* 52 (7): 536–547.

Butler, J., G. Wacker, and R. Balmer. 2008. *Religion in American Life: A Short History*. Oxford: Oxford University Press.

Carnegie, A. 1889. *The Gospel of Wealth*. New York: The Century Company.

Carstensen, L. L. 2006. The Influence of a Sense of Time on Human Development. *Science* 312 (5782): 1913–1915.

Carter, R. 1992. *The Gentle Legions*. Miami, FL: The Curtis Publishing Company.

Catholic Post. 1979. Archbishop Fulton Sheen Dies: Famed Radio and TV Preacher Native of Diocese. *Catholic Post* (of Peoria, IL) (December 16) Section 1, p.1 http://www.allendrake.com/elpasohistory/sheen/shncaps1.htm.

Center for Effective Philanthropy. 2009. Center for Effective Philanthropy. http://www.effectivephilanthropy.org/ (accessed July 9, 2009).

Chaplin, C. 1918. *The Bond*. Silent movie. The Internet Archive, April 15, 2009. http://www.archive.org/details/CC_1918_09_29_TheBond (accessed April 15, 2009).

Coppola, D. 2006. *Introduction to International Disaster Management*. Burlington, MA: Butterworth-Heinemann.

Cutlip, S. 1990. *Fund Raising in the United States*. New Brunswick, NJ: *Transaction Publishers*.

Eckel, C. C., and P. J. Grossman. 2003. Rebate versus Matching: Does How We Subsidize Charitable Contributions Matter? *Journal of Public Economics* 87 (3–4): 681–701.

Ehrenreich, J. H. 1993. Social Work and Philanthropy. *Encyclopedia of American Social History*, ed. M. K. Cayton, E. J. Gorn, and P. W. Williams, vol. 3, 2231–2239. New York: Charles Scribners' Sons.

Federal Trade Commission. 2009. Helping Family and Friends Avoid Charity Fraud. http://www.ftc.gov/bcp/edu/pubs/consumer/telemarketing/tel18.shtm   (accessed June 25, 2009).

Feldman, R. A., and S. B. Kamerman. 2001. *The Columbia School of Social Work: A Centennial Celebration*. New York: Columbia University Press.

Finkenbine, R. E. 2003. Law, Reconstruction, and African American Education in the Post-Emancipation South. In *Charity, Philanthropy, and Civility in American History*, ed. L. J. Friedman and M. MacGarvie, 161–178. Cambridge, UK: Cambridge University Press.

Flannery, H., R. Harris, and C. Rhine. 2009. *2008 donorCentrics™ Internet Giving Benchmarking Analysis*. Cambridge, MA: Target Analytics.

Franklin, B. 1766. *Memoirs of the Life and Writings of Benjamin Franklin: On the Price of Corn and the Management of the Poor*. Washington, DC: Library of Congress.

Franklin, B. 1999. *Wit and Wisdom from Poor Richard's Almanack*. Minneola, NY: Dover Publications.

Friedman, L. J., and M. MacGarvie, eds. 2003. *Charity, Philanthropy, and Civility in American History*. Cambridge, UK: Cambridge University Press.

Gary, T., and M. Kohner. 2002. *Inspired Philanthropy: Your Step-by-Step Guide to Creating a Giving Plan*. San Francisco: Jossey-Bass.

Gates, B. 2008. Making Capitalism More Creative. *Time, Inc.* (July 31).

Giving USA Foundation. 2008. Giving USA 2008: Annual Report on Philanthropy for the Year 2007. Indianapolis, IN: Center for Philanthropy at Indiana University.

Giving USA Foundation. 2009. Giving USA 2009: The Annual Report on Philanthropy for the Year 2008. Indianapolis, IN: Center for Philanthropy at Indiana University.

GuideStar. 2009. The State of Nonprofit Transparency, 2008: The State of Voluntary Disclosure Practices. Washington, DC: Guidestar.

Hart, T. R. 2002. ePhilanthropy: Using the Internet to Build Support. *International Journal of Nonprofit and Voluntary Sector Marketing* 7 (4): 353–360.

Hart, T. R., J. M. Greenfield, P. Gignac, and C. Carnie. 2006. *Major Donors: Finding Big Gifts in Your Database and Online*. John Wiley & Sons, Inc.

Hoffmann, C. 1970. *The Depression of the Nineties: An Economic History*. Westport, CT: Greenwood Publishing Corporation.

Independent Sector. 2003. *Deducting Generosity: The Effects of Charitable Tax Incentives*. Giving and Volunteering in the United States.

Internal Revenue Service. 2009. Ten Tips for Deducting Charitable Contributions. http://www.irs.gov/newsroom/article/0,id=106990,00.html (accessed June 6, 2009).

John, R. R. 1993. Communications and Information Processing. *Encyclopedia of American Social History*, ed. M. Cayton, E. Gorn, and P. Williams, vol. 3, 2349–2361. New York: Scribner.

Karlan, D., and J. A. List. 2007. Does Price Matter in Charitable Giving? Evidence from a Large-Scale Natural Field Experiment. *American Economic Review* 97 (5): 1774–1793.

The League. 2009. Learning to Give. http://learningtogive.org (accessed April 7, 2009).

Leonhardt, D. 2008. What Makes People Give. *New York Times Magazine* (March 9): 44.

Lindeman, E. 1936/1987. *Wealth and Culture*. Piscataway, NJ: Transaction Publishers.

List, J. A., and D. Lucking-Reiley. 2002. The Effects of Seed Money and Refunds on Charitable Giving: Experimental Evidence from a University Capital Campaign. *Journal of Political Economy* 110 (1): 215–233.

Mather, Cotton. 1966. *Bonifacius: An Essay upon the Good by Cotton Mather*, ed. with an introduction by David Levin. Cambridge, MA: Belknap Press of Harvard University Press.

McCarthy, K. D. 2003. Women and Political Culture. In *Charity, Philanthropy, and Civility in American History*, ed. L. J. Friedman and M. MacGarvie, 179–197. Cambridge, UK: Cambridge University Press.

Mosk, M. 2008. In Obama Fundraising, Signs of a Shift from Online to In-Person. *The Washington Post* (July 18): A06.

Network for Good. 2007. The Young and the Generous: A Study of $100 Million in Online Giving to 23,000 Charities. http://www.networkforgood.org/downloads/pdf/Whitepaper/20061009_young_and_generous.pdf (accessed June 26, 2009).

New York Times. 1936. S. Parkes Cadman Dies in Coma at 71. *New York Times* (July 13): 1.

Obama, B. 2009. Inaugural Address, January 20. Washington, DC.

Okasha, S. 2008. Biological Altruism. Stanford Encyclopedia of Philosophy. http://plato.stanford.edu/entries/altruism-biological (accessed June 29, 2009).

Orshinsky, D. 2005. *Polio: An American Story*. Oxford University Press.

Pirolli, P., and S. Card. 1999. Information Foraging. *Psychological Review* 106: 643–675.

Project Agape. 2009. Facebook Causes, Facebook. http://apps.facebook.com/causes/help?category=About+Us (accessed July 9, 2009).

Putnam, R. D. 2000. *Bowling Alone: The Collapse and Revival of American Community*. New York: Simon & Schuster.

Ragsdale, J. D. 1995. Quality Communication in Achieving Fundraising Excellence. *New Directions for Philanthropy Fundraising*, no. 10 (Winter), http://www.tacticalfundraising.com/uploads/Communication.pdf (accessed July 15, 2009).

Rooney, P. M. and H. K. Frederick. 2009. *2008 Bank of America Study of High Net Worth Philanthropy: Portraits of Donors*. Indianapolis, IN: The Center for Philanthropy at Indiana University.

Simon, H. A. 1971. Designing Information in an Information-Rich World. *Computers, Communications, and the Public Interest*, ed. M. Greenberger, 37–72. Baltimore, MD: Johns Hopkins University Press.

Slovic, P. 2007. Psychic Numbing and Genocide. *Judgment and Decision Making* 2 (2): 79–95.

Small, D. A., G. Loewenstein, and P. Slovic. 2007. Sympathy and Callousness: The Impact of Deliberative Thought on Donations to Identifiable and Statistical Victims. *Organizational Behavior and Human Decision Processes* 102 (2): 143–153.

Soetevent, A. R. 2005. Anonymity in Giving in a Natural Context—A Field Experiment in 30 Churches. *Journal of Public Economics* 89 (11–12): 2301–2323.

Solomon, L. 2009. *Tech Billionaires: Reshaping Philanthropy in a Quest for a Better World*. New Brunswick, NJ: Transaction Publishers.

Spink, A., and C. Cole. 2006. Human Information Behavior: Integrating Diverse Approaches and Information Use. *Journal of the American Society for Information Science and Technology* 57 (1): 25–35.

Thumma, S., and W. Bird. 2008. *Changes in American Megachurches: Tracing Eight Years of Growth and Innovation in the Nation's Largest-attendance Congregations*. Hartford, CT: Hartford Institute for Religion Research.

Toppe, C., A. D. Kirsch, and J. Michel. 2002. Giving and Volunteering in the United States 2001. In *Giving and Volunteering in the United States*. Waldorf, MD: The Independent Sector.

Toppe, C., et al. 2002. Faith and Philanthropy: The Connection between Charitable Behavior and Giving to Religion. In *Giving and Volunteering in the United States*. Waldorf, MD: The Independent Sector.

Trattner, W. I. 1998. *From Poor Law to Welfare State: A History of Social Welfare in America*. 6th ed. New York: Free Press.

Walters, R. G. 1978. *American Reformers 1815–1860*. New York: Hill and Wang.

Waters, R. D. 2007. Nonprofit Organizations' Use of the Internet: A Content Analysis of Communication Trends on the Internet Sites of the Philanthropy 400. *Nonprofit Management & Leadership* 18 (1): 59–76.

Wing, K. T., T. H. Pollak, and A. Blackwood. 2008. *The Nonprofit Almanac 2008*. Washington, DC: The Urban Institute Press.

World Vision. 2009. You've Chosen to Sponsor a Hope Child. http://donate .worldvision.org/OA_HTML/xxwv2DoChildSearch.jsp (accessed June 15, 2009).

4 Airline Travel: A History of Information-Seeking Behavior by Leisure and Business Passengers

Rachel D. Little, Cecilia D. Williams, and Jeffrey R. Yost

Introduction

Although scattered local airline companies began offering flights to passengers as early as 1913, scheduled domestic flights did not become widely available in the United States until the 1920s.[1] During the early years of commercial aviation, U.S. airline travel was limited to a small population of business travelers and wealthy individuals who could afford the high ticket prices. The majority of travelers relied instead on more affordable train services for their intercity transportation needs. Over ninety-five years later, the airlines have grown to be one of the most important and heavily used transportation options for American business and leisure travelers. Following deregulation of the airline industry by the U.S. government in 1978, airline routes increased, ticket fares decreased, and discount carriers prospered, thus making airline travel accessible to a much broader segment of the U.S. population. In 2008 alone, 649.9 million passengers traveled on domestic flights on U.S. airlines (Bureau of Transportation Statistics 2009).

Airline travel requires a significant level of information-seeking behavior on the part of passengers due to the inherent financial and emotional risks involved in travel planning (Pollock 1998, 237). Passengers must invest considerable funds in ticket purchases and may have various feelings and expectations about their airline trip in terms of customer service, airline security, and the purpose of their trip. Throughout the history of commercial aviation, airline passengers have responded to these risks by seeking information about airline carriers, ticket prices, schedules and routes, seat assignments, reservation policies, and in-flight services (e.g., meals, computing services, etc.). As the airline passenger population has diversified, special travel-related information needs have become increasingly important, especially for the elderly, individuals with disabilities, people with other medical issues (including food aller-gies), and families traveling with small children. To respond to these travel information needs, passengers have traditionally relied on such information sources as travel agents, airline ticketing and reservations agents, newspapers, magazines, airline publications, and trusted friends and family members. However, the introduction of the Internet in the mid-1990s,[2] and the growth of airline ticketing and booking Web sites, have given passengers a powerful new information source that is being widely embraced, but is at the same time creating new and more complex information needs.

The Internet as a current information source theoretically empowers travelers by giving them increased access to a broad range of airline ticketing and reservation information that was previously available only through intermediaries such as travel agents. However, this sense of Internet empowerment is illusory because the power to shape and trans-mit information remains under the control of the airlines. Furthermore, in this "disintermediated" world, passengers must now learn how to navigate the Web, how to conduct efficient searches online, how to use online ticket-booking programs, and how to ensure that the infor-mation source that they are accessing is thorough and trustworthy.[3] Most important, they generally wade through a large volume of infor-mation without the support and expertise of travel agents, who were a critical source of airline information in the years leading up to the introduction of the Internet. Airline passengers have therefore had to evolve as information users in response to the growth and evolution of the Internet, and to the airline industry's decision to make a wide range of information available through this media. While the Internet has reduced the role of travel agents as information intermediaries, and has led to a full-circle return to the direct information relationship between passengers and carriers that existed when commercial travel first began, it is not clear that the Internet represents a significant improvement in the airline industry's efforts to meet the information needs of travelers. Instead, the complexity of information online and loss of valued intermediaries have only increased the risks of airline travel decisions. Despite these challenges, the Internet has become one of the most important information sources for a large percentage of today's airline passengers. It is estimated that in 2003, over thirty-five million Americans used the Internet to make travel plans, and the Inter-net "generated nearly $53 billion in leisure travel bookings for 2004—nearly 22 percent of the travel industry's revenue" (Grubesic and Zook

2007, 420). As the Internet becomes increasingly ubiquitous, these numbers will continue to grow.

In addition to the role of the Internet, the information-seeking behavior of airline passengers and the sources that they have consulted have been historically influenced by various endogenous and exogenous factors. These social, industrial, economic, and technological factors are listed in table 4.1 and will be further explored in this chapter.

This chapter will provide a historical survey of the evolution of information users, information-seeking behavior, information needs, and information sources from the beginning of the commercial airline industry to

Table 4.1
Exogenous and endogenous factors that influence the information-seeking behavior of airline passengers

Exogenous factors	Endogenous factors
• Fuel prices	• Improvements to aviation technology (i.e., pressurized cabins, invention of jets)
• General state of the economy	
• The Internet and online ticketing websites (starting in the mid-1990s)	• Improvements to airline safety and security
• Corporate/business travel budgets	• Airline services—i.e., food and entertainment services, and in-flight phone, computer, and Internet access for business travelers.
• Increase in global nature of business operations and greater need for travel	
• Terrorism and safety fears (hijackings, plane bombings, plane crashes, 9/11)	• Incentive programs, such as frequent flyer programs and airline credit cards.
• Role of credit card companies in booking travel (starting in 1941)	• Creation of computer reservation systems (CRS) and global distribution systems (GDS) for airline ticketing and sales—e.g., SABRE and APOLLO (introduced in the 1960s, fully computerized in the 1980s)
• Societal attitudes toward travel (i.e. reaction to Charles Lindbergh's flight in 1927 and role of aviation in World War II)	
• Viability of alternative transportation (i.e. trains, buses, automobiles)	• Post-deregulation pricing and ticketing policies beginning in the 1980s (revenue management systems)
• Government deregulation of airline industry	
• General state of household economies and available money for leisure travel	• Cooperative relationships with other industries (hotels, car rental agencies) and other airlines (code-sharing programs)
• Changing demographics of airline travelers in terms of class, gender, and race	
• Industry use of celebrities as airline passengers and promoters—i.e., Shirley Temple, Eleanor Roosevelt	

the present. We will first discuss our methodology, including the limitations that we have imposed on our research, our approaches in formulating
our analysis, the time and topical boundaries of our study, and our
summary literature review. Four historical periods will be considered: (1)
the early years of airline travel from the 1920s through the early 1940s,
(2) the jet age of travel from the post-World War II era through federal
deregulation of the airlines in 1978, (3) post-deregulation travel from the
late 1970s through the mid-1990s, and (4) the Internet age of travel from
the mid-1990s through the present day. Finally, we will conclude with a
discussion of the dynamics of information-seeking behavior on the Internet, the challenging information needs facing airline passengers as Internet
users, and the limitations imposed on this information source by the
airline industry.

Methodology

This chapter focuses on the history of airline travelers as information
users from the 1920s, when regularly scheduled commercial airline travel
began, to the present. We will discuss the information-seeking behavior
of leisure and business travelers as individuals, but will not cover air
travel that is arranged by corporate travel offices. We are further focusing solely on U.S. airline passengers in order to create a research survey
that is manageable in scope. We are not applying a theoretical approach
to our analysis beyond using a historical survey to study how the
information-seeking behavior of airline passengers has changed over
time. We review the evolving information behaviors, needs, and sources
of everyday business and leisure travelers in this survey, with a particular
concentration on the use of such sources as airline ticketing and reservation agents, independent travel agents, and the Internet. We will
only explore travel by other means, including trains, as it pertains to
the larger discussion of the changing economic and social behaviors of
airline passengers.

The literature consulted for this chapter covers the fields of marketing and advertising, aviation and passenger social histories, airline industry
and travel/tourism research, sociology, psychology, information studies,
and women's studies. We studied historical travel advertisements in the
New York Times, as well as magazines such as *Travel* and *Holiday*. Finally,
we reviewed various airline and ticketing Web sites, as well as online
aviation archives, including those containing historic airplane menus,
photos, and printed flight schedules.[4]

Early Airline Passenger Travel (1920s–1941): An Era of Glamour, Speed, and Adventure

Regularly scheduled passenger airline travel began in the United States in the 1920s.[5] The U.S. Post Office had begun transporting mail by air in 1918, and the Air Mail Act of 1925 provided government contracts to private airline companies, leading to the establishment of air carriers such as American, United, and Eastern Airlines. The airplanes at this time were small and carried very few passengers. In 1926, an estimated 5,800 people purchased airline tickets (Heppenheimer 1995, 22). However, Charles Lindbergh's solo flight across the Atlantic Ocean in 1927 strengthened the public image of the aviation industry and "made aviation the most potent metaphor for progress in the United States" (Douglas 1995, 62).[6]

Although it was perceived as glamorous and adventurous, passenger airline travel during the 1920s was very grueling and expensive. It cost approximately $400 to travel across the country, an amount prohibitively expensive for working- and middle-class travelers. Trips were long, and in 1927, it took an average of thirty-two hours to fly across the country. Airplanes were noisy, lacked climate controls and pressurization, and could not fly at higher altitudes to avoid weather problems or turbulence. Flight delays were common, passengers were sometimes bumped by more profitable mail cargo, and passengers frequently suffered from air sickness en route.[7] Food services were limited and consisted of apples, sandwiches, and coffee from a thermos. If flights were canceled due to bad weather, passengers were forced to continue their trips by train, which was the dominant form of intercity transportation at that time. In 1929, Transcontinental Air Transport offered a combined rail and air service across the country whereby passengers would travel by rail at night, fly during the day, and arrive at their destination in forty-eight hours. While this service was considered to be very luxurious, the company failed after one year because of economic difficulties stemming from the Great Depression.

During the 1930s, the experiences of airline passengers significantly improved. By 1932, TWA was able to reduce cross-country trips to twenty-four hours and airline technology finally allowed planes to fly safely at night. The DC-3 plane was introduced in 1936. While these planes still flew at low altitudes of 5,000 to 6,000 feet, they could carry passengers at faster speeds and for longer distances, were better equipped to handle turbulent weather, and made airline passenger travel profitable for the first time. DC-3 planes had galleys, so the airlines were able to offer passengers improved meal options, including fruit cocktail, fried chicken, rolls, and tea or coffee. Meals were prepared and loaded on board prior to departure, and were heated and served to passengers in flight. The Douglas Sleeper Transport was also introduced in 1936, and could carry fourteen passengers on overnight flights while they slept. Meanwhile, progress was made in the late 1930s to pressurize airplanes so that they could fly at higher altitudes above the weather. Introduced in 1938, the Boeing 307 Stratoliner was the first plane to have a pressurized cabin and could fly at a maximum altitude of 26,000 feet.[8] Even though planes still had to make frequent stops on cross-country trips and delays were common, airline travel was considered to be fast, exclusive, and lavish, with only first-class services provided to passengers.

During the 1920s and 1930s, individuals seeking information about airline travel were predominately white businessmen (including traveling salesmen and company executives) and upper-class individuals. During the 1930s, approximately 90 percent of airline travel was completed for business purposes (Bilstein 1995, 110). Celebrities were also encouraged to fly for airline publicity purposes and did so more frequently after the Hollywood insurance companies lifted the bar against air travel in 1935. The airlines discriminated against African American passengers during the 1930s and would not sell tickets to anyone who was African American or who "sounded like a black person" on the phone (Ibid., 103). The few African American travelers who successfully purchased tickets were segregated on airplanes and in airports.[9] Despite the allure of airline travel, ticket prices and fears about airline safety kept most middle-class travelers from flying. Others were reluctant to fly due to frequent flight cancellations, their ignorance of airline travel and "its value in daily business matters," and the general discomfort involved in travel by plane (Bilstein 1995, 94). Wives of businessmen were particularly fearful about their husbands traveling by air. Some businessmen kept their flights a secret from their wives, but as one businessman said, "every one of us was proud to make a business appointment in a distant city, saying that we would fly to keep it. It gave us prestige" (Ibid., 93–94). To counter these fears, the airline industry actively promoted the speed, safety, and business value of its services and used celebrities such as Shirley Temple and Eleanor Roosevelt to support this message.

Despite marketing campaigns, Americans preferred at a ratio of fifteen to one to travel by train due to the perception that trains were safer and more comfortable, and the reality that rail travel was cheaper (Bilstein 1995, 97). However, improvements in aviation technology and the possibility of faster trips by plane than by train still led to a rapid increase in passenger travel, especially by businessmen. While airline flights in the 1930s were more expensive than Pullman or coach seats on the trains, business travelers were willing to pay higher prices because they could save time by flying overnight. In 1930, approximately 417,000 people traveled by air (Heppenheimer 1995, 22). However, by 1940, over two million passengers had traveled on domestic flights (Solberg 1979, 223). In 1941, this increased to over four million passengers. (Heppenheimer 1995, 124).

The information needs of this passenger population were focused on obtaining reliable information about airline routes and schedules, ticket pricing and purchasing, in-flight services (especially with regard to combating air sickness), transportation to the airports, and safety (given the frequency of accidents in the early aviation industry). Airline travelers obtained this information by contacting the airlines by phone, through airline ticket counters at the airports and hotels, through airline brochures and publications, and through newspaper and magazine articles and advertisements (such as those from *Travel Magazine, Holiday Magazine,* and the *Saturday Evening Post*), radio advertisements, and likely through business associates, friends, and family (despite the often negative public perceptions toward airline travel). The first *Official Aviation Guide of the Airways* was published in 1929, and provided a listing of flight schedules for thirty-five airlines for use by the public.[10] Passengers were required to contact the airlines directly to make ticket reservations, and in larger cities, callers were often connected to the airports because airline employees were available at these locations twenty-four hours a day. In the mid-1930s, the airlines set up reservation offices in large cities, where travelers could call and book travel for flights departing the following day. This served as an advantage over rail travel, as train passengers were usually required to purchase tickets in person at the train stations. Plane tickets were generally prepared by hand using carbon paper and were given to passengers upon their arrival at the airports. American Airlines created the first successful customer loyalty programs in the 1930s through scrip books and the Air Travel Card, which debuted in 1935 and allowed travelers to purchase airline tickets on credit and receive a 15 percent discount on flights. By the 1940s, all of the airlines had similar discount programs, and began allowing passengers to pay for tickets on credit in 1941. American Airlines additionally created the Admiral's Club in the late 1930s for "special patrons, celebrities, and politicians" (Bilstein 1995, 96).

Airline Travel during the Jet Age (1945–1978): The Beginning of General Leisure Travel

Passenger airline service was very limited during World War II, and many commercial airplanes were diverted to the war effort.[11] Passenger seating on existing flights was significantly reduced, and pursuant to government orders, planes could generally only carry military and government personnel. Pleasure trips were virtually nonexistent. Passengers who needed to travel for medical reasons or family emergencies had to compete for the few seats that were available by traveling on "standby" and risking being "bumped" by priority travelers (Solberg 1979, 274–275). During this time, Air Card Travel discounts were discontinued.

Nonetheless, the war increased the public's awareness of the importance of the airline industry to the war effort and the "idea of plane travel . . . captivated the American imagination" (Sickels 2004, 213). Once the war ended, passenger airline traffic increased dramatically. In 1945, 6.7 million passengers traveled by air, and this doubled to 12.5 million in 1946, as "suddenly, everyone wanted to fly" (Heppenheimer 1995, 124). This growth was primarily in response to a reduction in ticket prices, which finally made the airlines competitive with the railroads.[12] At the same time, the airline infrastructure was not prepared to handle this large increase in travelers. Airports in the mid-1940s were chaotic and crowded, telephone reservation lines were busy with callers, passengers often had to wait at least three weeks to get an available seat on a plane, and there were frequent problems with lost luggage. Not only was it difficult to book travel by phone, but it also was almost impossible to call the airlines and cancel a reservation. As a result, the airlines had many "no shows" and passengers who flew standby as "go shows" had a reasonably good chance getting seats (Solberg 1979, 333). Even those individuals who successfully booked a seat over the phone were not always guaranteed a seat. Airline agents did not consistently prepare formal passenger lists and flights were frequently overbooked. If too many passengers showed up for a flight, the airline would simply cancel the flight to avoid liability. To check in for flights, passengers were often required to report to downtown ticket offices and then take crowded buses to the airport.

Despite these problems, aviation technology continued to improve, and newly introduced airplanes, such as the Lockheed Constellation and Douglas DC-6, could fly nonstop across the country, over the ocean, and above turbulence and bad weather. To address the shortage of available seats, nonscheduled flights were initiated on surplus military aircraft.

Known as "fly-by-night operations," these planes operated at off-peak hours and only left the airport if they were full, thus requiring passengers to be flexible with their travel schedules (Heppenheimer 1995, 126–127).

During the 1950s, the number of airline passengers continued to increase dramatically, even though it was still relatively expensive to travel (Taaffe 1959; Young and Young 2004). Passengers were no longer given universal first-class treatment, and were rapidly beginning to represent a broader segment of the U.S. population. Coach services were established in the late 1940s and early 1950s, and by 1958, low-fare air travel constituted two-thirds of the North Atlantic airline business. Coach services were not as comfortable and meal services were limited, but the trade-off for passengers was cheaper airline tickets. It was not until the 1950s that coach and first-class services were put on the same planes. This trend toward passenger diversification continued into the 1960s and 1970s as flying began to be viewed as a regular part of life and less as a luxury, first-class experience. The first-class travel services that remained were not as luxurious as they had been prior to the war, though efforts were made to provide better food service in flight. Liquor service became widely available on flights in 1958 and in-flight movies were introduced in 1961.

At the same time, airlines began to actively and successfully market their services to railroad Pullman and coach passengers. As a result, by 1951, more individuals traveled by air than by Pullman train for the first time. In 1957, the airlines "surpassed both trains and intercity buses in passenger miles, making airlines the nation's leading intercity passenger carrier" (Bilstein 1995, 105). By 1959, fifty-five million individuals had traveled on scheduled domestic flights (Heppenheimer 1995, 170).

The late 1950s were the beginning of the jet age and the Boeing 707 was introduced in 1958, followed by the Boeing 747 in 1970, and the DC-10 and L-1011. Not only were the new jets pressurized, but also they could fly at even higher altitudes and speeds, and passenger discomfort related to turbulence was greatly reduced. Nonstop service was now being offered between the East and West Coasts, and in 1957, Pan Am introduced direct flights between New York City and London. It was now possible to travel anywhere in the world within twenty-four hours. The interiors of the new jets changed significantly, as the staterooms and sleeping berths of the past had disappeared. The public outcry to these changes led to the reintroduction of first-class lounges on jets a few years later. Food service in the form of hot meals continued to be important to travelers, with flight attendants delivering the meals via trolleys on board.

Airline travelers continued to grow in numbers. Passenger traffic reached 169 million in 1970, and by 1970, 54 percent of the U.S. population had traveled by plane (Bilstein 1995, 111; Heppenheimer 1995, 272). During this time period, airline travelers were predominantly businessmen. According to an article in 1964 in *Time* magazine, this group represented approximately 86 percent of domestic airline passengers, though other estimates placed this group at 66 percent of all travelers (Solberg 1979, 406). Nonetheless, the introduction of coach fares and vacation packages led to the growth of leisure travelers, and more families started to fly once or twice a year for holidays and vacations. By the 1970s, approximately 50 percent of passengers were traveling for leisure purposes (Bilstein 1995, 111). The elite "jet set" first-class travelers of the 1950s and 1960s were still important, but their numbers were overshadowed by economy-class travelers (Solberg 1979, 405). Women began to travel more by accompanying their husbands on business trips, and travel by women increased as family vacations became more frequent. College students became an important passenger population in the 1960s, and the airlines began to offer special youth standby fares. There was an increased interest in international travel as well. By 1970, five million Americans had traveled abroad (Bilstein 1995, 108).

Travelers during this time period needed information about pricing, routing, and safety. They were generally more concerned about the reliability and convenience of travel, and less about the traditional in-flight services that had been offered prior to World War II. Business travelers remained concerned about having regularly scheduled flights to accommodate their business schedules, while leisure passengers were focused on lower fares and flexibility in their flight itineraries.

Domestic airplane hijackings became a major problem by the late 1960s. During January 1969 alone, eight U.S. airliners were hijacked to Cuba. The Federal Aviation Administration (FAA) introduced new passenger screening procedures, which were first tested in 1970 and broadly deployed by 1972. On September 11, 1970, President Nixon announced a comprehensive antihijacking program. This included placement of federally trained armed personnel (sky marshals) on some domestic and international flights, accelerated efforts by federal agencies to enhance screening security, and an apparatus for consultation between the U.S. State Department and appropriate foreign agencies to prevent and combat hijacking (Krause 2008, 39–44).

While air travel by this time was safer than auto travel, passengers still worried about airline safety, particularly after two airplanes collided in

1960 over Staten Island and after the wave of hijackings—many by the Popular Front for the Liberation of Palestine—in the late 1960s.[13]

Airline travelers continued to obtain airline information and purchase airline tickets by telephoning the airlines. As information technology progressed, it took longer to book travel because the airlines were now using an electronic signaling system to check availability on flights and confirm seat reservations. During the 1960s, the airlines began using computer reservation systems to automate call centers and seat-reservation procedures for the airlines. Furthermore, tickets were starting to be purchased more frequently on credit, and many travelers were increasingly using travel agents to obtain information about airline tickets. Travel agencies existed in the interwar era, but grew rapidly in number and total air travel bookings from the 1960s through the 1980s.

Despite the growing travel agency industry, some air travelers continued to use airline ticket booths and offices that were available in many cities, and to rely on information sources such as newspapers, magazines (especially *Holiday Magazine*, the leading travel magazine during the 1960s), television, industry publications (such as "How to Travel by Air," which was published by airplane manufacturers and publicized in *Holiday Magazine* during the 1940s), airline route maps, schedules, and brochures, and word of mouth via colleagues, friends, and family. The *ABC (or Alphabetical) Air Guide* started being published in 1946, and this resource provided information about domestic and international flights. The *Official Airline Guide* began publication as an international edition in 1944, and absorbed the previous *Official Aviation Guide of the Airways*.[14] This guide was published monthly, was considered to be an authoritative source of airline travel information, and listed worldwide airline departures and arrivals "in a city pair format" (Lawrence 2004, 195). In 1970, Official Airline Guide (OAG) released the *OAG Pocket Flight Guide*, which was a portable listing of flight schedules targeted for business travelers.[15] Airline passengers additionally relied on the *OAG Travel Planner*, which was first published in 1958 and contained information about airline terminals, transportation to airports, locations of ticket offices, and other travel data.[16] This publication continued to be printed into the 1970s and competing publications were also available, such as the *Air Traveler's Guide* and the *Air Traveler's Destination Directory*.

The introduction of jets into the airline industry and increases in the size of aircraft meant that the airlines could transport more passengers more rapidly. However, airline ticket prices were still relatively high during this time due to federal regulations, and by the 1970s, the airlines were struggling to fill their seats. This "overcapacity" by the airlines would soon becoming a leading factor in the subsequent deregulation of the airlines, which would have far-reaching consequences for passengers in terms of increased flight schedules and reduced airline ticket prices (Gottdiener 2001, 165).

Deregulation and the Pre-Internet Era (1978 to Mid-1990s): A Time of Increased Travel Choices, Reduced Ticket Prices, and Advances in Computer Technology

In 1978, the Airline Deregulation Act was signed into law.[17] Starting in 1938, the airline industry had been managed as a public utility by the Civil Aeronautics Board (CAB). The CAB regulated virtually all aspects of the airline industry and any route or fare changes had to be approved by this agency. By the 1970s, the airlines were facing increased labor and fuel costs, and were having difficulty filling passenger seats due to high fares. American Airlines applied to the CAB and received approval to sell coach seats at discounts of up to 45 percent. These "Super Saver" fares were first offered in 1977, and increased passenger traffic by 60 percent on some routes (Pickrell 1991, 9). The CAB granted similar fare reductions to other carriers, and the success of these fare reductions was a key factor in support of the passage of airline deregulation during the following year.

Following deregulation, the airlines were allowed to set routes and ticket prices, and this increased and improved travel options and reduced fares for passengers (at least for those living in larger cities). During 1980, 57.5 percent of passengers traveled on a discounted fare and the average discount was 42.9 percent of the full-fare ticket price (Taneja 2002, 14). However, other commentators have noted that the average ticket price reduction following deregulation was closer to 15 percent (Pickrell 1991, 29). Airlines that served small communities were given subsidies to prevent these regions from losing airline services because of the expense of flight operations on less-utilized routes. Nonetheless, airline deregulation was a "mixed bag" for passengers and carriers, and consumers in larger markets benefited far more, in reduced prices and increased options, than those in smaller markets (Grubesic and Zook 2007, 418). The airlines created additional problems for passengers through their competitive models, which strove to keep other airline companies out of particular markets through "perimeter rules, long-term exclusive use gate leases, slot controls, and predatory pricing schedules" (Ibid., 419).

Two types of airline choices were available to passengers during the 1980s and 1990s. First, there were full service airlines, which provided more flights to destinations, in-flight and meal services, baggage handling, and direct reservation services. Passengers on these carriers were given more flexible ticketing options, compensation for flight delays, upgrades, free drinks in-flight, and the opportunity to join frequent flyer programs. These carriers appealed in particular to business and first-class travelers. The second option was to travel on a charter airline or discount carrier. These carriers, including People's Express and Southwest Airlines, offered basic travel with limited in-flight amenities (Freiberg and Freiberg 1996). Passengers on these carriers tended to be traveling for leisure. Flights were shorter hauls with more limited schedules. To compete with larger carriers, they tended to fly to "secondary airports" located within fifty miles of the primary airport serving the particular destination (Grubesic and Zook 2007, 420). The seats on these planes were less comfortable, the planes did not have first-class sections, limited compensation was offered for flight delays, there were greater restrictions on making ticket changes, and meal and beverage services were minimal. However, passengers tolerated these reductions in service because of the cheaper fares.

As a result of deregulation, low-cost carriers, and cheaper fares, airline travelers continued to increase in volume and diversity. During the 1980s, the number of domestic passengers increased by 62 percent, and the number of passenger miles traveled on domestic flights grew by 79 percent (Pickrell 1991, 15). In 1980, passenger traffic reached 297 million (Heppenheimer 1995, 314). At the same time, real personal income grew by 25 percent between 1978 and 1988, and passengers were now more willing and able to pay for airline travel (Pickrell 1991, 15). Individuals were also working fewer hours and had more time for leisure activities. Finally, the average size of families had begun to decrease, and it was easier for smaller families to travel by air. During the 1988 holiday season, a record twenty-two million passengers traveled by air (Batchelor and Stoddart 2007, 179). While there was still a mix of leisure and business travelers, the airlines during this time became a leading form of intercity travel for most individuals. In the 1990s, business travelers began to join leisure travelers on discount carriers in greater numbers.

Information needs of travelers during this time increased with the new deregulation pricing and routing schemes, including the "hub and spoke" system (Gottdiener 2001, 167). Prior to deregulation, passengers normally flew on direct flights to their destinations, and they were now being forced to travel through hubs and take connecting or commuter flights. Deregulation created confusion for travelers, who were frustrated with airline competition, fluctuating ticket prices, and the limited schedules of low-cost carriers. Travelers were also concerned about safety on flights as problems with air rage (disruptive behavior by passengers and crew while flying) and hijacking incidents increased. Additionally, a growing number of air travelers were interested in participating in frequent flyer programs that were being established by the airlines during this period. American Airlines started its current frequent flyer program and AAirpass corporate travel card in 1981, and other airlines soon followed suit. During the 1980s, approximately twelve million people joined these programs (Toh and Hu 1988, 11). Budget travelers tended to be less concerned about in-flight services than first-class and business travelers.

While the airline carriers continued to be an important information source through their internal reservation systems (IRS), airline passengers principally relied on travel agents during this time period—which helped counterbalance increasing needs for information. Travel agents were generally more accessible to passengers than airline representatives because of their convenient locations, their efficient services, and their ability to book hotels and rental cars in addition to airline tickets. By the mid-1980s, the airlines had introduced revenue management programs, in which airline computers would frequently change airline ticket prices and engage in discriminatory pricing in order to maximize their revenues. Ticket prices under this system generally depended on both the volume of unsold seats and the number of days until a particular flight. In charging higher fares, airlines took advantage of the greater price inelasticity of business travelers, who tended to book closer to flight dates than vacation travelers.

Travel agents were able to track pricing and schedule changes for passengers through computer reservation systems.[18] The CRS programs of the 1960s had evolved during this time period as computer technology improved. Starting in the mid-1970s, the airlines decided that it was more cost effective to contract out a percentage of their reservation and ticketing services to independent travel agents, and gave travel agents access to these CRS programs through rented computer terminals. In addition to using the *Official Airline Guide*, travel agents could now locate computerized information about flight schedules, fares, and seat availability. The OAG itself was later launched in 1983 as an electronic edition and booking capabilities were added in 1985.[19]

The new CRS programs now gave travel agents the ability to make reservations from terminals in their offices, and to access airline routing and pricing information on a "real-time" basis, which was critical because the airlines made constant changes to routes, service classes, and fares (Dombey 1998, 131). While fare information for each participating airline generally was listed on each of the individual CRS programs, the airline that owned the particular CRS would often highlight its flights on the system as a form of competition. Although government regulations were later introduced in 1984 to prevent this type of CRS owner bias, it was sometimes an issue that travel agents favored certain airlines, thereby reducing passenger choices. In 1976, the airline industry was served by approximately 12,000 travel agency locations with CRS, and this increased to approximately 25,000 locations by 1986 (Pickrell 1991, 23). By 1980, approximately 7,500 travel agent locations were equipped with at least one of the three computer reservation systems or global distribution systems available: SABRE (owned by American Airlines), APOLLO (owned by United Airlines), and PARS (owned by TWA) (Ibid., 24; Poon 1988, 541). By 1986, 95 percent of all travel agencies were equipped with CRS programs, and agents booked "more than half of the airline industry's total dollar value of ticket sales" (Pickrell 1991, 24). In return for booking travel, travel agents received a sales commission from the airline.

Because travel agents now had access to much of the same information as the airlines, they became the primary sales agents for flights and eventually represented approximately 75 percent of airline ticket purchases in the United States (Dombey 1998, 131). While the major airlines participated in CRS, Southwest Airlines did not do so and customers were often required to contact this airline directly in order to purchase their tickets. Given the complicated nature of CRS, reliance on travel agents relieved airline passengers of a considerable burden in terms of making travel reservations, and passengers benefitted from the expertise and personal service that travel agents could provide.

In addition to relying on travel agents, passengers continued to contact the airlines directly by phone and in person at airline ticket offices located away from airports, and used printed flight schedules to make travel decisions. It was often necessary to contact the airlines directly when making complicated ticket changes or addressing ticketing problems. Furthermore, passengers consulted newspapers, television advertisements, travel magazines (such as *Consumer Reports Travel Letter, Frequent Flyer, The Business Flyer, Traveler's Advisory*, and the various OAG guides),[20] airline in-flight magazines, friends, family and colleagues, and began to use CD-ROM and home software programs to research flight information.[21] Frequent flyer programs and publications additionally became more important, and tended to create brand loyalty in that passengers were encouraged to do business directly with one carrier in exchange for points or miles that could be redeemed for future travel or upgrades to first-class seats. Major changes to these information sources would arise as the Internet began to be more widely used in the late 1990s and the beginning of the twenty-first century.

Click and Fly (Mid-1990s–Present): The Age of the Internet

The 1990s were a time of considerable growth in the travel industry. [22] Tourism increased, as well as business travel, due to the greater distances between business locations and customers. Starting in the mid-1990s, the Internet became an important information tool for online commerce and consumers. The airline industry and the travel industry in general quickly took advantage of this technology. In 1996, the travel Web site Expedia. com (which was originally owned by Microsoft) was launched.[23] This Web site allowed passengers to research ticket prices and route schedules, and book airline tickets. During this same year, SABRE Holdings launched the competing Web site Travelocity.com.[24] Orbitz.com started online services in 2001 and other competing online companies soon followed.[25] These online travel agency Web sites brought CRS/GDS technology directly to consumers in a simplified Web-based format. At the same time, the individual airline carriers established their own Web sites where individuals could book travel directly and manage their frequent flyer accounts. In 1995, Alaska Airlines became the first airline to offer ticketing and booking services through its Web site.[26] In addition to airline ticketing programs, these Web sites eventually grew to allow consumers to make rental car, hotel, and vacation package reservations.[27] As online technology evolved, passengers were able to access flight information from their computers and their cell phones. The Internet as an information source sparked a revolution in information-seeking behavior that had important consequences for the airline industry and for independent travel agents—impacts that have become increasingly profound in the past decade with the growing ubiquity of the Internet. In fact, it was online travel purchases that kept the airlines viable during the economic downturn that occurred following September 11, 2001.

At the same time as technology was changing, the airline industry struggled to remain competitive in the face of post-deregulation competition, economic decline, and fluctuations in jet fuel prices. Following the tragic events of September 11, airline travelers faced increased security and fears about future terrorist acts every time that they traveled. Passenger travel dropped immediately after the attacks, imperiling the already pre-carious financial situation of many airlines. This drop resulted from a combination of passenger fears and the declining business travel resulting from the economic downturn that was accentuated by the events of September 11. Although travel has since rebounded, the airline industry has continued to lose money. In response, the airlines have periodically dropped cities as hubs, reduced flight schedules, added fuel surcharges to airline tickets, merged with other carriers, or completely gone out of business.[28] The airlines have also begun using smaller regional jets in an effort to reduce costs and remain competitive.

Airline travelers as information users during this time period have continued to fall into one of two major categories: those desiring luxury or specialized services available on full-service airlines, and those most concerned about low fares available on full-service and budget airlines. Luxury or niche travelers have proven to be as price conscious as other travelers, though unusual destinations and special menus and in-flight services were of greater importance to this group. In contrast to prior historical periods, business travelers have become a smaller segment of the airline passenger community, though they still continue to represent the majority of ticket revenue for most airlines. Leisure travel has grown as household incomes increased and ticket prices were reduced, and airline travel has become common for families and individuals. To remain competitive with business travelers, the airlines have begun to resurrect some of the luxury services that were offered to passengers in the 1930s (such as sleeper seats), and combined these services with options that appeal to business travelers, such as telephone and in-flight Internet services, business lounges, early boarding, extra baggage allowances, and work stations that were built into airline seats. To accommodate the dining needs of passengers, the airlines have begun to provide specialized meals, including those that are vegetarian, kosher, low-sodium, or gluten-free. The airlines have continued to use code-sharing agreements, which benefit consumers by allowing them to book with a larger range of airlines and redeem frequent flyer miles with other airlines as well. At the same time, in response to rising costs, the airlines have begun to charge coach passengers for services that were previously complimentary, including meal services and checked baggage.

The informational needs of leisure travelers have continued to evolve in response to changes in the airline industry. Pricing and routing information have remained critical to most airline travelers, while continuing diversification of the passenger population has meant that travelers needed additional information, such as how to travel if one has medical issues, dietary concerns related to in-flight meal services, and ticketing options for small children and infants. Since September 11, 2001, airline travelers have been concerned with the new security procedures in airports and on board airplanes. The federal government has responded to this need by providing online information about these security procedures through the Transportation Security Administration Web site and advertisements. Passengers have also relied on the news media and individual airline carriers for this type of information. Because of the financial instability of the airline industry, travelers have often required information to ensure that their booked travel would take place and to obtain some type of warning if it appeared that a carrier was planning to cut services or declare bankruptcy. Considering that these changes could result in less convenient and more expensive travel, access to this information has become critical and often comes from airline Web sites, airline personnel, the news, travel agents, or online travel agency Web sites.

As the Internet has grown in importance as an information source, the role of traditional bricks-and-mortar travel agents has declined significantly. This change has been partially driven by the airlines. By the late 1990s, the airlines were spending $6.4 billion per year on travel agent commissions (Lewis, Semeijn, and Talalayevsky 1998, 22). To reduce ticketing costs and make more efficient use of the Internet, the airlines made CRS/GDS programs directly available to passengers through online travel sites such as Expedia, Travelocity, and Orbitz. These online travel agencies were joined by low-cost booking services such as Priceline.com and Cheaptickets.com; by general information sources such as kayak.com that collect ticket-pricing information from other Web sites and make it available to consumers; and by specialty airline information Web sites such as seatguru.com (which provides information about airplane layouts and seating). In 2003, the GDS program Worldspan processed more than 65 percent of all online travel agency bookings (Grubesic and Zook 2007, 421). At the same time, the airlines capped and later eliminated commissions paid to independent travel agents, and this caused many agencies to close or merge with larger companies.[29] To further reduce costs, the airlines introduced e-ticketing. It had traditionally cost the airline an average of $35 to $45 dollars to issue a paper ticket through a travel agent, while e-tickets issued through an airline's Web site cost between $2 and $5 dollars (Lewis, Semeijn, and Talalayevsky 1998, 23). Southwest Airlines went a step further and began offering ticketless travel for all flights in 1995.

Through the Internet, the airlines were able to continue to adjust fares relative to their competitors in real-time as they had done before, to reduce ticket costs to consumers, and now had the option to personalize ticket prices based on private information and travel preferences that were shared by an individual consumer online. The Internet as a new information source therefore broke the control that travel agents held over the information options available to passengers, and passengers now had access to a disintermediated information model, allowing users to bypass retail travel agencies and deal directly with airline suppliers and the airlines themselves through the Internet. This direct access in turn required airlines to create user-friendly interfaces to the CRS/GDS technology so that passengers could successfully negotiate this technology online. As an incentive to encourage Internet use, airlines lured customers to their Web sites by offering cheaper fares, frequent flyer discounts, and extra frequent flyer miles for booking online at the airline's proprietary Web site. Southwest Airlines, which had never used a GDS program, continued to reach customers directly through its own Web site.

Nonetheless, there continued to be a limited role for traditional travel agencies in the information-seeking strategies of travelers, in particular for those who were older, less experienced with the Internet, or wished to supplement the information that they had found online with personal advice.[30] Airline passengers who were planning costly trips (on which they would likely spend on average 58 percent more than the typical leisure traveler) were more likely to consult travel agents to make final arrangements after gathering initial information online (The Economist 2005, 67).[31] In-person or telephone contact with a travel agent or airline representative generally tended to be necessary for itinerary changes, booking frequent flyer travel, and making emergency travel arrangements (such as to attend a funeral). Airlines have been reluctant to make these services accessible online, though some airline Web sites have improved their capacity to book frequent flyer rewards travel. Meanwhile many airlines imposed fees for using airline customer service representatives on the phone rather than making purchases or itinerary adjustments online. Thus, segments of the population—such as some elderly individuals, who continue to engage in transactions by phone (with airlines or intermediaries)—were penalized with higher airfare costs.

In addition to the Internet and travel agents, passengers have continued to use traditional information sources to locate flight information, such as airline reservation agents, television, newspapers, travel magazines, airline publications, and friends and family via word of mouth. Advertisements in magazines, newspapers, radio, and television since the 1990s have largely focused on e-travel companies, and there has been an increase in advertising on the Web. Friends and family remained an important resource, but electronic word-of-mouth became significant as well.[32] Internet users could now read and rate reviews of airlines, airports, or destinations, create blogs devoted to a particular location or airline, or even share photographs with other travelers. The rise of the Internet has virtually eliminated airline ticket offices beyond those available at airports, and printed airline schedules have been replaced with electronic versions that are readily available online via airline Web sites.

The Internet and Current Information Issues

While Internet access has dramatically changed the way in which airline passengers access travel information, this information source has not necessarily empowered travelers and is still primarily under the control of the airline industry and CRS/GDS programs.[33] Information users further face various concerns and problems with the Internet related to the security of the personal information that they transmit across the Internet when purchasing tickets, the reliability of online travel information, and the constant fluctuations in travel information. The Internet is a convenient and efficient source of information, but today's passengers still face the same historical challenges in terms of their ability to seek useful information from the airlines. They also face difficulties given the growing sophistication and volume of information on the Internet and the frequent lack (or growing cost) of useful intermediaries to guide them through this information. In reaction to these issues, many airline passengers have used the Internet to create educational blogs and information-sharing programs that rate the services airlines provide.

The demographic profile of airline travelers who use the Internet to gather travel information and buy airline tickets has changed over the past decade. Early Internet users in the late 1990s were predominantly well-educated males whose yearly salary was over $50,000 (Dombey 1998, 132). That men were the principal online shoppers of airline travel during the early years is not surprising, as this corresponds with the preceding statistic for Internet users generally. However, over time women have gained more familiarity with both computers and the Internet, and an AcuPoll survey in 2004 found that 54 percent of female travelers used the Internet to make travel arrangements (Marketing to Women 2004, 1). Today's users are drawn to the Internet because it is convenient and user friendly. As users have become more comfortable and familiar with computers and the Internet, they have embraced the Internet as a significant source of travel information. In 2005 54 percent of Internet users started their travel information search with an online travel agent (The Economist 2005, 65).

In terms of search patterns, leisure travelers generally have more time to book travel and search for deals online than business travelers. Some leisure travelers "love the `thrill of the hunt' . . . when individually booking through online brokers; while others may be easily frustrated by the seemingly endless choices of travel combinations" (Smith and Rupp 2004, 87). Leisure travelers online have generally been interested in "a convenient way to access information such as airline schedules and fares, and obtain personalized advice or share past experiences" (Lewis, Semeijn, and Talalayevsky 1998, 22) Business travelers are more focused on getting reasonable fares, but "trip planning is not high on their daily agenda—in fact, for many business travelers, it is a last minute detail that is reluctantly attended to" (Smith and Rupp 2004, 88). For business travelers, their greater concern with online travel Web sites is to have the ability to "control travel costs and enter into long-term pricing arrangements with travel suppliers. They also want to have repeat information (such as travel preferences and corporate travel policies) stored for later use" (Lewis, Semeijn, and Talalayevsky 1998, 23).

Despite the convenience of the Internet, online shopping is "characterized by a struggle for control between producers and users" (Pan and Fesenmaier 2006, 826). This is the dominant theme for travelers using the Internet as an information source. Access to travel information has always been asymmetrical. Travel agents and the airline carriers controlled privileged information during the pre-Internet era, and now the airlines and the major travel sites manage the information available to travelers. For many travelers, the perceived difficulty of online transactions influences their decision whether to use the Internet in addition to other sources of information. However, as Internet use has become increasingly common, shopping or seeking information online is no longer seen as quite as complicated.

The single most important concern for the majority of travelers shopping online is the security of sensitive information. Many travelers who do not purchase travel products online cite their feelings of discomfort with sharing private information online, such as their address or credit card number.[34] Users are often unable to evaluate the trustworthiness of online vendors, and this issue of increasing user confidence in sharing personal or financial information online has led retailers to make their Web sites more comfortable and transparent by posting privacy policies.

Reliability and authenticity of online travel vendors is another important concern of travelers seeking information online. Much of the language of travel information is in the industry's language, which may be confusing to many Internet users. Travelers using the Internet are starting to seek authenticity of information by establishing an open dialogue with other users, allowing for a democratization of travel information. As will be discussed further, this open dialogue includes access to travel blogs, comments and reviews on industry or cyberintermediary travel Web sites, and the ability to share individual opinions with the airline industry.

The quality and quantity of information choices on the Internet has become a problem for many travelers. Travelers are often overwhelmed by the amount of airline information available online and are unsure how to locate the answers to their questions. This overwhelming quantity stems in part from the numerous Web sites produced both by the airlines themselves and by cyberintermediaries. In response, travelers have sought to identify and use authoritative information "hubs or clusters" as a way to avoid collecting invalid or inaccurate information.[35] The independent online travel sites, such as Travelocity, Kayak, and Expedia, were created through partnerships of GDS/CRS technology and commercial entrepreneurs. Some cyberintermediaries are owned by airlines as well.

The number of "players" affects the validity of online content, with many sites providing conflicting information about ticket prices (Yost 2008, 320–326). Some of this disparity is due to differences in ticket commissions of the various online vendors. Some of the major cyberintermediary companies, such as Expedia, have recently eliminated online booking fees.[36] Travelers are expected to negotiate varying price structures and conflicting information on their own, and this displacement of work onto the consumer is one of the notable changes in the Internet era. Today's travelers must spend considerable time doing the work they used to pay a travel agent to perform. And while some travelers have been able to successfully use travel Web sites to strategically purchase airline tickets, airlines are hardly ignorant of users' stratagems and constantly revise pricing models to ensure maximized revenue. Furthermore, while some travelers may enjoy the independence and freedom to make their own arrangements, all travelers face significant financial and emotional risks in seeking the "right" ticket.[37]

Travelers generally begin an online search for airline information with some background knowledge or perception of the airline, destination, or source. They conduct an internal search based on their own parameters, and an external search using information gathered from family and friends, and the marketplace. Previous experiences are weighted heavily, including satisfaction or frustration with specific airlines, airports, or destinations. When searching online, travelers sometimes use search strategies and semantics that differ not only from each other but from the airline industry and retailers themselves. Users' searches are often frustrated by the marketing and promotional language dominating travel Web sites; such sites could be more beneficial to passengers if they used personalized, or non-marketing, language. However, frustration with the language of Web sites does not prevent users from readily searching or purchasing airline products online, as users often perceive Web sites as having the best prices available.

It is difficult for travelers to become strategic buyers. As buyers make their choices from the wide selection of travel times and ticket prices offered by airlines and independent retailers, they believe the Internet is giving them good insight into the comparative pricing of tickets by competitors. However, the systems are not as transparent as they appear. Most users are unaware of the airline industry's use of sophisticated revenue management systems designed to benefit the airlines, not passengers. The pricing algorithms these systems use enable airlines to anticipate user behavior and prevent travelers from strategic buying. For example, the Web permits airlines to continually adjust the number of seats available for various fare rates—ensuring maximum profit. In 1997 United Airlines, Inc. alone made ten million fare changes, and with the Internet today, dynamic pricing strategies have become more popular with airline carriers (Ibid.).[38]

To the detriment of airline passengers, some price scamming operations have recently been detected on the Internet. Since airlines can change their fare rates within seconds, users may find that "slippery price displays" result in ticket prices that appear in one amount and then increase by the time the passenger is ready to book his ticket reservation (Elliott 2008). Other purchasing models use equally deceptive and confusing practices. Hotwire.com requires users to complete an online purchase before it shares details about the travel product with the user, including which airline the user will be traveling. Some users understandably are uncomfortable with this approach. The purchasing model on Priceline.com allows customers to define the prices they are comfortable with paying for specific travel products, but this model has not been embraced by the majority of e-travelers, who prefer using online intermediaries such as Orbitz or Expedia that provide some customer support and value-added information. These inter-mediaries do not simply "dump" all available data on users; instead the user-friendly interface is designed to allow users to walk through the selection process. For example, rather than respond to a user's search by listing *all* available flights on a specific date, an intermediary will ask the user to sort by her preferences, including total flight time, price limit, time of day for departure, and preferred number of connections (nonstop vs. one or more plane changes).

Internet users are able to browse information and address information needs as soon as they know they have to travel. While many travelers access the Internet seeking specific airline information, the convenience of the Internet encourages casual browsing. Many Internet users visit travel information sites without an immediate intent to purchase airline tickets, but rather for long-term information gathering (The Economist 2005, 66).

The popularity of the personal computer has allowed airlines to offset many of their traditional expenses. Airlines since the 1990s have increasingly relied on electronic tickets that in many instances travelers can print out at home or at hotels. Airlines have embraced e-tickets for two main reasons: they facilitate a direct connection with the user in his home, and they are cheaper to produce than traditional tickets. E-ticketing appeals to users because it reduces the overall cost of airline tickets. As users have become more familiar with this technology, e-ticketing has reduced their need to consult travel agents to handle ticket purchasing. While many travelers may find the convenient, low-price tickets available online satisfying, this should be balanced with the understanding that skilled and knowledgeable people, such as travel agents, can provide uniquely specialized information based on their considerable time invested in learning the intricacies of airline systems and travel product vendors.

Finally, airline information on the Internet is often complex and voluminous. Without trusted intermediaries, consumers are left to depend on the airlines and online booking programs in order to understand this information. Rather than rely on travel agents, the Internet has forced many travelers to depend on each other for information and empowerment. As a result, many airline passengers have started using the social networking technology that is available on the Web to create chat groups, blogs, and advice-based Web sites. These sites provide word-of-mouth advice and support for travelers, who can post complaints, ratings, or useful airline information. Some noteworthy examples include http://www.smartertravel.com/, http://www.tripadvisor.com, and http://www.thirtythousandfeet.com/travel.htm. Blogs specifically targeted to business

travelers have additionally grown in popularity, such as http://www
.inflighthq.com/ and http://crankyflier.com/. Web sites created by travel
experts and trusted travel publications have also proliferated in an effort
to educate and help airline passengers find useful information online.
These include the "Travel Troubleshooter" Web site at www.elliot.org,
Budget Travel Magazine at http://www.budgettravel.com/, and National
Geographic *Traveler* magazine at http://traveler.nationalgeographic.com/.
Airline passengers further have the option to post complaints with an
airline directly on that carrier's Web site. While independent Web sites do
not necessarily change the level of control that airlines exercise over the
creation of ticketing and routing information, they do have the potential
to improve the information-seeking efforts of airline passengers, to force
airline ticketing policies to become more transparent, and to offer the
airline industry a clearer picture of the information needs of airline
passengers.

Conclusion

Information-seeking behavior has come full circle, as information intermediaries such as brick-and-mortar travel agents have diminished in importance, and travelers today seek ticketing information directly from the
airlines via the Web sites of online travel agencies and of individual carriers.[39] The Internet is the new distributer of airline information, and in a real
sense users can interact directly with airlines and reduce the time and cost
of consulting human intermediaries. While it can be argued that through
the Internet the "individual becomes his/her own travel agent," information seekers are being lured into a false sense of empowerment as the airline
industry continues to control the prices, services, and destinations that
travelers can choose from (Smith 2004, 296–297). The complexity and
volume of information that passengers must process on the Internet further
creates significant challenges for them. Internet users can band together to
post complaints about airline policies and offer each other advice on how
to navigate travel Web sites safely and efficiently, but these users still must
interact within the confines of the airline industry's CRS technologies, and
within the constraints of the Internet itself. Certain airline services such as
frequent flyer programs have been kept under close industry control and
Internet intermediaries are only able to provide limited information about
these services. There is a growing assumption that this information should
be freely and easily available online, and this may eventually force the
airline industry to share information without restraint.

Table 4.2
Summary of information users, information needs, and information sources as related to airline ticketing and travel

Time period	Information users	Information needs	Information sources
1920s–1941	Predominately businessmen (including traveling salesmen), upper-class individuals, and celebrities. Relatively few women and middle-class travelers traveled by air during this time.	Airline routes and schedules, ticket pricing and purchasing, in-flight services, transportation to the airports, and aviation safety	Telephone calls to airline reservation offices; visits to airline ticket counters in airports and hotels; airline publications and brochures; newspaper and magazine articles and advertisements (such as those from *Travel Magazine*, *Holiday Magazine*, and the *Saturday Evening Post*); radio advertisements; the *Official Aviation Guide of the Airways*; and business associates, family and friends
1945–1978	Predominantly businessmen and wealthy leisure travelers. Overall increase in economy-class leisure travel. Diversification of passenger population to include families, women, and college students	General: ticket pricing, airline routing and safety (especially regarding airline hijackings), and reliability and convenience of travel. Business travelers: locating flights that accommodate business schedules. Leisure travelers: identifying low fares and flexible flight schedules	Telephone calls to airports and airline reservation centers (which were using CRS); visits to airline ticket offices and airports; independent travel agents; newspapers, magazines (especially *Holiday Magazine*), and television; airline industry publications (e.g., "How to Travel by Air,"); printed route maps and flight schedules; publications such as the *ABC Air Guide*, the *Official Airline Guide*, the *OAG Pocket Flight Guide*, and the *OAG Travel Planner*; and business associates, family, and friends

Table 4.2
(continued)

Time period	Information users	Information needs	Information sources
1978–mid-1990s	Mix of business and leisure travelers, with overall increase in volume and diversity of passengers	Post-deregulation airline pricing and routing schemes (e.g., "hub and spoke" system), airline safety (as related to hijackings and passenger "air rage" incidents), frequent flyer and incentive programs. Overall focus on locating affordable tickets and feasible flight schedules over in-flight services	Principal information source: independent travel agents with CRS access. Other sources: airline ticketing and reservation agents using internal reservation systems (contacted by phone or by visiting airline offices); airline publications and printed flight schedules; newspapers, television advertisements, travel magazines (such as *Consumer Reports Travel Letter*, *Frequent Flyer*, *The Business Flyer*, *Traveler's Advisory*, and the various OAG guides); frequent flyer publications; CD-ROM and home software travel programs; and word of mouth
Mid-1990s–present	Mix of business and leisure travelers flying on full service and discount carriers. Continued diversity of travelers to include broad cross section of population	Ticket pricing and routing information, airline security and safety (especially post-9/11), diversification of information needs (including information about travel with medical issues, dietary concerns related to in-flight meal services, and issues with traveling with small children and infants), frequent flyer information, fees for in-flight services, and overall financial stability of airline carriers	Internet, including airline carrier websites, online travel agencies (e.g., Travelocity and Expedia), blogs, and independent information/advice Web sites; brick-and-mortar travel agents (to a significantly lesser degree); phone contact with airline ticketing and reservation agents; television, newspapers, and travel magazines; airline publications; and friends and family/word of mouth

Today's travelers have the ability to consult multiple Internet and traditional information sources when making travel plans, and many users are taking advantage of the variety of sources and forms of travel information that are available. Because of existing limitations on the Internet as an information source, such as differences in semantics and language between online travel vendors and users, it is unlikely to become the sole source of travel information for airline passengers. The Internet has forced traditional sources of airline travel information to change the way they conduct business. However, some traditional sources of information, such as family and friends, will continue to be valuable to travelers.

Finally, the role of women as information users has evolved and this represents an important trend in the field of information seeking behavior by airline passengers. Women in the 1920s and 1930s were generally frightened by the thought of their husbands traveling by air and rarely flew themselves. In the new millennium, airline tickets are one of the most common items that women purchase online, and six out of ten married women are responsible for purchasing airline tickets (Marketing to Women 2004, 1). The airlines are actively marketing to this population of information users, and American Airlines recently created AA.com/Women to reach female travelers, who make up 48 percent of its customer base (Marketing to Women 2007, 1). Overall, the average air traveler has changed tremendously since the 1920s, as have the information sources that travelers consult when planning and booking airline trips. While the Internet has made the information needs of today's travelers more complex, the basic information that travelers seek regarding ticket prices, travel routes, airline schedules, and security and safety remain very much in line with those of the original pioneer airline passengers of the 1920s (table 4.2).

Notes

Please note that all Internet citations and URLs listed in this chapter are current as of November 24, 2009.

1. This section relies heavily on Bilstein 1995, Freeman n.d., Pollock 1998, and Solberg 1979.

2. Throughout this chapter, the term *Internet* is used to refer the commercial Internet—made possible in the early to mid-1990s by the World Wide Web and browser software. The Internet, a network of networks based on the adoption of standard protocols (TCP/IP), was in place by the early 1980s (though only used by limited segments of the population, particularly individuals employed in government and academia).

3. See Dombey 1998 and Lewis and Talalayevsky 1997.

4. See http://www.departedflights.com/, http://www.centennialofflight.gov/, and http://www.library.northwestern.edu/transportation/digital-collections/menus/.

5. Although a few small airline companies began offering limited flights in California and Florida, regularly scheduled commercial passenger service did not begin until the mid-1920s. One historian sets this date at 1926 (Hudson 1972, 31).

6. This section relies heavily on Bilstein 1995, Cruz and Papadopoulos 2003, Douglas 1995, Freeman n.d., Gottdiener 2001, Heppenheimer 1995, Hruschka and Mazanec 1990, Hudson 1972, Lyth 1993, O'Hara and Strugnell 1997, and Solberg 1979.

7. On Western Air flights, passengers were given "burp containers" for this purpose (Solberg 1979, 115). Airplane windows could also be opened as well and "many simply leaned out and upchucked" (Ibid.). In some cases, the interiors of airplanes had to be hosed down after landing because so many passengers had gotten sick during flights (Heppenheimer 1995, 25).

8. See Solberg 1979, and The Aviation History Online Museum, http://www.aviation-history.com/boeing/307.html.

9. This segregation was eliminated in airplanes and terminals following the passage of civil rights legislation in the 1960s. See further discussion in Bilstein 1995.

10. See http://www.oag.com/oagcorporate/history_of_OAG.html.

11. This section relies heavily on Bilstein 1995, Cruz and Papadopoulos 2003, Gottdiener 2001, Heppenheimer 1995, Hruschka and Mazanec 1990, Hudson 1972, Lyth 1993, Reilly 2003, and Solberg 1979.

12. In 1941, passenger miles on trains were six times greater than those on airplanes. This was reduced to a three to one ratio by 1946 (Heppenheimer 1995, 125).

13. The FAA has long compiled and publicly disseminated information on air travel safety and accident and incident statistics—these data are now provided on the FAA Web site, http://www.faa.gov.

14. We had difficulty locating historical information about this publication. This search was complicated by the fact that at least four different publishers have owned this publication at different times throughout its history, and that different geographic editions have been issued. The best information we could locate was from the Library of Congress Web site, at http://lccn.loc.gov/sv%2088085741; http://lccn.loc.gov/sv%2088085641. This publication is also referred to as the *OAG Tariff Book* (Brown 1989, 69).

15. See http://www.oag.com/oagcorporate/history_of_OAG.html.

16. See http://www.usps.com/judicial/1975deci/2-78.htm.

17. This section relies heavily on Boberg and Collison 1985, Brown 1989, Cruz and Papadopoulos 2003, Dombey 1998, Gottdiener 2001, Grubesic and Zook 2007, Hruschka and Mazanec 1990, Kleit 1992, Pickrell 1991, Poon 1988, Schulz 1997, and Taneja 2002.

18. CRS programs that handle ticket sales and bookings for multiple airlines are also known as global distribution systems (GDS). Examples of GDS programs include SABRE, Worldspan, and Galileo (further discussed in the text). Various works consulted for this chapter used the abbreviations CRS and GDS interchangeably. See Dombey 1998, 129.

19. See http://www.oag.com/oagcorporate/history_of_OAG.html. The print edition of the OAG continued to be important for travel agents who made bookings on airlines that did not participate in CRS/GDS programs (O'Connor 2001, 121). The *Official Airline Guide* is still published today, and OAG continues to release various travel planning products online that are geared toward corporate travel planners.

20. During the late 1980s through the mid-1990s, the airlines spent approximately $574 million per year on travel advertising (Crotts 1999, 149–150).

21. See Bruce 1987 for a thorough retrospective discussion of technology and travel information in the 1980s, including a description of Prestel's "Travel Desk," an early electronic travel agency that mimics today's Internet travel sites.

22. This section relies heavily on The Age 2005, Beach and Lockwood 2003, BRW 2008, Dietrich, Snowdow, and Washam 1997, Dombey 1998, The Economist 2004, Frank 2008, Grant 1996, Grubesic and Zook 2007, Horrigan (2008), Hruschka and Mazanec 1990, Lewis, Semeijn, and Talalayevsky 1998, Oxoby 2003, Pae 2008, Raine 2001, Reklaitis 2008, Smith 2004, Smith and Rupp 2004, Swarbrooke and Horner 2001, Transportation Group International, LC 2002, and Weber and Roehl 1999.

23. See http://overview.expediainc.com/phoenix.zhtml?c=190013&p=overview.

24. See http://www.sabre-holdings.com/ourBrands/travelocity.html.

25. See http://pressroom.orbitz.com/index.php?s=43&item=302.

26. See http://www.alaskasworld.com/newsroom/ASNews/AS-Fact-Sheet_Innovation.asp.

27. For example, Expedia currently owns hotels.com, Hotwire.com, and TripAdvisor.com.

28. Following 9/11, United Airlines announced that a 20 percent fleet reduction would occur by 2005 (Grubesic and Zook 2007, 417).

29. Traditional travel agents that have survived the Internet revolution tend to be those that handle travel for large corporations, though further discussion of this dynamic is beyond the scope of this chapter. See Lewis, Semeijn, and Talalayevsky 1998.

30. See Smith and Rupp (2004, 84), noting that "regardless of the amount of information and control consumers are given over their travel plans, there is always going to be a portion of the population that will still [relies] on travel agents. Regardless of the expanding power of the Internet, many people prefer to deal with people, not computer screens."

31. Bennett 1993 in particular discusses the danger travel agents face, noting that the future employment prospects of travel agents depend not only on their ability to work with the Internet and CRS technology, but also on their exceptional personal-service skills.

32. See Crotts 1999 for a further discussion of the power of word of mouth advertising among friends and family as it relates to the travel industry.

33. This section relies heavily on Anderson and Wilson 2003, Chen and Gursoy 2000, Clemons, Hann, and Hitt 2002, Dombey 1998, Elliot 2008, Horrigan 2008, Jang and Feng 2007, O'Connor 2004, Pan and Fesenmaier 2006, Pröll et al. 1998, Smith 2004, Smith and Rupp 2004, Transportation Group International, LC 2002, Weber and Roehl 1999, Wellman and Haythornthwaite 2002, and Yost 2008.

34. One study noted that 92 percent of consumers were worried about online privacy, "with 61 percent concerned enough to refuse to shop online" (O'Connor 2004, 403).

35. See Pan and Fesenmaier 2006.

36. See Nassauer 2009, D2.

37. Cooper 1993 offers a description of the way airline carriers compete with each other over ticket prices and often sell each other's products. Travelers are for the most part unaware of the effort made by airlines to control price structures, and these efforts further complicate travelers' efforts to find the right ticket to purchase.

38. This paragraph additionally draws from Taneja 2002.

39. This section relies heavily on Dombey 1998, Pan and Fesenmaier 2006, Reklaitis 2008, The Economist 2005, Transportation Group International, LC 2002, and Weber and Roehl 1999.

Bibliography

The Age. 2005. Two Big Airlines File for Bankruptcy. TheAge.com, September 15 (accessed November 20, 2005).

Anderson, C. K., and J. G. Wilson. 2003. Wait or Buy? The Strategic Consumer: Pricing and Profit Implications. *Journal of the Operational Research Society* 54 (3): 299–306.

Anthes, G. H. 1998. The Price Had Better Be Right: New Technologies Let Companies Peg Prices to the Marketplace. *Computerworld* 21 (December): 65–66.

Batchelor, B., and S. Stoddart. 2007. *The 1980s*. Westport, CT: Greenwood Press.

Beach, W. A., and A. S. Lockwood. 2003. Making the Case for Airline Compassion Fares: The Serial Organization of Problem Narratives during a Family Crisis. *Research on Language and Social Interaction* 36 (4): 351–393.

Bennett, M. M. 1993. Information Technology and Travel Agency: A Customer Service Perspective. *Tourism Management* 14 (4): 259–266.

Bilstein, R. E. 1995. Air Travel and the Traveling Public: The American experience, 1920–1970. In *From Airships to Airbus: The History of Civil and Commercial Aviation, Volume 2—Pioneers and Operations*, ed. W. Trimble, 91–111. Washington, DC: Smithsonian Institution Press.

Boberg, K. B., and F. M. Collison. 1985. Computer Reservation Systems and Airline Competition. *Tourism Management* 6 (3): 174–183.

Brown, G. A. 1989. *The Airline Passenger's Guerilla Handbook: Strategies & Tactics for Beating the Air Travel System*. Washington, DC: The Blakes Publishing Group.

Bruce, M. 1987. New Technology and the Future of Tourism. *Tourism Management* 8 (2): 115–120.

BRW. 2008. One Step Beyond. *BRW* 30, no. 19 (May 15): 38.

Bureau of Transportation Statistics. 2009. December 2008 Airline Traffic Data: System Traffic Down 5.7 Percent in December from 2007 and Down 3.7 Percent in 2008. US DOT Publication No. BTS 12–09, June 1. Washington, DC: U.S. Government Printing Office.

Chen, J. S., and D. Gursoy. 2000. Cross-Cultural Comparison of the Information Sources Used by First-time and Repeat Travelers and Its Marketing Implications. *International Journal of Hospitality Management* 19 (2): 191–203.

Clemons, E. K., I. Hann, and L. M. Hitt. 2002. Price Dispersion and Differentiation in Online Travel: An Empirical Investigation. *Management Science* 48 (4): 534–549.

Cooper, R. E. 1993. Communication and Cooperation among Competitors: The Case of the Airline Industry. *Antitrust Law Journal* 61 (2): 549–557.

Crotts, J. C. 1999. Consumer Decision Making and Prepurchase Information Search. In *Consumer Behavior in Travel and Tourism*, ed. A. Pizam and Y. Mansfeld, 149–168. New York: Haworth Hospitality Press.

Cruz, A., and L. Papadopoulos. 2003. The Evolution of the Airline Industry and Impact on Passenger Behaviour. In *Passenger Behaviour*, ed. R. Bor, 32–44. Burlington, VT: Ashgate Publishing Company.

Dietrich, B. L., J. L. Snowdow, and J. B. Washam. 1997. The Promise of Information Technology in the Travel Industry. In *Information and Communication Technologies in Tourism 1997*, ed. A. Tjoa, 129–139. New York: Springer-Verlag/Wien.

Dombey, A. 1998. Separating the Emotion from the Fact: The Effects of New Intermediaries on Electronic Travel Distribution. In *Information and Communication Technologies in Tourism 1998*, ed. D. Buhalis, A. Tjoa, and J. Jafari, 129–138. New York: Springer-Verlag/Wien.

Douglas, D. 1995. Airports as Systems and Systems of Airports: Airports and Urban Development in America Before World War II. In From Airships to Airbus: The History of Civil and Commercial Aviation, Volume 1—Infrastructure and Environment, ed. W. Trimble, 55–84. Washington, DC: Smithsonian Institution Press.

The Economist. 2004. Click to Fly. *The Economist* 371, no. 8375 (May 15): 8–11.

The Economist. 2005. Flying from the Computer. *The Economist* 377, no. 8446 (October 1): 65–67.

Elliott, C. 2008. Unfair Fares: 5 Shopping Strategies. http://www.tripso.com/columns/unfair-fares-5-secrets-for-avoiding-the-bait-and-switch/ (accessed November 26, 2008).

Frank, T. 2008. Ads Try to Change Fliers' Attitudes Toward Screening. *USA Today*, November 25, p. 01a.

Freeman, R. n.d. The Pioneering Years: Commercial Aviation 1920–1930. U.S. Centennial of Flight Commission. http://www.centennialofflight.gov/essay/Commercial_Aviation/1920s/Tran1.htm (accessed December 8, 2008).

Freiberg, K., and J. Freiberg. 1996. *Nuts! Southwest Airlines' Crazy Recipe for Business and Personal Success*. Austin, TX: Bard Press, Inc.

Gottdiener, M. 2001. *Life in the Air: Surviving the New Culture of Air Travel*. Lanham, MD: Rowman & Littlefield Publishers, Inc.

Grant, B. J. 1996. Trends in US Airline Ticket Distribution. *The McKinsey Quarterly* (4): 179–184.

Grubesic, T., and M. Zook. 2007. A Ticket to Ride: Evolving Landscapes of Air Travel Accessibility in the United States. *Journal of Transport Geography* 15: 417–430.

Heppenheimer, T. A. 1995. *Turbulent Skies: The History of Commercial Aviation*. New York: John Wiley & Sons, Inc.

Horrigan, J. B. 2008. Online Shopping: Internet Users Like the Convenience But Worry about the Security of Their Financial Information. Pew Internet & American Life Project, February 13. http://www.pewinternet.org/~/media//Files/Reports/2008/PIP_Online%20Shopping.pdf.pdf (accessed November 17, 2008).

Hruschka, H., and J. Mazanec. 1990. Computer-Assisted Travel Counseling. *Annals of Tourism Research* 17: 208–227.

Hudson, K. 1972. *Air Travel: A Social History. Bath.* Somerset, England: Adams & Dart.

Jang, S., and R. Feng. 2007. Temporal Destination Revisit Intention: The Effects of Novelty Seeking and Satisfaction. *Tourism Management* 28 (2): 580–590.

Kleit, A. N. 1992. Computer Reservations Systems: Competition Misunderstood. *Antitrust Bulletin* 37 (4): 833–861.

Krause, T. L. 2008. *The Federal Aviation Administration: A Historical Perspective, 1903–2008.* Washington, DC: U.S. Department of Transportation.

Lawrence, H. 2004. *Aviation and the Role of Government.* Dubuque, IA: Kendall Hunt Publishing.

Lewis, I., and Talalayevsky, A. 1997. Travel Agents: Threatened Intermediaries? *Transportation Journal* 36 (3): 26–30.

Lewis, I., J. Semeijn, and A. Talalayevsky. 1998. The Impact of Information Technology on Travel Agents. *Transportation Journal* 37 (4): 20–25.

Lyth, P. J. 1993. The History of Commercial Air Transport: A Progress Report, 1953–1993. *Journal of Transport History* 14 (2): 166–203.

Marketing to Women. 2000. Women Look for Smooth Online Transactions. *Marketing to Women* 13 (8): 1–2.

Marketing to Women. 2004. Women Handle Most Travel Arrangements: Their Top Tool Is the Internet. *Marketing to Women* 17 (11): 1–3.

Marketing to Women. 2007. Women Are Clicking: The Online World Becomes a Vital Platform for Reaching Female Consumers of All Ages. *Marketing to Women* 20 (6): 1–2.

Nassauer, S. 2009. Expedia Eliminates Fees For Booking Flights Online. *Wall Street Journal*, May 28, p. D2.

O'Connor, P. 2004. Privacy and the Online Travel Customer: An Analysis of Privacy Policy Content, Use and Compliance by Online Travel Agencies, 401–412. In *Information and Communication Technologies in Tourism 2004*, ed. A. Frew. New York: Springer-Verlag/Wien.

O'Connor, W. E. 2001. *An Introduction to Airline Economics.* Westport, CT: Praeger.

O'Hara, L., and C. Strugnell. 1997. Developments in In-Flight Catering. *Nutrition & Food Science* 3: 105–106.

Oxoby, M. 2003. *The 1990s.* Westport, CT: Greenwood Press.

Pae, P. 2008. American Airlines to Charge for Checked Baggage. Los Angeles Times, May 21. http://travel.latimes.com/articles/la-trw-american22-2008may22 (accessed December 6, 2008).

Pan, B., and D. R. Fesenmaier. 2006. Online Information Search: Vacation Planning Process. *Annals of Tourism Research* 33 (3): 809–832.

Pickrell, D. 1991. The Regulation and Deregulation of U.S. airlines. In *Airline Deregulation: International Experiences*, ed. K. Button, 5–47. London: David Fulton Publishers Ltd.

Pollock, A. 1998. Creating Intelligent Destinations for Wired Customers. In *Information and Communication Technologies in Tourism 1998*, ed. D. Buhalis, A. Tjoa, and J. Jafari, 235–247. New York: Springer-Verlag/Wien.

Poon, A. 1988. Tourism and Information Technologies. *Annals of Tourism Research* 15: 531–549.

Pröll, B., W. Retschitzegger, P. Kroiß, and R. R. Wagner (1998). Online Booking on the Net—Problems, Issues and Solutions. In *Proceedings of the 5th International Conference on Information Technology in Tourism (ENTER)*, ed. D. Buhalis et al., 268–277. New York: Springer.

Raine, L. 2001. How Americans Used the Internet after the Terror Attack. Pew Internet & American Life Project, September 15. http://www.pewinternet.org/PPF/r/45/report_display.asp (accessed November 17, 2008).

Reilly, E. J. 2003. *The 1960s*. Westport, CT: Greenwood Press.

Reklaitis, V. 2008. E-Travel Outfits Packing More Treats. http://beta.investors.com/NewsAndAnalysis/Article.aspx?id=450555 (accessed November 20, 2008).

Schulz, A. 1997. Electronic Market Coordination in the Travel Industry: The Role of Global Computer Reservation Systems. In *Information and Communication Technologies in Tourism 1997*, ed. A. Tjoa, 67–74. New York: Springer-Verlag/Wien.

Sickels, R. 2004. *The 1940s*. Westport, CT: Greenwood Press.

Smith, A. D. 2004. Information Exchanges Associated with Internet Travel Marketplaces. *Online Information Review* 28 (4): 292–300.

Smith, A. D., and W. T. Rupp. 2004. E-Traveling Via Information Technology: An Inspection of Possible Trends. *Services Marketing Quarterly* 25 (4): 71–94.

Solberg, C. 1979. *Conquest of the Skies: A History of Commercial Aviation in America*. Boston: Little, Brown and Company.

Swarbrooke, J., and S. Horner. 2001. *Business Travel and Tourism*. Boston: Butterworth-Heinemann.

Taaffe, E. 1959. Trends in Airline Passenger Traffic: A geographic case study. *Annals of the Association of American Geographers. Association of American Geographers* 49 (4): 393–408.

Taneja, N. K. 2002. *Driving Airline Business Strategies through Emerging Technology.* Burlington, VT: Ashgate Publishing Company.

Toh, R. S., and M. Y. Hu. 1988. Frequent-Flier Programs: Passenger Attributes and Attitudes. *Transportation Journal* 28 (2): 11–22.

Transportation Group International, LC. 2002. *Consumer Attitudes and Use of the Internet and Traditional Travel Agents.* Washington, DC: National Commission to Ensure Consumer Information and Choice in the Airline Industry.

Weber, K., and W. Roehl. 1999. Profiling People Searching for and Purchasing Travel Products on the World Wide Web. *Journal of Travel Research* 37 (3): 291–298.

Wellman, B., and C. Haythornthwaite, eds. 2002. *The Internet in Everyday Life.* Malden, MA: Blackwell Publishers Ltd.

Yost, J. R. 2008. Internet Challenges for Non-media Industries, Firms, and Workers: Travel Agencies, Realtors, Mortgage Brokers, Personal Computer Manufacturers, and Information Technology Services Professionals. In *The Internet and American Business*, ed. W. Aspray and P. Ceruzzi, 315–349. Cambridge, MA: MIT Press.

Young, W. H., and N. K. Young. 2004. *The 1950s.* Westport, CT: Greenwood Press.

Travel Web Sites Cited
Alaska Airlines—http://www.alaskasworld.com/newsroom/ASNews/AS-Fact-Sheet_Innovation.asp

Expedia, Inc.—http://overview.expediainc.com/phoenix.zhtml?c=190013&p=overview

Official Airline Guide (OAG) Corporate History—http://www.oag.com/oagcorporate/history_of_OAG.html

Orbitz, LLC—http://www.orbitzworldwide.com/SABREHoldings/Travelocity—http://www.sabre-holdings.com/ourBrands/travelocity.html

5   Genealogy as a Hobby

James W. Cortada

Many Americans have discovered that working out a genealogy table can be as much fun as solving a crossword puzzle.
—Peter Andrews (Andrews 1982, 17)

Introduction

The most widely read book in nineteenth-century America was "the family Bible." It was not called just "the Bible," but rather, "the *family* Bible." Between the seventeenth century and the 1960s, nearly every Christian resident of the United States became familiar with the Bible. It provided the only comprehensive set of perspectives on religious instruction, life, and worldview for many native-born Christian Americans. Much of the Old and New Testaments were genealogical, about generations of people, beginning (literally) with the story of Adam and Eve on the first page of the Book of Genesis.

Most Americans living in the nineteenth and first half of the twentieth centuries would have been familiar with passages in the Gospels of Matthew and Luke, which record the genealogy of Jesus Christ, and of his parents Joseph and Mary. Family linkages were discussed either through the daily routine reading of the Bible or in other forums. Bibles published in the United States even included several blank or preprinted pages placed between the Old and New Testaments on which families routinely recorded births, weddings, and deaths of relatives, often replacing their Bibles when these pages no longer had room for more names. It was not uncommon for a family Bible to be used for over a century to record such family events, and in some families, the practice continues. For those curious about their family history, these Bibles often provided a convenient starting point for identifying ancestors. Indeed, not until the early years of the twentieth century did state and federal agencies begin replacing this form of record keeping with forms, most notably birth and death certificates, as true evidence of familial standing, necessary for documenting proof of identity, age, and details needed for passport applications.

When the U.S. Constitution went into effect in 1787, it mandated that a census be conducted every ten years by the federal government in order to apportion representatives to the lower house of Congress. Officials have conducted this census every decade since 1790. The census is conducted for political and operational reasons, not out of some genealogical consideration, but in the process the government has amassed a treasure trove of reliable genealogical information for over two centuries. These data are unique in the world and their value as a source of family information has increased over time.

In addition to counting people, from the very first census the government collected information about families: names of the heads of households, ages of all members of a household, and their gender. In subsequent censuses, enumerators collected additional information tied to families: names of children, occupations of all members of a household, relationships (parents versus children), and place of birth of parents. By the late nineteenth century, even American passport applications required the applicant's parents' names and ages, the applicant's place of birth, and both the applicant and parents' community of residence. Family and family history were tied together, even though the amount of data collected remained limited.

The family as an institution in society is part of one's worldview, a component of each individual's self-identity. This chapter explores how that awareness and, therefore, interest in one's family's past manifested itself in what today is often acknowledged as the nation's most widely indulged hobby—genealogy. But first some distinctions are needed. Genealogy is largely accepted to be the identification of one's ancestors: their names, their lifetime dates, their specific generational and familial relationships to living members of a family, and often where they lived and how they made their living. Family history is often intertwined with the concept of genealogy in that it too seeks to document previous members of a family, but also includes stories and evidence that describe the collective biography and handed-down values of all who came before us as part of the "family tree." Often the two terms are used interchangeably. The word hobbyist routinely is used to describe one who pursues genealogy, in part because so few of the many millions of people who engage in it have had any academic training in the subject. That fact is ironic, however, because genealogy is also a body of highly defined, formal best practices developed over many hundreds of years into a disciplined profession and line of intellectual activity. The amateurs—hobbyists—and the professionals, and the discipline they share have had increased interaction from the start of the Gilded Age in post-Civil War America to the present, and are now also joined by some of the most sophisticated uses of information technology available today. How did all of that happen?

Who Are These People?

There is a growing body of evidence suggesting that genealogy has become an important pastime engaged in by many people. A few statistics suggest the extent of that engagement. As of late 2008, nearly three million people, the majority from the United States, had become active users of Ancestry. com, one of the largest Internet sites for genealogical research. Over nine hundred thousand have subscribed to its extended services, paying a monthly fee.

The number of Americans expressing interest in genealogy nearly doubled from the late 1970s to 2008. Interest surged when genealogical information became available through the Internet, with interest expanding by over a third between 1995 and 2001 (Bernstein 2001). These were not passive players. The Church of Jesus Christ of Latter-day Saints, which operates familysearch.org, one of the largest and most active online genealogical sites, reported that in the early 2000s it received in excess of eight million inquiries per day accessing its database of over one billion names (Kilborn 2001).

The profile of the person who engages in genealogy has changed over the decades. Formal study of one's family history dates to the early 1800s in North America with the formation of genealogical societies. These societies were located largely in New England, but subsequently spread, over the century, to the Mississippi River, and by the early 1900s across the continent. Early interest in the subject grew out of normal curiosity about one's family history, but also as a manifestation of social status through membership in patriotic societies that required documentation of one's genealogy. Wealthy and upper-middle class Americans of the nineteenth and twentieth centuries hunted distinguished relatives and even royal relations. Organizations that played important roles in encouraging such research, patriotism, and historical preservation included the New England Historical Genealogical Society (founded 1845), Daughters of the American Revolution (founded 1890), Colonial Dames of America (founded 1891), the General Society of Mayflower Descendants (founded 1897), the

National Genealogical Society (founded 1903), and the American Society of Genealogists (founded 1940 for professional genealogists only).

Throughout the twentieth century, the white middle class in the United States joined the wealthy and status-conscious in pursuit of their ancestors. They were curious about their origins and increasingly able to devote time and funds required to conduct such research. The amount of genealogical data increased during the last third of the twentieth century and was easily accessed in libraries and on the Internet. Ethnic groups, most notably African Americans, Hispanics, Jewish Americans, and Native Americans, then became involved. By late 2008, the number of genealogical sites devoted to ethnic groups had expanded rapidly: over seven hundred devoted to Hispanics and African Americans, some five hundred to Jewish families, and even some six hundred concentrating on Native Americans. In short, by the end of the twentieth century all major social, economic, and ethnic cohorts were actively participating in genealogical research.

Genealogy is considered a hobby. The price of admission is interest, time, and activity, much as with any hobby. Because one is not required to be licensed or trained (credentialed) to participate, there are many problems with factual errors in what data are collected. To be sure, formal training and resources are available—and there are professionally trained (even licensed) genealogists who conduct research for free or for fees, particularly in other countries where Americans need help tracking down information without having to travel overseas.

There are the inevitable questions about why genealogy is attractive to Americans. Answers stem from a variety of endogenous and exogenous forces at work. Personal reasons fall into four essential categories. First, there is the normal curiosity about one's heritage. "Where did I come from?" "Who were my people?" "Am I related to any famous individuals in history, or to some horse thief?"

Second, genealogical research has served as an excellent way to meet other members of one's family and to maintain connections with them. Genealogy provides a way to connect with relatives given the history of social and physical mobility that has characterized ever-migrating American families. The Internet's social networking and e-mail tools have made this even more possible in recent years.

Third, as individuals achieve a certain age they begin to think about their legacy and relations, about the past and future of their families. Increasingly, we read that Baby Boomers are particularly interested in genealogy, but the historical record suggests that in all decades elder members of a family were often the most interested in passing on stories of their families to younger members.

A fourth motivation to engage in genealogy concerns the inspiration that emerges from family incidents and stories: the grandfather who fought at Normandy in 1944 telling his family about his Italian roots first in New York and earlier in Italy; African Americans attempting to identify their ancestral home in Africa; or others with tales of adventures in Europe, Latin America, or elsewhere in the world all serve as grist for the hobby. The discovery of documentation in an elderly relative's home is frequently an additional spur to investigation. One survey done in 2005 reported that 65 percent of those doing genealogical research became interested because of stories heard within their families, a percentage that seems low given the much higher percentage of people (85 percent) who have reported talking to relatives as their initial source of information (Market Strategies, Inc. 2005).

Exogenous forces, too, have long stimulated interest in genealogy and influenced how Americans approached the subject. There are three sets of such forces. The first comprise activities and influences immediately outside the family. Genealogical activity is largely centered within a family. Therefore, it is reasonable to consider influences outside the family as exogenous in order to understand this activity as a hobby. An essential collection of such forces are events that stimulate interest among those family members who conduct genealogical research, such as the anniversary of a national historic event.

A second category of exogenous forces influencing genealogical activity are various actions taken by governments, libraries, and other institutions either to make available or to deny access to information considered essential (or useful) to someone conducting genealogical research. The availability of a new decade's worth of census data in the United States is an example of this kind of force, as is the continued refusal of a government agency to make available certain types of data, such as personnel records for living relatives or tax returns.

A third force—and one that will receive more attention in this chapter than the first two—is the innovation in information technology (IT) that has profoundly affected genealogical activities in the United States, to such an extent that IT has become the most influential of the three exogenous forces.

Within the category of the first class of exogenous forces, the centennials of the American Revolution in 1876 and the Civil War from 1861 to 1865 led individuals to explore the roles of their families in those wars. In the years following the Revolutionary War centennial, genealogical and historical societies came into being that served as major centers of research about family history. The Civil War's centennial in the 1960s led to a sharp increase in research concerning military records of soldiers fighting on both sides; the bicentennial of the American Revolution in the 1970s did the same. Both world wars of the twentieth century also stimulated interest in family history, although evidence is anecdotal.

A more direct influence was publication of Alex Haley's 1976 novel, *Roots*, which subsequently became a television series watched by some of the largest TV audiences up to that time (over 130 million viewers). One archivist at the National Archives reported increases in genealogical research of some 300 to 400 percent in its various centers following broadcast of the series (Haley 1976). The novel traced the history of an African American family, generating enormous interest on the part of people of all races and ethnic backgrounds to explore their ancestry, particularly African Americans, and often for the first time. Librarians all over the United States in the 1970s reported a substantial increase in interest in family history across all ethnic and racial lines, not just among African Americans, so much so that the *Roots* broadcast on television is widely considered by genealogists, historians, and librarians as a turning point in the history of American genealogy.

Where interest had long existed in genealogy, *Roots* did not necessarily encourage dramatically new efforts. For example, in the South, where interest in familial history and in the role of ancestors in the Civil War was and is widespread, *Roots* had less impact among white residents. However, it did have an enormous effect on African Americans. In short, each national event had varying influences on people doing genealogical research, depending on their ethnic background, local culture, and social status, suggesting that one must be cautious about broad generalizations of effects on amateur and professional genealogists alike.

Various events in American life continue to supply reasons for pursuing genealogical research: religious requirements (Mormons), patriotism (historical anniversaries), difficult times (world wars, the Great Depression), media events, and the possibility of conducting research in new ways. Exogenous forces of the second type—availability of new sources of information—continue to stimulate interest independent of some national event such as an anniversary or publication of a book. In the nineteenth century these included establishment of genealogical research centers and publication of the first "how to" books on the subject, followed by the creation of massive research collections, such as those of the Library of Congress (over 140,000 publications), the creation of online collections, and most recently the study of DNA. Sometimes the surge in interest became substantial. For example, as a byproduct of the *Roots* phenomenon and in the same decade as the bicentennial of the American Revolution, the Library of Congress reported an 80 percent increase in the use of its genealogical resources between 1972 and 1977, largely by first-time researchers (New York Times 1977).

In 2001, twenty-five million records of immigrants and travelers entering the United States through Ellis Island at New York City between 1892 and 1924 became available online through the American Family Immigration History Center's Web site (http://www.EllisIsland.org). The site became an instant success, with well over eight million visits on the first day of its availability. The center possesses documentation on the names of people arriving in the country, as well as their countries of origin and when they arrived. Even the names of the ships that transported them have been preserved.

In 2008, the American government began making available a few income tax records from the early years of the twentieth century, which in time will probably lead to another surge of interest in family history.

The influence of newly available information in sparking research grew sharply once online genealogical services appeared, because they could rapidly spread news about additional accessible records to targeted audiences certain to want these kinds of announcements. Prior to such portals and services on the Internet, news about the availability of additional records spread more slowly, largely through word of mouth, when mentioned in a how-to manual, or when a researcher showed up at a library or archive that happened to have these materials. In the arcane language of the business manager, use of the Internet reduced the "mean time" to news and hence, the "cycle time" to process new data. In other words, researchers worked faster with the most currently available sources.

The Information that People Seek in Genealogy

What questions about genealogy do researchers seek to answer? Most are interested in knowing by name their blood relatives, going back in time as far as possible. They are keen to examine documentation that lists names of parents (e.g., birth certificates, passport applications, census records) and who married whom (e.g., marriage certificates and census records). They want to know birth, marriage, and death dates for each person and how relatives are related to each other. There has long been interest in military records of ancestors and, to a lesser extent, in wills and land records to

establish an ancestor's wealth and profession. Ties to churches, social organizations, and roles in communities also prove interesting. Of particular
interest—and concern—is the changing spelling of a family's name over
the centuries, which can either link families together or demonstrate that
they are not related.

Once curiosity about direct lineage is satisfied, many individuals
researching their family history become interested in acquiring stories of
families associated through marriage. After this basic information is accumulated, researchers typically want to begin formulating a narrative of
their family's past, such as what their ancestors did in Europe or Africa
before coming to the New World and their subsequent roles after immigration to the United States. Weaving a personal history of a family into the
much larger tale of the nation's past becomes the ultimate objective.

The epigraph at the start of this chapter strikes at the heart of the search
for information, in that it can be quite varied, difficult to accomplish, and
dispersed. Table 5.1 lists many types of widely consulted records essential
to the genealogist. This list is by no means complete. Yet, note how varied
and how many sources of information are available. Repositories of genealogical data can be voluminous, yet there might be few citations or even

Table 5.1
Variety of records consulted by genealogists

Birth certificates	Biographies
Adoption papers	Obituaries
Marriage certificates	Census reports
Divorce records	Military records
Death certificates	Church sacramental
Telephone directories	Association membership roles
School and university grades	Criminal records
Wills	Civil law suits
Family letters	Diaries
Emigration	Passport applications
Land deeds	Tax statements
Lineage organization files	Medical records
Newspapers	News and trade magazines
Oral histories	Professional
Photographs	Ship passenger lists
Tombstones	Voter registrations
Probate records	Family trees
Local history books	Histories of major events, battles

only one citation for a specific ancestor in literally thousands of pages of records. Genealogical inquiry is normally a tedious process that benefits from the use of finding aids, such as indices, and, most recently, software tools and the Internet to mine for specific data. Table 5.2 lists examples of the size and scope of the records various organizations hold that are essential for genealogical research, suggesting the mountain of material one might encounter in conducting such research. Table 5.3 presents a list of major Web sites accessed routinely by researchers today while table 5.4 lists widely used software tools.

Table 5.2
Size of genealogical archive, library, and online holdings, circa early 2000s

Library of Congress (LC)	40,000 genealogies, 100,000 local histories
National Archives and Records Administration	4 billion pages
Church of Jesus Christ of Latter-day Saints (LDS)	3.2 million microfilm rolls and microfiche; 356,000 books, serials, and other formats; over 4,500 periodicals; 3,725 electronic resources
Ancestry.com	1 billion names online
Cyndi's List	260,000 links to family history

Table 5.3
Widely used genealogical Web sites, circa early 2000s

Cyndi's List (http://www.cyndislist.com)
FamilySearch (LDS) (http:/www.familysearch.org)
Ancestry.com (http://www.ancestry.com)
Ellis Island (http://www.EllisIsland.org)
Genealogy.com (http://www.genealogy.com)
RootsWeb.com (http://www.rootsweb.ancestry.com)

Table 5.4
Widely used software tools

Family Roots (popular in 1970s to 1980s)
Ancestral Quest (early genealogical tool, still available in early 2000s)
Personal Ancestral File (PAF) (developed by LDS)
Family Origins (available in 2000s)
Family Treemaker (links to email)
Ultimate Family Tree

As of the late 1990s, *Family Tree Maker* had become the bestselling software tool of the decade, operating on PCs using Microsoft systems; it has proved relatively easy to use (Roberts 1998). Furthermore, all major libraries and archives also have Web sites and are increasingly making available online reference guides to their collections and providing access to key documents. One could just as easily have also cited the thousands of user manuals, guides, and articles published over the past two centuries that have been invaluable to researchers.

In most instances, the results from research that people need and want include five core categories of documents. These categories of information have remained remarkably consistent for over a century. The first category includes family trees, because they illustrate relationships among people within a family over time, usually including birth and death dates. These records, the minimum that researchers want, often prove so difficult to determine accurately that the effort to document them can consume a lifetime or more of work. A second set of desirable records consists of photographs of family members going back as far into the nineteenth century as possible; or prints, engravings, and portrait paintings, which can extend the visual record even further back in time. A third collection consists of myriad family records (e.g., legal and religious documents such as birth certificates, wills, and family letters). Fourth, equally prized, are collections of military records. These are well preserved in the United States, from the Revolutionary War to the present. Fifth, recordings of family voices, motion pictures, and videos of familial events and people, dating back to as early as the 1930s in many cases, have long been another source of documentation about the family. Families have been willing to spend considerable time and money to preserve these memories by moving film to video and most recently to DVDs, and transferring recordings from reel-to-reel tape to diskettes, then to CDs and DVDs. Businesses exist specifically to move this kind of content from one medium to another and to capture images off old photographs and slides and convert them into digital formats. Table 5.5 lists some examples of these businesses.

The most advanced form of family documentation is a written narrative of a family's history from its origins to the present. These narratives provide biographies of key family members, family stories and myths, handed-down practices within the clan, family members' roles in great and small events of a nation's past, and their trials and successes over time. Some family histories are informal: typed or even handwritten manuscripts that circulate within the family. Others are privately printed, and still

Table 5.5
Examples of U.S. media conversion firms, circa early 2000s

DigMyPics (http://www.digmypics.com) [largely photo and film scanning, conversion]

FotoBridge (http://www.fotobridge.com) [largely photo and film scanning, conversion]

ScanCafe (http://www.scancafe.com) [largely photo restorations]

QA Video Services (http://www.quavideoservices.com) [specializes in preserving home movies]

Digital Transfer Services (http://www.dtsav.com) [preserves video, film, recordings, and still shots]

Source: Online directory of digital restoration and preservation services, http://www.dmoz.org/Shopping/Entertainment/Recordings/Media_Preservation_and_Transfer/Video/ (last accessed March 15, 2009).

others are preserved in the form of fictionalized novels and history books published by established publishing houses, such as Alex Haley's *Roots* by Doubleday and various academic studies of plantation families in the pre–Civil War South, published by university presses. The Library of Congress, long a collector of family histories, houses more than forty thousand such narratives in its collection. The ultimate goal of this kind of documentation is to tell "the story" of a family.

How People Conducted Genealogical Research before Personal Computers

While issues that genealogists and family historians investigate have remained essentially the same over time, the arrival of digital tools, beginning in the 1980s and assisted by the availability of personal computers (PCs) and early online files, began to change the way researchers worked to such an extent that one can reasonably divide the modern history of genealogical research into two periods, pre- and post-digital. However, this chronological division still requires us to acknowledge that techniques from the pre-digital era are still used today—such as discussions with family members and research visits to European churches—in part because much information remains undigitized.

In the United States then—and now—research begins with family members telling each other stories about the activities and life experiences of their own, then of their parents, and then of prior generations. This storytelling remains both the initial source of information about a family's past and often the key source of early inspiration for a member to pursue genealogical research. Research often proceeds to books about a community's history, which are typically maintained in local libraries. Trips to ever-larger city and state libraries that hold larger collections of genealogical records, books, and periodicals, became standard practice by the late 1800s, and continued throughout the twentieth century.

Beginning in the 1930s, primarily federal institutions, but others as well, began to microfilm records such as census materials, making them widely available in regional libraries and at large universities. The widespread microfilming of newspapers, undertaken by the U.S. Library of Congress and many state university libraries, dating from the eighteenth century to the present, provided researchers with a new source. In fact, consulting census and newspaper microfilm (later microfiche) became two core activities of genealogical researchers. Those investigators who had the financial means might also hire a professional genealogist to conduct research, although extant evidence suggests that this practice remained quite limited (New York Times 1894). Thirty-four state genealogical societies that people consulted existed before Haley published *Roots*; eight more were formed afterward. But the number of professional genealogists remained low relative to the whole population of people doing genealogical research; indeed, most members of genealogical societies were hobbyists. To be sure, many were as highly skilled in the ways of genealogy as credentialed researchers.

Finally, one should acknowledge an old American practice, dating back to the eighteenth century: the how-to book for all manner of information and all kinds of work and play. Beginning in the nineteenth century and extending to the present, numerous books and articles appeared that guided researchers through various genealogical issues and sources of data, and advised on how to collect and organize information. Researchers in North America routinely turned to such publications to teach themselves the process, while classes and seminars sponsored by universities and local libraries provided a relatively minor source of guidance. An entire book could be written about genealogical guides, but suffice it to note that they had common features:

• Descriptions of what kind of information constituted genealogical data
• Lists of physical repositories of such information
• Issues related to accuracy of information from various sources, such as members of one's family
• How to create and manage a research agenda
• How to organize physical records and information
• How to write/tell a family's story

• Extensive lists of publications, libraries, and archives
• Discussions of special challenges, sources, and issues of specific ethnic groups, such as Catholics, African Americans, and Jews

These manuals often provided the only training researchers had unless, of course, they were professional genealogists.

Collecting information in pre-computer times proved to be a slow, tedious, time-consuming exercise. A piece of data would be picked up at a church or courthouse, off a tombstone, or in a family Bible or genealogical publication only to become an isolated fact that had to be matched to others so as to extend the narrative. A researcher's trips to various libraries to read microfilm consumed many weekend afternoons, while the volume of correspondence with librarians, archivists, and family members could prove massive. In fact, such correspondence was so great for so many researchers that authors of how-to books routinely devoted attention to its organization and use as part of the research process. In addition, a research-er's correspondence often became the occasion to exchange with family members or other interested parties records such as photographs, letters written by ancestors, and wills and other legal documents. Nothing was more exciting to a researcher than the arrival of a fat brown envelope in the mail containing several photographs of early forebears or a stack of old documents. Such a correspondence could go on for decades.

At the risk of oversimplifying the research process before the advent of computers, (which usually took hundreds of pages to describe in a user manual, varying by family and by issues pursued), a summary of what took place suggests how all these various tools were and continue to be used, serving as useful context for understanding the profound influence that information technologies have had on such activities. Initial research often started with relatives describing ancestors. That dialogue normally contin-ued for the life of the project, usually for months, but often over many years and decades. Novices focused on identifying basic data, such as names of grandparents. More advanced researchers pursued answers to very specific questions that clarify existing information. There was always the race to query the oldest relatives before their insights and memories were lost. The next stage was often to acquire a user manual, study it, and then collect or consult the kinds of records already discussed (for instance, census or military records). Documentation proceeded to family trees, then added profiles of individuals with their birth and death dates, followed by details on individuals, such as their careers, civic activities, personalities, where they lived, and so forth. As progress was made, record keeping became important; that is when individuals begin creating voluminous records stored in file cabinets, collecting material into three-ring binders, and writing to libraries and archives to build and verify the family narrative. Simultaneously, genealogists communicated with relatives to collect specific pieces of information and copies of documents, and to solicit help in visiting nearby libraries, archives, and courthouses to collect more data.

The process was highly iterative as one moved back in time, beginning normally with one's parents, moving next to investigating one's grandparents, then great-grandparents, and so forth. Along the way decisions were made about how broad to make the search (e.g., whether to build trees of aunts and uncles and their descendants) and whether to explore the roots of families that married into the family of the genealogist. New facts necessitated retracing steps to fill in gaps—just as one does in filling out a crossword puzzle—and to validate data now of questionable accuracy or to answer new questions.

This routine remained the unbroken practice from at least the mid-nineteenth century until information technologies began to emerge. While the IT tools described in the section that follows have affected profoundly how research is done and what kind of data can be collected, the core tasks and purposes remain unchanged. One still needs family trees, names, and relationships identified, birth and death dates, information about family events, and so forth. In the age of the Internet word of mouth remains as profoundly important as before, because much tacit knowledge lies locked only in the minds of elderly relatives, only to be revealed when asked the right question. People have traveled to libraries and archives all over the world from the 1800s on into the early 2000s, because the Internet has not eliminated distance from the life of a committed genealogist and family historian.

Research after the Arrival of Computers and the Internet

The emergence of information technologies made possible several fundamental changes in genealogical practices that represent the third category of exogenous forces at work. These changes affect research methods rather than the nature of issues addressed by researchers. A number of specific technological changes made computing accessible and affordable to researchers. Beginning largely in the 1960s, computer data storage expanded in size and dropped in cost by about 20 percent compounded each year, with the result that by the early 1980s, storing data on a computer was inexpensive enough for individuals to afford. That long-term trend continued to the present. One megabyte of memory on a computer in 1979 cost about $110,000; today it is less than $25. This evolution made storing massive quantities of information in computers a possibility for the even the most casual researcher.

A second development, beginning in the late 1970s but not widespread until the mid-1980s, was the repackaging of computing power into ever smaller and more affordable devices. The multimillion-dollar mainframe computers of the 1950s and 1960s made way for PCs in the 1970s and 1980s that could be acquired initially for up to $5,000 each; then came the $600 computers and laptops of the early 2000s. By 2000, PCs and other computer devices were small enough to be perched on a table or carried into an archive. They were also relatively inexpensive and many more individuals could afford to use computers to collect, organize, and display data in an effective and speedy manner.

A third and crucial technological development was the emergence of telecommunications linked to computers. Beginning in the 1960s commercial organizations began sending data over telephone lines from one computer to another, from terminals to mainframes and minicomputers. By the early 1980s, one could receive or send information from a personal computer to another PC, or access information from a large mainframe. When that function became reliable and practical, genealogists could conduct research through networks tapping into information in other computers without having to travel. At first they used dial-up telephone lines (in the 1970s and 1980s), then subscribed to consumer-oriented private networks (in the1980s and early 1990s), and then accessed the Internet directly through telephone or cable communications (beginning in the mid-1990s).

We have already alluded to the emergence of software designed to help genealogists. Three functions proved essential. First, software could be used to scan multiple sources of data available on various computers. This functionality was very limited in the 1970s and 1980s, for two reasons: technological limitations of the software and limited availability of information in computers. Google and other search engines did not become widely accessible until the mid-1990s, but nonetheless, searching was an early application of the software.

Second, software made it possible to organize information into databases (files, spreadsheets), beginning in the early 1980s on PCs and using software algorithms and methods developed originally for large mainframe files as far back as the 1960s. One could now store thousands of pieces of information without filling up rooms with paper, using methods that were effective in accessing this information.

A third development in genealogical software involved the display of data. Before computers, family trees were often messy, hand drawn paper representations, plagued by erasures and bad handwriting. In contrast, even the early software created family trees in neatly printed formats that could either be populated by hand, with people typing in the information, or eventually by using software to pull in data from online files stored in one's PC or available through such service Web sites as Ancestry.com. Over time these software-created trees became more detailed, moving from just names to names and dates of birth and death to each name being, in effect, linked to an electronic folder containing whatever information, pictures, and scanned documents related to that individual was available.

As with commercial computing in companies and government agencies, often technologies evolved faster than the content that went into them. However, beginning in the 1970s and extending to the present, the amount of information that governments, libraries, and researchers converted from paper and microfilm to digital formats became nothing less than a data tsunami. Key records in the United States, such as census materials, and other economic data moved to the Internet; churches set up Web sites and began populating these with birth, marriage, death, and burial records, as did state and county offices. The Church of Jesus Christ of Latter-day Saints began posting hundreds of millions of records in online formats. Hardly a week goes by without an announcement by some agency or organization in the United States (or Europe) that it has now made available new records over the Internet. The volume of electronically available material is so massive that it is nearly impossible to estimate the quantity. Attempts to do so have led many observers to provide statistical sound bites about how many online files now exist. That availability of online data has transformed most genealogical how-to books largely into IT user manuals, focusing on researching over the Internet.

By the end of the 1990s, therefore, Americans doing research had available the necessary IT tools and increasing amounts of digitized data they could use to construct family histories, along with essential infrastructure. In 1990 approximately 20 percent of American households had a PC; by 2000 deployment had exceeded 40 percent (and these PCs were connected to the Internet). As of early 2009, over 60 percent were Internet linked. However, if one takes into account the percentage of people online daily from any source—home, work, mobile device—then the statistics are higher: 50 percent in 2000, and nearly 80 percent in late 2008 (Pew Foundation 2009).

Simultaneously, the speed with which researchers could access information increased as they moved from slow dial-up lines to high-speed broadband in the early 2000s. At the start of 2009, well over half of the American public had access to broadband computing. Broadband connections made quick transmission of large volumes of information orders of magnitude greater than that which could be transmitted over a dial-up line. For example, in the 1990s one could download a photograph of a relative to one's PC over a dial-up line in about five to six minutes. In the early 2000s, the same photograph, along with several dozen others, could be transmitted over a broadband connection in less than half a minute. The convenience and ease with which this important activity could take place encouraged more such transactions, reducing the need to wait for the U.S. Postal Service to deliver those brown envelopes so anticipated in the past. Table 5.6 lists major uses of the Internet for genealogical research as of the early 2000s. Table 5.7 shows types of information that became available over the Internet since approximately 2003, reflecting the current pattern of ever-more varied materials becoming available.

As information became more accessible, behavior by researchers began to shift, moving from activities of lone genealogists to more family-centered projects. To be sure, in the 1980s and early 1990s, individual researchers worked in the same ways they had before, simply grafting use of PCs onto their existing practices. As email usage spread in the early 1990s, the amount of communication one could engage in with relatives increased, leading to expanded collaboration within families. The extent of that expanded dialogue has yet to be studied. Mounting evidence suggests

Table 5.6
Major uses of the Internet for genealogical research, circa 2003

Search indexes, vital records, and statistics of individuals (birth, death, marriage, census, immigration)

Search classified directories and collections of Web genealogical resources

Create personal family-history pages

Explore searchable databases of personal family-history pages, commercially collected data products, indexes of government-collected data sets

Participate in genealogical discussion groups

Use genealogical freeware and shareware

Access city, county, and state genealogical society Web sites

Access individual library Web sites and online catalogues

Source: Diane K. Kovacs, "Family Trees on the Web," *American Libraries* 34 (7) (August 2003): 44.

Table 5.7
New types of data available over the Internet, since 2003

Maps and other GIS data sets
Ship passenger lists
Cemetery transcriptions
Wills, deeds
Tax files
Newspaper runs from earlier decades
Genealogical journals
Personal family histories
Business directories published by cities and professional associations
Discussion groups

---

that a new activity grew in the 2000s: social networking. Web sites such as Ancestry.com and others that used social networking tools added MySpace-like community functions and bulletin boards that made it possible, for example, for multiple family members to populate simultaneously a family-wide tree database, working from various locations, updating the tree in real time. In recent years, social networking has facilitated the sharing of family trees, photographs, and other documents, as well as the creation of birthday calendars.

As Americans began to use commercially available Web sites, a Wiki-type of behavior emerged, recommended by almost all the Internet-oriented how-to manuals of the 1990s and 2000s. These provided the capability to create online files housed at such sites as Ancestry.com. Others could add content, correct errors, and make material available to other individuals. Tasks were performed by Web site users, not by site providers. These tools resulted in significant user adoption. For example, between July 2006 and October 2008, management at Ancestry.com recorded:

• More than 7.5 million family trees created
• More than 725 million profiles added to those family trees
• More than 6.5 million Ancestry.com users created family trees
• More than 10 million photos uploaded into Ancestry.com
• Nearly 150 million Ancestry Hints accepted (Bonner 2008)

These statistics are useful indicators of this Internet-facilitated behavior, because it occurred within the largest genealogical Web site in the United States. Rival services could reasonably be expected to report similar increases in activity, since the ability to post data at these sites is a core function offered. It is a short step to yet another community-oriented action: creation of family Web sites.

Families increasingly set up their own Web sites, a practice that began around 2005, within existing social networking tools such as MySpace and more frequently, Facebook. Often these family sites were created to facilitate communications among relatives and to share photographs. Younger tech-savvy family members created these sites for such activities as weddings (e.g., www.eWedding.com, www.Wedsite.com) and other family projects; some then evolved into ongoing family Web sites. As Americans over the age of 45 (traditionally the age group most interested in genealogy) began interacting with these social networking based sites, they began transforming into genealogical Web sites. More members of a family could participate in a collaborative way in conducting genealogical research, learning more family history and even organizing family reunions in the process.

In the period following wide availability of Web browsers (post-1995), which made use of the Internet dramatically easier for most people, genealogical research led to an unanticipated consequence: people discovering lost friends and relatives. Both white and yellow page telephone directories in the United States became increasingly available online in the early 1990s. Over time, various local directories were interconnected so that by the end of the 1990s one could look up, for example, everyone in the United States whose last name was "Cortada," or everyone with the name of one's lost college roommate.

Services appeared that performed data mining searches, for instance, of high school classes by year of graduation, for a fee. It did not take long for family historians to use the same tools to find lost or new relatives. As one observer noted in 1999, with online genealogical tools, including telephone directories, "many who became separated from their families, and thus their family histories, have discovered that they can find themselves reunited with long-lost cousins and long-forgotten lore" (Howells 1999). Nearly a decade later, observers still reported expanded use of the Internet to thicken connections within families and to bring new participants into family discussions, so that the "digital age family history has become a sociable, co-operative movement" (Hudson and Barratt 2007, 20–21).

To summarize, research methods involving use of the Internet extended the range of pre-Internet work: digitization of records, extensive access to noncommercial and commercial Web sites, establishment of family social networking sites, extended communications among family members, friends, and other hobbyists for sharing and exchanging research, collaboration among larger groups of people, discussion and information requests from a larger pool of potential informants, and socialization in general. The most immediate results that the emerging Internet use had on genealogical research can also be summarized. Most obvious is the increased speed with which information could be accumulated, using the power of software search engines and telecommunications. Mail delivery speeds became increasingly irrelevant. The amount of personal travel to libraries and archives declined as online researching reduced the need to physically visit sources of information. Internet users could prepare for a research trip far better than before, because so much preparatory information could be examined in advance, such as the online indices of records to be examined.

Yet other behavior did not change. For example, hobbyists still bought books on how to use the Internet, teaching themselves how to conduct genealogical research. A search of Barnesandnoble.com, the online bookstore, in 2009 indicated there were several thousand books for sale on or about genealogical matters, while Amazon.com listed over one hundred titles available on its e-book, Kindle. Emblematic of Americans' interest in relying on books and other instructional materials was Terri Stephens Lamb's book, *E-Genealogy: Finding Your Family Roots Online,* published in 2000, which addressed the reader in the first person and speaks of "easy-to-read steps" intended to "show everything you need to know to bring your heritage home," personalizing the obligation to learn the techniques through use of the manual (Lamb 2000). This book is a useful example of what had been learned in the previous half-dozen years about using Internet-based genealogical tools and is a good representation of practices that have continued a decade after publication.

The Internet Did Not Change Everything

As important as the Internet became in augmenting genealogical research, it has not proven to be the be-all and end-all for family historians. One major problem that defies resolution is the spread of misinformation. Beginning in the 1800s, experienced genealogists warned about sloppy research and emphasized the need to verify data, even if from official sources or family Bibles, because records are collected inaccurately, people lie, information is lost, and bad data survives and circulates perniciously. The Internet facilitates dissemination of false facts in much the same way as paper records.

Elizabeth Powell Crowe, author of one of the most widely respected guides about the use of the Internet in genealogical research, still felt compelled to comment extensively about trusted sources as late as 2002. She almost could have written the same way about published sources a century earlier: "Most serious genealogists who discuss online sources want to know if you can 'trust' what you find on the Internet. Many professional genealogists I know simply don't accept what's found on the Internet as proof of genealogy, period. Their attitude is this: A source isn't a primary source unless you've held the original document in your hand" (Crowe 2002, 12–13). Mimicking standard good practice of historians, she commented that even a paper source should not to be fully trusted unless its data are confirmed by another paper source. "In my opinion, you must evaluate what you find on the Internet, just as you evaluate what you find in a library, courthouse, or archive" (Ibid., 14).

Genealogical scams, such as companies offering to sell family history, certificates of authenticity, and family heraldry have been active for decades and continue to be a problem for novice researchers. So, normal discretion and a healthy skepticism remain essential best practices for a family historian—behavior that is infrequently evident. Thus, for example, the millions of contributions of content to Ancestry.com's databases are not all guaranteed to be accurate.

A second problem that persists is the existence of material not yet on the Internet, which means one still must travel to U.S. courthouses around the country, to church archives across Europe, and to libraries all over the world. Only a fraction of all relevant materials are online. For example, the Roman Catholic Church has kept records of who received the sacraments (birth, first communion, confirmation, marriage, and death) for nearly two thousand years, and made it a formal responsibility of parishes since the late Middle Ages. To be sure, wars, fires, sloppy management or workmanship by a semi-illiterate priest in one church or another compromised the process from time to time. Nonetheless, billions of records exist in Catholic parishes and libraries all over the world that have yet to move onto the Internet. Catholic Church records are absolutely an essential source of genealogical information for Americans probing their families prior to their arrival in the New World, and even for early migrants for hundreds of years in Latin America and the Caribbean. Table 5.8, therefore, is the flipside of table 5.7, showing what is not yet widely available on the Internet.

This leads to a related discussion, of what the Internet still cannot do. The Internet cannot be used to communicate directly with a source or

Table 5.8
Sample types of data not widely available over the Internet, circa 2008

Most Catholic Church records
Most local European newspapers
Most U.S. county court records, pre-1980
Post-1930 U.S. Census records
Vast majority of documents in state and national archives
Municipal records from Eastern Europe and Asia
Colonial records from European country archives

---

family member who does not use the Web or even anyone who does not use email. Finding information on many family names, which may have various spellings, remains tacit knowledge (that artificial intelligence has yet to grapple with). Some names cannot be found on the Internet; they have not migrated there. It is difficult to effectively verify the accuracy of information found on the Internet; that remains a very important human activity of discernment and critical judgment, informed richly by the context of known family history.

Nearly as frustrating is the fact that there is cost associated with placing accurate copies of original documents on the Internet, so relatively few accurate copies of original documents are available. Further, original documents will only be obtained via online application forms made available by reliable sites, such as online records of applications for passport originally submitted as an e-service by a government (such records are not available today).

Other aspects of genealogical research that the Internet only partially facilitates are the contextual conversations and visitations so essential to understanding a family's history. Talking to elderly residents of a village about life in the community in which one's ancestors lived can only be done by visiting the town and sitting down with the "old men in the park" to understand what it was like during a critical period, such as in war, famine, and economic boom times. For example, visiting a rural locality gives one an enormous sense about lifestyles there, including how far home was from work, school, and church; and the effects that weather and soil conditions had on a family's prosperity and worldview. Just as military historians universally insist that one must walk a battlefield to understand how a battle was fought, won, or lost, experienced family historians walk where ancestors trod to understand more deeply their prior experiences. On such trips one can photograph "ancestral homes," have a face-to-face meeting with a distant relative who gives the genealogist the gift of some old papers and photographs, discover a school textbook used by an ancestor in the 1800s, and buy local histories not listed by online book dealers such as Amazon.com. We still live in a time when digital means are used only as part of a much larger set of tools to do genealogical research.

A New Frontier: Research after the Internet

In 1998, a group of African American descendants of slaves belonging to President Thomas Jefferson proved through DNA testing that they were his blood relatives, tracing their ancestry back to Jefferson's slave mistress Sally Hemings. This topic became the first major historical subject of debate that turned on the issue of DNA testing in genealogical research. The effect of this finding proved electrifying among African Americans interested in their genealogy, much as *Roots* had stimulated their interest in the 1970s. This discovery opened a new chapter in genealogical research on the possibility of using of DNA to link families together through biological proof made possible by advances in genetics research.

During the 1990s various groups took early steps to conduct genographic research, communicating largely through the Internet with interested researchers. However, it was not until the early 2000s—when a world-wide study was initiated to determine the DNA relationship of all humans and, more specifically, to prove that all people were related and had come out of Africa sixty thousand years ago—that general interest in the subject expanded (Behar et al. 2007). IBM and the National Geographic Society collaborated with Spencer Wells, a geneticist and anthropologist at National Geographic, to collect specimens of DNA from various groups of humans all over the world. To offset partially the project costs, National Geographic publicly invited everyone worldwide to donate a small fee and their DNA to the effort and receive in return a summary of migratory patterns of other and earlier people who shared their types of DNA.

The project was accompanied by television publicity, press coverage, and extensive discussion over the Internet; hundreds of thousands of individuals participated almost immediately. Various other organizations began offering similar services. For example, Genetic Genealogy, DNA Tribes, and Familybuilder all merchandize their services primarily over the Internet. At the moment, the premier project is National Geographic's. More to the point of our story, this is a new use of technology and science in support of genealogical research and, as with earlier computer tools, is quickly being embraced by family historians.

Conclusions

This book is devoted to a discussion of how Americans find the information to support their everyday activities. The experience of genealogists reflects a long-standing practice evident across many facets of American life, of incorporating each new form of information-handling technology that has appeared in support of work, war, play, religion, politics, sports, and hobbies. It is a theme that the author and others have argued is an important feature of American society in evidence since colonial times (Chandler and Cortada 2000). Internet users rely on the Web at the same time that they rely on all manner of preexisting tools and sources of information. Thus, it is not unusual for American genealogists to go to the Web for some sources of material, visit their bookstore for how-to manuals, and make pilgrimages to courthouses, archives, and libraries to look at books and actual documents, talk to experts face to face, and spend many hours reading microfilm and microfiche. Researchers take their notes on 8½ × 11-inch pads, 3 × 5-inch index cards, notebooks and three-ring binders as well as using laptops, personal computers, digital cameras, and photographic functions on their cell phones. They scan photographs, store jpg files of these pictures, and go to their neighborhood Walgreen's pharmacy to pick up paper copies of images emailed to them by distant relatives. One other piece of evidence in support of the notion that genealogists simultaneously use multiple technologies: The vast majority of how-to books on genealogy published from about 1995 to the present in the United States include either a CD or DVD with various software tools and forms. Not even technical manuals published for computer programmers were adorned as comprehensively as those designed for genealogists.

Americans value speed of execution, and it is one of the enduring features of the Internet that individuals can gather large amounts of information about their families in a very short period of time, accomplishing in hours what used to take years. This can be accomplished from the comfort of their home offices, using a laptop connected wirelessly to well-stocked Internet sites such as Ancestry.com or that of the Church of Jesus Christ of Latter-day Saints. Convenience increased activity, as the reports from commercial Web sites, archivists, and librarians suggest. Sales and contents of how-to manuals also reflect this growing use of the Internet by which individuals and families explore their past. One could argue that this speed of execution and do-it-yourself approach to genealogy across all social, economic, and racial groups reflects the larger American practice of applying "sweat equity," similar to the way in which they personally remodel their homes or restore old automobiles and boats.

We live in a data-hungry world. At one time it was enough to have the family tree decorated with the names of ancestors, and the years of their birth and death. The rest of the information was largely stored in people's minds and passed down through oral traditions and narratives told by grandparents to grandchildren, occasionally aided by the fact that someone forgot to clean out a box or trunk in an attic containing papers and photographs. Even the most casual viewer of the popular public television program *Antiques Road Show* can be amazed at how much material is saved accidently by seemingly everyone. Serious genealogical researchers, indeed often the professional genealogists, began arguing in their professional publications, in presentations, and in the press, for more context and information about individuals, beginning in the late 1800s; not until the last third of the twentieth century did this imperative become possible for the masses.

One of the features of Internet-based genealogical research is the massive diversity of information people want to explore. Hobbyists want important collections put online. When the Internal Revenue Service announced in early 2009 that it would begin to make a few tax records available to the public largely from pre–World War I, reaction was less "wonderful news!"— although it was news—and more "how can we access it over the Internet?" The thought of having to go to some regional National Archives library to read microfilm now seemed positively primitive. As in other areas of their lives, Americans became accustomed to searching myriad databases for information at work and play, so they applied the same practice to their study of family history. The literature on how to conduct that kind of research and studies about genealogical practices in the United States, many cited in this chapter, leave behind a documented trail of evidence that clearly shows an expanding appetite for varied information that can inform family history. Indeed, the whole notion of moving from genealogical or family trees to family history is symptomatic of the change made possible by various types of information technology, including, but limited to the availability of the Internet.

Another feature of how genealogists use the technology—and one not discussed by observers—is the impact users of Internet sources are having on the technology itself. So many people are using the Internet to do genealogical research that enterprises serving these consumers can justify the cost of creating new software tools and augmenting search engines' mathematical underpinnings and capabilities, as evidenced by advances in major genealogical Web sites. Economies of scale and the volume of demand provide markets for new tools, such as the three-dimensional family trees that include, within each name on a tree, paths to sub-data-bases that hold images of photographs and original documents, as well as video and audio files. Functions associated with family trees are far more advanced (as of 2009) than many of the data management tools and portals available in more traditional business, teaching, and government applications. The functionality provided has more in common with the data-intensive research projects one finds in pharmaceutical genomic research, the hunt for new medical compounds, or modeling of weather.

As soon as new functions become available to the genealogist they are rapidly adopted. New requests made for further advances. Like other multifaceted data-intensive applications, those of genealogists are pushing information technology forward across a broad band of activities: database management, data mining, data visualization, graphical representation, memory systems, computer cycle speeds, and, of course, demand for access to an ever-larger capacity of broadband communications across the entire country. The only other corner of American society where so much innovation is routinely possible exists within the U.S. Department of Defense, where, for example, requirements to train annually several million people, most of whom are under the age of thirty and accustomed to video games and screens, makes it affordable to do ground-breaking research and innovation in IT-based teaching and real-time automated high-performance systems, such as battle-field management. Other civilian applications described in this book are also contributing to the evolution of the Internet itself, and it is the confluence of multiple types of uses and transformations that affect genealogical work. One final point about the power of economies of scale: the monthly fee to use such advanced tools at a major genealogical site is as little as $12 to $30 dollars per month (as of 2009).

Since many of their work activities involve use of computers, it should be no surprise that users would bring over to genealogical activities the same tools and methods to other aspects of their lives including genealogy. For example, a genealogist might use the laptop assigned by their employer to access a genealogical Web site and perhaps write up search results using Microsoft Word that he or she learned to use at school. Organizing materials or conducting research in the manner of a management consulting project, as many how-to books suggest, grew out of practices in wide use in manufacturing, military, and research functions and industries. Given what we know about the demographics of those conducting genealogical research today (for example, employed Americans age forty-five and older) we can assume that the majority learned to use the Internet not by doing genealogical research but by using computers at work. Over 80 percent of all workers interact with computers in one fashion or another, and we are rapidly reaching the point where over 90 percent of all people between the ages of ten and forty-five interact with the Internet.

Acknowledgments

This chapter was built upon the research of Michelle Bogart and Holly Green for which the author is deeply grateful.

Bibliography

Andrews, Peter. 1982. Genealogy, the Search for a Personal Past. *American Heritage* 33 (August–September): 17–21.

Behar, Doron M., et al. 2007. The Genographic Project Public Participation Mitochondrial DNA Database. *PLOS Genetics* 3 (6): 1083–1095.

Bernstein, Elizabeth. 2001. Genealogy Gone Haywire. *Wall Street Journal*, June 15, W-1.

Bonner, Susanne M. 2008. Email to author, October 21.

Chandler, Alfred D., Jr., and James W. Cortada, eds. 2000. *A Nation Transformed by Information: How Information Has Shaped the United States from Colonial Times to the Present*. New York: Oxford University Press.

Crowe, Elizabeth Powell. 2002. *Genealogy Online*. New York: Osborne/McGraw Hill.

Haley, Alex. 1976. *Roots: The Saga of An American Family*. New York: Doubleday.

Howells, Cyndi. 1999. Tracking Your Family through Time and Technology. *American Heritage* 50 (1). http://www.americanheritage.com (accessed May 1, 2010).

Hudson, Jules, and Nick Barratt. 2007. The Rise of Family History. *History Today* 57 (4): 20–21.

Kilborn, Peter. 2001. In Libraries and Cemeteries, Vacationing with Ancestors. *New York Times*, August 19, 1, 24.

Lamb, Terri Stephens. 2000. *E-Genealogy: Finding Your Family Roots Online*. Indianapolis, IN: Sams Publishing.

Market Strategies, Inc. 2005. Market Strategies Survey. Unpublished, in author's possession.

New York Times. 1894. Its Twenty-Fifth Birthday: New York's Genealogical and Biographical Society. *New York Times*, February 19.

New York Times. 1977. Tracing Their Roots. *New York Times*, February 20.

Pew Foundation. 2009. Pew Internet & American Life Project, Pew Research Center. http://www.pewinternet.org (accessed May 1, 2010).

Roberts, Ralph. 1998. *Genealogy via the Internet: Computerized Genealogy*. Alexander, NC: Alexander Books.

6 Sports Fans and Their Information-Gathering Habits: How Media Technologies Have Brought Fans Closer to Their Teams over Time

Jameson Otto, Sara Metz, and Nathan Ensmenger

Introduction

On any given Monday night during the fall National Football League (NFL) football season, roughly eight and a half million Americans will watch a professional football game on television. In the face of declining television viewership, ratings for televised football games continue to increase. In 2008, ESPN's *Monday Night Football* represented cable television's most watched television series, with one game in particular (the Eagles/Cowboys matchup), attracting an audience of almost thirteen million, cable television's largest household audience ever.[1]

But while the experience of watching football on television remains the dominant and most familiar form of sports consumption in the United States, an increasing number of football fans are choosing to engage in less conventional, less passive forms of sports entertainment. During the same 2008 season in which televised football viewership topped out at thirteen million, for example, more than twenty-seven million football fans participated in the sports statistics-based game known as fantasy football. Fantasy football is a game based entirely on the manipulation of sports information. Fantasy football players/managers use data drawn from real-world competitions as a means of evaluating the performance of their own imaginary football teams. In a process more reminiscent of financial accounting than of traditional sports fandom, they construct spreadsheets, analyze scenarios, and compare outcomes.

Fantasy football, along with the more general fantasy sports phenomenon, represents an enormous economic opportunity for the sports industry. For most fantasy football participants, watching a specific game (whether in real life or on television) is only the beginning of a long process of evaluating its information content: the average fantasy football player spends an additional nine hours a week managing his or her teams,

tracking statistics, and organizing his rosters and strategies. Entire industries have emerged to feed fantasy football players' growing demand for
sports information: Web sites, magazines, books, software, and even dedicated television channels. These represent not simply more of the same
kinds of sports coverage that have existed in earlier periods, but rather
an entirely new form of sports consumption based almost wholly on the
communication of abstract information.

What strikes many observers as being unique, and perhaps inexplicable,
about fantasy football is how divorced it seems from the reality of actually
playing the game of football. All sports consumers are, of course, experiencing sports only at a distance. But traditional sports media have always
attempted to create a sense of vicarious participation, an illusion of reality
(or perhaps hyperreality) meant to mimic the real-world experience of
being at an athletic contest. Fantasy sports seem so obviously intangible
as to destroy this illusion altogether.

To a certain degree the emergence of fantasy sports is a novel development that can only be understood in the context of very recent innovations in information technology. But in other respects, it is only the
most recent expression of an informational turn in sports consumption
that began at least as early as the nineteenth century. The very existence
of a professional sports industry has always been dependent on the existence of technologies that allow local events to be readily described,
encapsulated, and delivered to mass-market audiences, the vast majority
of whom would never be able to experience the actual competition firsthand. Box scores in newspapers, radio play-by-play commentary, and
even television broadcasts are all highly technologically mediated forms
of experiencing sporting events. In many ways, sports remain the primary
venue in which many Americans consume information, or at least sophisticated statistical information. Even watching a game on television exposes
us to some of the most sophisticated forms of information management
and presentation available today.

This chapter explores the central role of information in the everyday
life of American sports consumers. It considers the many ways in which
Americans have traditionally consumed sports information—attending
live sporting events, watching matches on television, listening to them on
the radio, reading about them in the newspaper or online, wagering bets
on them—as well as more modern, technologically driven forms of sports
consumption, such as playing video games based on sports or participating
in fantasy sports leagues. The goal is to critically examine the ways in
which American sports fans have related not only to their favorite teams

and athletes, but also to the various forms of media that serve as conduits between the fans and their teams. The central argument is that recent advances in the media landscape have allowed fans to develop a more and more active relationship with their fantasy teams, enabling them to act as virtual general managers and sports correspondents through online fantasy sports teams and sports blogs.

Sports and the American Experience

Sports and the sports fandom have long played a central role in the formation of American Identity. Millions of Americans every year attend sporting events, watch matches on television or listen to them on the radio, read about them in the newspaper or online, wager bets on them, play video games based on them, play fantasy games that are rooted in the actual games, and, of course, participate in live sports competitions. The focus of this chapter is not on the history of playing sports, however, but on how Americans consume them. The focus therefore, will be on sports fans, and not athletes.

Many sports fans take an unusually strong interest in the performance of their teams, acting like the successes and failures of their teams are in fact *their* successes and failures. If one were to talk to a loyal fan about a recent win they would likely say "we won" rather than "they won." What causes these deep-seated connections to these teams? What motivates a fan to devote so much time and effort following groups of men and women in flashy uniforms who toss and kick balls around for a living?

The motivations that drive sports fans are based in many factors. One of the more obvious is geography. Fans show pride in their particular locality or nation, and they make it a matter of personal pride if the home sports team is successful.[2] Examples of this include the emergence of nationalism during the Olympics and the fervent regional followings that are rejuvenated during the playoffs of professional sports leagues. These regional rivalries are essential for sports' survival. Try and imagine professional baseball without the Boston Red Sox and New York Yankees rivalry, basketball without the Los Angeles Lakers vs. the Boston Celtics, college football without Ohio State vs. Michigan. The entire foundation of American sport would be weakened without regional pride to motivate fans.

Sports provide a form of social currency, a basis for conversation between friends and coworkers and a simple icebreaker among strangers on the bus or riding together in the elevator.[3] Unlike politics or music tastes, one is not likely to offend by bringing up the topic of sports unless one person purposely goads the other by expressing enthusiasm for the local team's rival. A lesser-known connection between people and sports is due to advances in technology. Players of sports-related video games and people who dabble in fantasy sports leagues can develop a strong interest in the source of those outlets. Those who happen to play a sports video game at a friend's house or participate in a fantasy league are subjected to extensive information about nonlocal teams and players and may thus gain perspective (and perhaps appreciation) of the sport as a whole. One fascinating example involves the popular *Football Manager* video game, in which a video game player chooses a club soccer team and assumes the role of manager of the club. Sports games originally allowed for the player to play virtual matches, but this particular version lets the player make major roster decisions, set scouting budgets, and bid for free agents within the game. The immersion into this fantasy sports world is so engaging that video games that follow this model have been cited as having demolished players' relationships and careers.[4]

Fantasy sports bring a new dimension to being a fan, requiring previously unseen levels of commitment in following various sports teams and athletes. In addition to analyzing the connections between American sports fans and the media sources that bring information to them, this chapter will also take a look into the aggressive information-gathering habits of fantasy sports players who seek out hidden nuggets of information for a competitive advantage against their opponents, showing that good fantasy sports players are uncommon, even among other sports aficionados, in their voraciousness for information.

A Brief History of Sports Media

By and large, fans have always been able to connect to their favorite teams simply by attending the games. One would gather one's friends and family, enjoy an afternoon at the ball game, and talk about it later that night, repeating this process from time to time. The relationship stopped there however. Despite their enthusiasm at the game, unless they personally knew a player or coach on the team, fans were effectively cut out of knowing the interesting details surrounding their favorite team and its players. The force that changed all this was the media. For well over a century, media sources including the newspaper, radio, television, and the Internet have brought sports games, interviews, and expert analyses into the homes of everyday Americans on a daily basis, providing them with the means to collect information about their favorite athletes and teams.

Newspapers

Eighteenth-century American accounts of sporting events were often reported as part of the newspaper's observation of the community at large, and were published days or weeks after the event without detailed coverage. This was due largely to America's Puritan heritage, in which sports were seen as promulgating idleness and ungodly behavior.[5] While individuals may have felt no fear in eschewing these directives, through-out the eighteenth century newspapers were still regulated in thought, if not in law, by these Puritan prohibitions.[6] The target audience of early American newspapers—the wealthy, literate elite—also limited the amount of sports coverage that existed during this period.[7] Horse racing and boxing were in most cases the only sports reported on, due to both their immense popularity with all social classes and the regional or national social contexts that could be engendered by these competitions.[8]

Like earlier newspapers, the larger papers of the mid-nineteenth century, such as the *New York Times*, covered sports only occasionally and offered little commentary; their coverage often included an expression of "regret at the role of sport in society."[9] On the other end of the spectrum, "penny press" papers, aimed at working- and middle-class readers, played up the sensational aspects of sporting events to draw in readers.[10] Overall, the news media limited their coverage to what was either socially appropriate or openly sensational; there was little interest in the kinds of information their readers wanted to know.

This would soon begin to change, as industrialization and increased urban population growth led to an increased popularity in a variety of sports and the development of organized and often commercialized groups. The first professional baseball team was founded in 1869, with the development of the National League occurring less than a decade later; amateur players also began forming semiprofessional and club teams with the sponsorship of industrial employers or local communities.[11] Within this new social order, sports recreation began to become more conventional and was increasingly promoted for its health benefits to city dwellers.[12] As sports became acceptable, local teams began to organize and grow into commercial entities. This was especially true for baseball, whose popularity spread following the return of Civil War veterans to their hometowns, bringing knowledge of and passion for the game with them.[13]

As teams and leagues began to form and a professional veneer developed on what had previously been a community affair, sports fans demanded that newspapers cover these organizations and provide in-depth coverage of games and players.[14] Newspaper publishers responded by covering the most popular sports, realizing that this coverage could generate readership. Their coverage also served to legitimate these sports, standardize the rules of the games, and create in-depth information about game play and players similar to what readers demand today.[15] Finally, improvements in technology, such as the development of long-distance telegraph networks, allowed newspapers to provide sports information from across the country in a reasonable amount of time, creating an immediacy to their reporting that readers rewarded.

As the twentieth century dawned, sports coverage created not only celebrity players but also celebrity sportswriters, who combined their enthusiasm for the sport with colorful, descriptive writing that drew in readers and increased circulation—devoted fans would purchase a specific paper solely to read the columns by their favorite sportswriter.[16] Henry Chadwick, considered the father of professional sports journalism, was not only a popular early writer, but also the creator of the box score and statistical measures such as batting average and earned run average.[17] During this period, there also grew an increasing trend of self-identification by users of these statistics as fans of a specific sport or team away from the stadium—one no longer needed to be an athlete or be present at a game to participate or display pride in a certain team. This movement was especially pronounced among newer immigrants, who often tended to create distinct groups of fans within certain cities and gave their chosen teams a loud and local personality.[18]

Following World War I, newspapers remained dependent on sports coverage not only to draw in increasing numbers of readers, but also as a source of steady advertising income. Unfortunately, this led newspapers toward a tendency to promote only "safe" and community friendly articles, to avoid antagonizing local teams and community leaders who desired the promotion of their sporting events and players without any investigative reporting.[19] This close relationship between newspapers and the sporting industry declined somewhat with the advent of radio and film as mediums for advertising, but this still remains a strong component of the modern sports page.[20]

Sports remain a popular and prosperous area of coverage in the modern newspaper industry. The sports pages are often considered to be the escapist center of the traditional newspaper, although the lighthearted, insider style common in this section provides little critical analysis of the sports industry and often toes the party line in regard to major league sports teams and players.[21] Some sports fans are savvier information users and continue to demand that sports writers undertake a higher standard of reporting within their industry, and in response some newspapers attempt to include stories of a more serious nature.[22] But the close relationships between the press and the professional sports teams continue to cause problems with readers who expect the same journalistic integrity found in other sections of the newspaper.

Newspapers are still the standard source for users gathering sports information. According to a 2004 poll by ESPN, sports fans are more likely to turn to newspapers to fulfill their information needs than to radio or television. Over 20 percent of fans surveyed obtained information from newspapers eight or more times a week. According to the survey, fans prefer newspapers since they are better than other broadcast media at providing in-depth statistics and analysis.[23]

It remains an open question how newspapers will deal with users' increasing ability to gather the same information on the Internet. Currently, some papers choose to offer all of their sports information, from feature articles to statistical analysis, for free on their Web sites, hoping to capitalize on user traffic to increase in-site advertising revenue. Others papers are choosing to restrict their data to print forms or subscription sites, which require the payment of fees for access. As newspapers decide on their future direction, there is no doubt that sports fans will be an important part of any decision.

Radio

In the 1920s, radio was a new medium with limited coverage. While it was possible for the average American to purchase a radio receiver, there wasn't much to listen to. In 1922, only 1 in 400 households owned a radio but between 1923 and 1930, 60 percent of American families purchased one.[24] Sports helped to popularize the new device, while at the same time, radio broadcasters helped promote sports to a wider audience of people who may never have attended a major sporting event.[25]

A key factor in the acceptance of radio over the newspapers was its immediacy, which was seen as a key advantage for the sports industry.[26] Radio networks were quick to build a relationship with sports; the first broadcast on NBC's radio network was the heavyweight championship fight between Jack Dempsey and Gene Tunney.[27] If the networks needed any hint that they were on the right track, the fact that this one prizefight in 1926 led to the sale of over $90,000 dollars worth of radio receivers from a single New York store would certainly point to the confluence of sports fans and the new medium of radio.[28]

Throughout the 1920s and 1930s, sporting events became increasingly popular on the airwaves due to the relatively inexpensive cost of broadcasting a preexisting event compared to the price of a scripted radio drama. The innate dramatic nature of certain dynamic sports, notably boxing and horse racing, allowed them to translate well to the new medium and entice listeners across the nation.[29] Fans of team sports, particularly baseball, increased radically; stations learned to "chain-feed" important games, particularly the World Series, along long-distance lines so as to satisfy users in distant cities. This also increased the number of listeners with a stake in the team's victory, creating a greater sense of community and a larger fan base for specific teams and players.[30]

Radios soon became the quickest way to receive information about sporting events since the nature of the medium allowed for live local coverage or live reporting of scores as the information rolled in from the wire services or were reported by telephone. While the immediacy of information and ability to hear the game live was a boon to fans and broadcasters alike, teams began to worry about a decline in attendance as families chose the radio over a trip to the ballpark, especially on game days with inclement weather.[31] Some college football teams were concerned about the increase in "Home Radio Parties," where fans would listen to the games and diagram the action on a blackboard at home, which some considered "almost as good as being at the game."[32]

To make listeners really feel they were attending the game, radio broadcasters—stations generally were unwilling to pay for expensive long-distance broadcasts—would often "ad-lib" game play for away games based on wire reports, creating a lively atmosphere. This not only made new fans for specific sports or teams, but also for the exuberant sportscasters.[33] Often these game recreations weren't wholly faithful to the real game; James Michener, the famed author, reported listening to a college football broadcast in the 1930s during which the home team was portrayed as having players who could perform miracles, only to learn at the conclusion of the game that they had lost without scoring a single point.[34] Former President Ronald Reagan, who began his career in the public eye as a sports announcer for a Chicago radio station, fielded a similar situation, when the wire went silent, forcing the high-spirited broadcaster to find new ways to describe a single player fouling repeatedly for almost seven minutes until the connection was restored.

Beginning in the 1960s, television began to eat into the audience for sports broadcasts, leading stations to eventually turn to talk radio and sports commentary to turn the tide and lure listeners back. In 1987, the first major all-sports radio station, WFAN (660 AM) out of New York, provided a sea change, incorporating shows from sports broadcasting legends with listener phone calls and up-to-the-minute sports updates.[35] Talk radio also changed the radio landscape by creating informal networks of fans who were able to communicate with the commentators and other fans from their home or work, in what has been regarded as a democratizing way of sharing and spreading information, with the opinion of fans considered on an equal level with that of the commentators.[36]

As with newspapers, the future of sports on radio seems to be heading toward a subscription model, with satellite radio companies providing the latest and best of sports commentating along with extended coverage of multiple sports.[37] At the same time, local stations still remain a cost-effective way for broadcasters and advertisers to communicate with their core demographic and create a community of fans who feel free to express themselves and share information.[38]

Television

Sports and television weren't an immediate match made in heaven. The first televised sporting event was a Columbia–Princeton baseball game held on May 17, 1939, followed by the taping of several professional baseball games.[39] Team sports were not the best choice for televised events; the rudimentary equipment could not accurately capture the fast pace of play in football and baseball, while wide views of the field made it impossible to discern one team—or individual players—from one another, as they appeared like "white flies" on the television screen.[40]

In the early years of television, due to both technological and budgetary issues, other sports such as boxing and roller derby were more popular on broadcast networks. Filming these sporting events could be accomplished with a single camera and small crew, requiring little payout for broadcast to a huge audience.[41] It was not until the 1950s and 1960s that television began to overtake radio for sports supremacy, in part due to the increased market penetration of television, as television receivers became more affordable and new technology improved the quality of sports broadcasts. Videotape technology allowed for recording and replaying of sports events from around the globe, while the development of instant replay in 1963 made sports events look more cinematic and increasingly attractive to fans.[42]

In the early 1960s, growing cooperation between professional sports leagues and newly experienced broadcast networks led to the advent of regularly televised professional sports games. With vast amounts of money on the line as broadcast rights and advertising revenues came into play, viewers at home were introduced to the products of the NFL, Major League Baseball (MLB), and the National Basketball Association (NBA), forcing smaller sports onto the sidelines and pushing the networks' idea of sports programming onto the public.[43]

While networks were busy making money, broadcasters and producers were delivering new ways to attract and keep viewer interest by staging broadcasts not from a bird's eye view of the game, but rather from the perspective of the fans in the stands, allowing viewers at home to feel as though they were part of the action.[44] Beyond the action on the field, broadcasters focused on the sidelines, cheerleaders, celebrities in attendance and, most of all, the fellow spectator, giving the viewers at home the feel that they were part of the event.[45] This coverage also helped develop the water-cooler culture of sports, allowing viewers to discuss and dissect a game together even though they watched it separately.[46]

In the modern era, televised coverage and commentary on sports caters to active, sophisticated users. ESPN was created in 1979 to appeal to users who desired more information than "a score and a winner," from detailed coverage of major sporting events to an exploration of different venues and kinds of sports.[47] Soon, user-driven coverage led to an increase in specialized networks, from regional networks covering every kind of sport, from professional to college to amateur, to "power" stations such as WGN and WTBS, sending local sports coverage out to the national audience.[48] Suddenly, the Atlanta Braves were America's team. Spanish language networks also stepped in to fulfill the needs of minority and immigrant groups, offering worldwide sport specific coverage, particularly *futból* and World Cup reporting.[49] These specialty stations and twenty-four-hour coverage networks often have higher ratings than national networks, proving that listening to user demands can improve television's business model.[50]

Television provides a bridge from the analog to the digital. Television users who watch sports are active participants in their viewing habits: they search for content with intent and intentionally choose to watch sporting events.[51] This is opposite to other genres of television viewers, who arbitrarily find content and "settle" on something to watch.[52] With technological advancements, viewers have also moved beyond the couch to take an active role in choosing what they want to see, especially with the growing popularity of DVR and TiVo use and an increase in the number of premium cable and satellite stations available. Major League Baseball now offers a service called MLB.tv, which allows fans to route live-game broadcasts through their computer onto the television to ensure that users are viewing exactly the content they desire.[53]

Internet

Unlike other media, the Internet did not immediately pick up on the importance of sports. In its earliest days, the only information that could generally be accessed, due to the slow speed and limited access of the Internet, was basic sports scores.[54] Eventually, access became faster and more readily available, leading to the creation of the first generation of sports-related Web sites, among them ESPN.com, Yahoo!, and AOL Sports. The initial offerings at these Web sites were not all that sports fans may have wanted; ESPN.com's original offerings consisted of regularly updated scores and stories from various wire services, with very little original or critical content.[55] The focus of these sites was also limited to the most popular sports, leaving followers of other, less-followed sports, such as soccer or tennis, without any information.

As the Internet continued to grow in popularity, fans became a bell-wether for the kinds of data that sports-oriented Web sites needed to provide to their users. Both larger sports Web sites and sports teams and organizations became aware that they needed to create an encompassing Web presence or risk being bypassed in favor of specialty or fan-created Web sites.[56] This realization accentuated a major aspect of the transition to the Internet, what Michael Real, an expert in the intersection of mass media and popular culture, calls the "death of distance," or the fact that users are no longer separated from their favorite team or sport by distance or local disinclination.[57] No longer do fans have to subscribe to specialty publications or wait for newspaper or television reports; now the action is available live around the globe, along with real-time communication with the press and other fans.[58]

The Internet has caused an essential change in the nature of being a sports fan. Before the development of the Internet and its vast capa-bilities, sports fans were seen as passive users of information, often portrayed as couch potatoes dutifully ingesting the information trans-mitted to them via their media of choice.[59] The constructive and inter-active elements of the Internet have revolutionized that image, allowing sports fans to have an increasing amount of control over the informa-tion they want to consume and allowing them at the same time to produce their own information for other fans to utilize. The uses of this new technology in sports information production range from the creation of specialized blogs and forums for specific sports or sports teams, to the uploading of audio or video from local sporting events, to providing an unofficial regulatory presence for teams, players, and coaches.[60]

While the Web has rapidly become the primary resource for disseminating sports information, what makes this form of media particularly novel is that sports fans, as information users, have devised new ways to harness information to create new and unique forms of use and interaction. A few of these new vehicles for sports information will be discussed later in the chapter, but first we turn to an in-depth analysis of one of the wildly popular new markets made more viable due to the nature of the Internet: fantasy sports.

From Fact to Fantasy

Sports are most obviously enjoyed for the physical drama of the game: the crack of a wooden bat on a baseball; the grunting of two lines grasping for leverage after a football is hiked; the dichotomy of complete silence on a putting green save for the birds and cicadas and the clamorous uproar when the putt is made. There are legions of sports fans, however, who find just as much exhilaration and pleasure from looking at columns of numbers on newsprint or on a Web page. There has been a demand for sports statistics in the United States almost as long as there have been organized sports. *Sporting News* first published the *Official Baseball Register*, a collection of baseball statistics, in 1886.[61] Baseball enthusiast Bill James began writing his *Baseball Abstract* in 1977. That year he sold only seventy-five copies for $4 apiece.[62] Today his *Abstracts* and related books sell thousands of copies. The demand for books such as these have grown only stronger, especially with the exploding popularity of fantasy sports in the mid-1980s and the expansion of fantasy sports into the Internet over the past decade.

While there are many fans out there who enjoy reading statistics for statistics' sake, fantasy sports have emerged as an outlet for putting these statistics to use, rather than having them serve only as fodder for discussion. Fantasy sports is based on using real statistics to play in virtual sports leagues. They have drastically changed the character of the U.S. sports market. Fantasy sports have sparked fan interest in all of the major American sports and created another outlet for fans of less popular sports to pursue their interests. As of 2004, fifteen million people were participating in some form of fantasy sports.[63] By 2008, this number had increased to 29.9 million people playing in the United States and Canada, spending $800 million directly on fantasy sports. If products that appeal to fantasy players "such as DirecTV's NFL Sunday Ticket and XM Radio's coverage of all MLB baseball games" are included, fantasy sports potentially create a market of up to 4.48 billion dollars.[64] Given that "fantasy players on average watch three hours more football a week than nonplayers,"[65] the major television networks have programmed their televised games to cater to this large body of users. It is not unusual for the television networks to deliver statistical information directly to the fans via a scrolling ticker at the bottom of the television screen. If one were to watch an NFL football game on television, it is likely that the commentators will wrap up the show with a recap of the fantasy player of the game, the athlete who amassed the greatest number of statistical points even though that player might not be the one most responsible for the successful outcome of the actual football event. If a baseball pitcher gives up a towering grand slam, thousands of people across the country are liable to utter a simultaneous groan—not only because the pitcher just lost the game for his baseball team, but also because the fantasy "owners" of that pitcher saw their fantasy ERA (earned run average) skyrocket, hurting their ranking in their fantasy league.

Broadcasting companies are not the only ones cashing in on these interests. Cell phone companies are also finding ways to make money off of fantasy sports. More and more sports fans are using their cell phones for the purpose of looking up sports scores. But one does not need the latest browsing technology on a cellular phone to do so—companies such as 4Info specialize in providing programs that alert sports fans about final scores and other statistics via text messages, making it easier than ever for sports fans to satisfy their interests.[66]

What kinds of people are playing these fantasy sports? Games are normally played by kids and teenagers—so one might expect this to be an activity dominated by young people. Actually, according to a study conducted by a University of Connecticut sociologist, the types of people playing fantasy baseball are remarkably homogeneous and not as young as one might expect. The demographics of fantasy baseball participants are 94 percent white, 98 percent male, 63 percent married, and overwhelmingly college educated.[67] When one looks beyond baseball to all fantasy sports, the hobby is not as overwhelmingly male. Across the board, 70 percent of all fantasy sports players are male, but female participation is increasing rapidly. 77 percent of female players have begun playing in the past five years.[68] Thirteen percent of all American teens had participated in fantasy sports in 2003, slightly higher than the national average of 11 percent. This percentage has most likely increased since 2003, given that fantasy sports have become more popular in the United States since then.[69]

While the activity of playing fantasy sports has become far easier through the use of the Internet, numerous varieties of the game have existed for decades. One form in which it existed was game simulation. For example, when computers came along in the 1950s, mainframe computers were utilized for sporting purposes. People created software that allowed them to "play" baseball by typing in actions and counterplays. The computer would then simulate a baseball game based on the commands the player chose.[70]

Another popular kind of fantasy game was tabletop baseball. Created in 1961 by the company Strat-O-Matic Baseball, this game was enjoyed by many as a way to "play" a game of baseball in twenty minutes from the comfort of one's home.[71] It consisted of a deck of special cards, three dice, and a collection of charts. Each card represented an athlete and the three dice were used to guide the player through pitching, hitting, and defense. On the back of each player card one would find several columns containing possible outcomes (baseball "plays") based on what a player rolls. A die would be cast to determine which column to use, and then the remaining two dice would be cast to determine the action of the pitcher, hitter, or fielder.[72]

Simulation technology has become much more sophisticated over the past four decades. One of the most complex and arguably the most interesting examples of professional sports simulation is a Web site called *What If Sports*.[73] At this site the sports fan can choose to simulate games from the four major American professional sports (baseball, basketball, football, and ice hockey) as well as college-level football and basketball. What makes it so fascinating to the site's sports fan users, however, is not only being able to simulate games between any two current teams, but also between teams of different eras. Who would have won a pitching duel between Roger "The Rocket" Clemens and the legendary Cy Young? Which undefeated regular season team would come out on top in a faceoff, the 1972 Miami Dolphins or the 2007 New England Patriots? The program utilizes collected statistics to simulate not only the winner, but also a play-by-play account of the entire game, giving just a little bit more substance to those timeless arguments that fans like to make about which teams from different eras are superior.

Although fantasy sports systems can be used to compare historic teams and players, they most commonly simulate current sports activities, and tend to rely on current, usually live statistics. Today leagues exist for nearly any sport imaginable. While the most popular versions center on the MLB, the NFL, and the NBA, one can find fantasy leagues for hockey, tennis, cross-country running, swimming, and even one for curling.

There is debate as to when fantasy sports started. Many cite the 1950s, when Wilfred Winkenbach created a form of fantasy golf.[74] Some argue that fantasy sports began with baseball enthusiast Bill Gamson in the 1960s, while others point to the Greater Oakland Professional Pigskin Prognosticators League, a fantasy league that tracked the American Football League in the 1960s. The idea of fantasy sports as we know it today, most agree, originated with Daniel Okrent's Rotisserie baseball league in 1980.

Okrent, an author and editor for the *New York Times*, had many arguments with his friends over who knew the most about baseball, so they started the first modern fantasy league in order to settle which of them would be the best general manager. They drafted players, and Okrent kept up with the statistics throughout the course of the season. Since these were the days before the Internet and convenient statistical spreadsheets, "Okrent found himself driving to a distant newsstand to obtain *The Sporting News* two days before it would otherwise arrive in his mailbox."[75] When the proto-fantasy season was over, Okrent published a two-page article about the league in the May 1981 issue of *Inside Sports*.[76] After that, the game grew rapidly in popularity, which Okrent named Rotisserie baseball, after *Rotisserie Francaise*, the diner in which he and his friends met.

By 1988, *USA Today* estimated that five hundred thousand people were playing fantasy baseball.[77] While this certainly reflects impressive growth in the hobby since it became public knowledge seven years earlier, the number of participants today is much larger, with almost thirty million players in the United States.[78] Fantasy baseball is still the most commonly played fantasy sport, but American football has seen a rapid rise in interest in recent years.

Before going into greater detail about the history of fantasy sports and how its participants gather pertinent information in order to play, it would be beneficial to have a working description about how the two most popular versions of fantasy sports are set up and scored.

Fantasy Rotisserie Baseball

Like most versions of fantasy sports, Rotisserie baseball begins with a "draft," a session during which the league's participants get together, either physically or online, and choose the players they want for their team. Some leagues set it up so that participants can choose any athlete they want in any round they want, as long as the athlete is still available. However there are more complex versions of fantasy drafts—one of the more popular alternatives involves consulting a preset spreadsheet with dollar values assigned to each athlete. Each fantasy participant starts out with a set number of dollars and is free to choose any available player as long as they do not exceed their dollar limit.[79]

The draft in itself is easily the most sociable portion of the fantasy season and possibly the highlight of the year, even more so than the playoffs. Everyone's spirits are still high about the possibility for the future and the fatigue of a long season has not yet set in. Though there are many ways that a draft can be completed, there are primarily two methods practiced. The first is best practiced in a league composed of friends—all of the fantasy players gather in one room with lists of available athletes, a wish list of desired athletes, and any notes, books, and laptops the players might have. They then set about making picks and arguing over draft choices. Besides having the utilitarian benefit of laying the groundwork for everyone's fantasy roster for the rest of the year, this type of manual draft allows for the spreading of both camaraderie and rivalry among the participants, two great motivating factors that can keep the league going during a long season.

The more modern and automated version of fantasy drafting involves using programs that are provided by the Web site through which the fantasy players are hosting their league. To use Yahoo! Sports as an example, at a set date and time the league's participants will meet in a special Yahoo! chat room to make their picks. The chat room includes a list of all the athletes available to draft as well as lists of all the athletes the various fantasy league participants have drafted. A draft order is preset and league participants draft their picks in order. Once everyone has made their first pick, everyone then picks again in reverse order (in the interest of fairness). This is repeated until everyone has a full roster and several backup athletes.

An advantage to drafting online is that it avoids the scheduling difficulty of getting everybody together in one place. League participants can go online and create wish lists of athletes and the computer will make picks automatically for the fantasy player based on those lists or preseason rankings of athletes. Most players choose to make their own selections, however, because the live draft is one of the more entertaining parts of the fantasy season and a computer cannot react to unexpected turns of events during the draft, such as unusual picks by rivals.

Once the draft is complete, the fantasy player then chooses his or her starting lineup, placing athletes in their respective positions on the baseball roster and pitching rotation. Assuming that the fantasy player is participating in the league through an online provider, the fantasy roster can be updated at any time, the only exception being that once the athlete's game (in real life) begins, that athlete is locked in his particular position in the fantasy roster until the following day. Athletes can be traded between fantasy players throughout most of the season, and athletes can also be chosen from the "waiver-wire," or pool of undrafted players.

Scoring is a complicated process, although it has been simplified through online automation. The ultimate goal is to accumulate the most fantasy points, which are acquired by accumulating points based on statistics in the real world. Before the season begins, the league participants choose which statistics they want to use in competing. Usually it includes ten to twelve hitting and pitching statistics such as runs, home runs, RBIs (runs batted in), batting average, stolen bases, wins, saves, ERAs, strikeouts, and WHIP (walks and hits per innings pitched), though other stats are also available for inclusion.

If an athlete makes a play that affects one of the chosen statistics *and* he is on the fantasy player's active roster, the statistic is added to the fantasy player's running total and is then compared to the other fantasy players' statistics. The fantasy player who most dominates a particular statistical category earns the most points. Assuming there are ten fantasy players in a league, the person with the most runs would acquire ten fantasy points, the person with the second-most runs would acquire nine fantasy points, and so on. The important thing to remember is that in Rotisserie baseball, fantasy points are not cumulative. If the person with the most runs has a slump and does not acquire any over a period of time, it is possible and likely that someone will pass the player in that statistical category, causing the player to lose the fantasy points he enjoyed while leading the category. The fantasy player with the most total fantasy points on the last day of Major League Baseball's regular season is the winner of the league.

Basic Information Gathering in Fantasy Sports

For the act of gathering useful predraft information about the available athletes, it used to be that the average fantasy player would have little choice but to rely on old newspaper box scores, their own gut instincts and memories of the previous season's statistics, or a purchased copy of the *Bill James Baseball Abstract*, a collection of baseball statistics and analyses. Today there are innumerable high-quality services that a fantasy player can obtain for free or for purchase in order to prepare for a fantasy draft or maintain a team during the course of a season.

Options abound for a fantasy player aiming to make a splash in his league and run the table for a season. There are dozens of books on sports statistics and fantasy tips as well as regularly updated Web sites to which fantasy players can subscribe. One anecdote involves Sam Walker, the author of *Fantasyland*. Over the course of a baseball season Walker spent roughly $19,500 in his ultimately unsuccessful quest to win a fantasy league playoff against the most elite players in the United States. He spent $7,400 on scouting trips, $1,800 on computer components, and $895 on encyclopedias and Web subscriptions, as well as salaries for two individuals who helped him plan his draft and make other personnel decisions throughout the season.[80] Not surprisingly, this statistic is not typical of the normal American fantasy sports player. Walker is a sportswriter who was pulling out all the stops to win in an elite league (he finished eighth—not a win but nevertheless a respectable finish). It is not unreasonable to assume that between book, magazine, and Web site purchases a fantasy sports player who participates in two to three different leagues could spend around $175 per year on this activity.[81] To stick with the example of baseball, books such as the *Bill James Baseball Abstract* can easily be purchased by a group of players who split the costs, while major sports magazines such as *Sports Illustrated* are an inexpensive source of current information about which athletes are expected to perform well in the upcoming season. Baseball Prospectus (http://www.baseballprospectus.com) is a highly touted online option for sports fans wishing to subscribe to regular updates about everything of value happening in Major League Baseball.

There are many Web sites (e.g., espn.com, cbssports.com) that allow people to play online fantasy sports, but by and large the most popular fantasy sports Web site is Yahoo! Sports (http://sports.yahoo.com/fantasy).[82] The popularity of Yahoo! Sports with fantasy players can be attributed to its abundance of sports statistics, its graphical StatTracker™ that allows fantasy players to "follow" both real sporting matches and fantasy matches in real time, and the fact that all this is free of charge. For a nominal fee, Yahoo! Sports also provides a draft kit that includes projected lineups of professional teams, in-depth analysis of what to expect from the best players, and expert predictions of where each player will likely be picked during a typical draft. Such information can be invaluable for fantasy players seeking an edge against their competitors.

It is not necessary to pay for reliable, up-to-date sports information. There are many good fantasy sports players who never spend a dime. The official Web sites of the professional sports organizations (e.g., mlb. com, nfl.com, nba.com, nhl.com) offer a plethora of team and player information. On the official Major League Baseball site one can find standings, team schedules, news articles, audio and video clips, interviews, and complete statistical histories of any player's professional career.[83] Beyond the official league Web sites, hundreds of privately run sites, wikis, and blogs exist. These are discussed in greater detail later.

Fantasy sports companies have been in the news lately because of a lawsuit involving Major League Baseball Advanced Media and the intellectual property rights to the names and statistics of its athletes.[84] CBC Distribution and Marketing, Inc., the parent company of CDM Sports, argued that all sports statistics are public information and that CDM Sports' use of the statistics is no different from the local newspaper publishing box scores in the sports section. MLB countered that the issue at hand is the use of the players' names for competitive advantage, and that the hundreds of fantasy sports companies are violating intellectual property laws by putting a player's name next to a statistic without a license. CBC won the case in 2006 and the 8th Circuit Court of Appeals upheld the ruling in October 2007, with the panel stating, "It would be strange law that a person would not have a First Amendment right to use information that is available to everyone."[85] This suggests that fantasy sports companies will continue to remain competitive and prosper so long as the concept itself is popular.

Information-Gathering Strategies of Fantasy Sports Players

It is difficult to do well in fantasy sports without conducting research into the sport in question. Complex decisions have to be made in order to put up the highest numbers possible; multiple variables must be taken into consideration. But how do the fantasy sports players come by the information to make these decisions?

The information-gathering habits of fantasy sports players differ by skill and experience level. Generally speaking, when attributing information-gathering habits, fantasy players are cleanly split into one of two groups: beginners and experts. The difference lies not in whether players do research before making their roster decisions—most fantasy players who put forth a good-faith effort to win in their leagues conduct at least a rudimentary amount of research when drafting and making roster changes. Instead, the fantasy players differ in the sources they use to gain their sports information. Expert players are more likely to think of fantasy sports as an investment and seek out highly researched, quality information from a source such as the *Bill James Baseball Abstract*. Novice players, however, are more likely to be satisfied with the mainstream news that is broadcast through television, newspapers, and major Web sites, where they would get their normal headline sporting information (e.g., www.si.com, www

.espn.com). These are good sources of sporting information, but this information is readily available to everyone in the league. While one will learn that one's ace pitcher is not going to be starting tonight due to a damaged rotator cuff by watching SportsCenter on ESPN, one is far less likely to learn about an up-and-coming player in the minor leagues who might be worth obtaining. Information in fantasy sports becomes more valuable if it is received quickly, implemented quickly, and is esoteric. Anyone who has watched a football game knows that Peyton Manning is an excellent quarterback to have on a fantasy team; it takes a shrewder fantasy player to realize that Eddie Royal is a more valuable athlete to pick up because he produces points yet can be obtained inexpensively (i.e., one would not have to use an early draft pick or give up a valuable player in a trade in order to obtain him).

Nuances are what lead to fantasy championships, and that is why expert fantasy players are more likely to win. It is not unusual for an expert player to consult numerous specialty sources in order to gain an edge. Though specialty sites exist for most professional sports, baseball has the highest quality of analysis due to both the sport's popularity and its identity as one of the most statistics-heavy sports played in the United States. One of the most popular sites utilized by fantasy baseball players is Baseball Prospectus, mentioned earlier in the chapter. This Web site gains much of its respectability from having been created by a group of sabermetricians—members of the Society for American Baseball Research (SABR). Sabermetricians attempt to answer objective baseball questions that can be solved by analyzing statistics such as, "Which batter most contributed to his team's success?"[86] An expert fantasy player who knew what she was looking for could learn which types of athletes she needs by looking at the site's statistics page that breaks down the information by category.[87] Which statistic is most useful for a hitter to dominate? Are pitchers who have a tendency to pitch ground balls more likely to win than those who pitch fly balls? Answers to questions like these can be found on this site.

MLB Trade Rumors is a Web site of incalculable worth to both fantasy sports players and regular sports fans. It functions as a hub for information regarding the personnel of all Major League Baseball teams. A user of this site has the option of looking up his favorite team and reading information about who the team might draft, trade for, or bring up from the minor leagues.[88] Fans find great value in learning what players their favorite team is seeking. Fantasy players could gain an advantage by having access to this type of information. For example, if it were rumored that an above-average pitcher for a bad team was going to be traded to the New York

Yankees, a historically successful baseball team, an astute player would attempt to acquire him for his fantasy team before the other fantasy players found out, as a decent pitcher playing for a good baseball team will be much more successful at racking up points in the "Wins" fantasy statistic.

Beyond being more skilled at making smart moves after hearing the latest news stories, the expert players are also capable of making informed decisions from raw statistical evidence.[89] Based on the information they glean from specialty Web sites, the best fantasy players are able to take into consideration, to use fantasy baseball as an example, the following information: Pitcher A's vital statistics such as ERAs and strikeouts-per-nine-innings (K's/9) ratio, the pitcher's historical rate of success against a particular team, the pitcher's rate of success over the past few games, the pitcher's general rate of success at that point in the season, the dominant arm of the pitcher, the latest injury reports relating to the pitcher, the size of the ballpark, and the weather conditions at the ballpark the pitcher is playing in. Using that information and more, a dedicated fantasy player can make the decision of whether to play Pitcher A over Pitcher B, generally with a respectable level of success.[90]

Beginner players, on the other hand, have their own methods of researching and choosing athletes. As mentioned earlier, they are far more likely to rely on mainstream sources for their information, but they are also more likely to make their fantasy decisions in a less statistics-based manner. They have a tendency to rely heavily on the reputations of athletes and teams when making their decisions. To use professional football as an example, there is generally little reason not to play a statistical juggernaut such as Tom Brady. The odds are good that if he plays, he will put up respectable numbers. However, a beginner may possibly elect to bench Brady if his team were to compete against a team with a reputation of playing good defense such as the Baltimore Ravens. No statistics would be taken into consideration, only the reputation of the opponent.[91]

Another approach that beginners take is to choose their players based on team loyalty. Rather than researching the best possible athlete to obtain, the beginner fantasy player might select athletes from his local or favorite sports team (e.g., a New York Yankee fan will choose the best players from that particular team, rather than the best players from around the league).[92]

In many instances, information is also gathered in a less direct way. Most fantasy Web sites include a community message board so that the players can communicate with each other. While this is most often used as an outlet for boasting about the superiority of one fantasy team or another, studies have shown that fantasy players sometimes utilize the message board to administer friendly advice or offer esoteric statistical information that does not work to their personal benefit.[93] This implies that although following sports and participating in a competitive league are important, the ultimate goal of sports fans that participate in these leagues is to have fun and foster friendships through the medium of sports.

Fantasy sports remain a popular pastime for sports fans of all varieties, from those with passing interests to the serious sports enthusiast. It would not be nearly so popular, however, were it not maintained on the Internet. Fantasy players existed before the activity moved to the Internet, but it was a laborious pastime requiring extensive research of statistics and hand calculations of athlete rankings in order to play. Now, the process has been streamlined and made more accessible on the Internet, with statistics automatically updated online and league rankings generated without any action needed by the user. As a result, the fan can concentrate solely on researching the athletic information that is out there for the taking and enjoy the camaraderie with his or her peers while playing.

Sports Message Boards and Blogs

People are increasingly turning to the Internet to obtain sporting information and meet with other fans. Based on a study by Woo conducted as part of his master's thesis at Indiana State University, fans would flock to sports teams' official message boards in order to satisfy those desires.[94] Message boards are online Web sites where individuals can log in and post threads about topics of interest: ravings about how good/bad the team played that day, information about a player, tirades against the team's management, and so on. Fans find a certain solace in congregating in a virtual arena where most of the connected individuals are supporters of the same organization. Woo measured the optimism and pessimism levels of the various threads for eight different team sites (four teams with winning records, four with losing ones). The tone on the message boards for the four winning teams was overwhelmingly more positive than on the four losing message boards.[95] This mirrors the exultations and lamentations that one could expect to find in a typical verbal communication about the home team, suggesting that the Internet has not fundamentally changed the desires of sports fans, just created an additional venue where they can express those desires.

Sports blogs are a valuable conduit of athletic information, coming not from the teams' organizations or professional journalists, but instead from highly motivated fans. Many of these sports blogs are run by individuals who start up their own Web site, write opinion pieces, and (sometimes) grow a base of supporters. Sports Blogs Nation (www.sbnation.com) has acquired a retinue of inspired fans to blog and offers a centralized location to which these bloggers can link their respective blogs. What this hub does is recruit the best bloggers for every individual team for the most popular sports in the United States, then prominently provides links to their team-specific blog alongside the blogs of other teams in the league's division. Sports Blogs Nation boasts nearly two hundred blogs, managed by some of the most professionally minded fans of baseball, football, basketball, hockey, and college sports in the United States.[96]

Because they are not official members of the press, bloggers generally do not enjoy access to professional teams' clubhouses or high-ranking members of the front offices. Therefore, the work of a sports blogger is largely steeped in research. Not unlike a highly competitive fantasy sports player, a blogger who wants to be taken seriously consults numerous other sporting Web sites and statistical collections to support their impassioned arguments about predictions for the season's future success or changes that the team should make.[97] One advantage that bloggers do have is the connection to their readers. While professional sportswriters are almost certainly also sports fans, sports bloggers generally are not paid a living wage for their postings. They most often conduct their research and talk sports because they love it. So when a reader comments on one of their posts with a new opinion or criticism, they can usually be assured that the sports blogger has a deep understanding of where the fan's impassioned argument is coming from.

Conclusions

One argument that is often made about technological improvements in sports media is that they strengthen the ties between fans and athletes. On the one hand, it is certainly true that, since the late nineteenth century at least, no matter where one resides in the United States, one can get access to one's favorite team. A person can "watch" an entire baseball game simply by reading a box score in a newspaper. Catching a ball game does not necessarily have to mean purchasing tickets, fighting traffic, paying exorbitant parking fees, and pushing through hordes of people solely to see faceless athletes play from a distance—it can just as easily mean watching on television from the comfort of one's home. Many would argue this is even better than going to the game as it provides the fan with cushioned seating, cheaper food, live play-by-play commentary, instant replay, multiple camera views with close-up shots of athletes, and a wealth of statistics.

On the other hand, the increasing amount of information available about athletes also highlights the differences between their world and the world experienced by most of their fans. Many professional athletes can earn more in a single game than most Americans could hope to make in a year: Alex Rodriquez, for example, made approximately $28 million for the 2008 baseball season.[98] He had 510 at-bats that season, meaning that he made almost $55,000 every time he stepped to the plate.[99] The cost of attending games (and to a lesser extent watching games on television) has become so expensive that it is less viable as a family activity. The average ticket price for an MLB game in 2008 was $25.43 and is expected to rise in 2009.[100] Reading tweets about how an athlete shops at Target or having an athlete enter the living room through a television show for an hour a week only highlights the absence of a legitimate connection between athlete and fan.

In addition to high salaries, the introduction of free agency and excessive trades has also damaged the average fan's loyalty to his or her team of choice. Following a team is an investment in time, money, and emotion, so it is surprisingly stressful when a fan follows a team religiously (watches the games, purchases a player's merchandise, etc.) only to find that her favorite player is traded or voluntarily flees for a bigger paycheck. Shaquille O'Neal, one of the icons of the NBA, has already in his ongoing career played for five teams (Orlando, Los Angeles Lakers, Miami, Phoenix, and Cleveland). Kenny Lofton was a true cosmopolitan among his MLB peers, playing for eleven different teams in seventeen seasons. The 2009 Boston Red Sox retained only three players (David Ortiz, Jason Varitek, and Tim Wakefield) from its 2004 World Series-winning team.[101]

The concept behind trades and free agency is not new, however. Players, even superstars, have been traded for decades. Jerry Rice, Wayne Gretzky, even Babe Ruth were all traded during their careers (only the last one thought it fitting to curse his previous organization however). What is changing, however, is the seemingly decreasing likelihood that a star player (such as Larry Bird, David Robinson, or Cal Ripken Jr.) who a team builds its success around stays with one organization for his entire career. Should a fan choose to maintain loyalty to an individual athlete rather than one particular team, there is no reason why that fan cannot continue to watch or read about their favorite athlete even if they get shipped halfway across the country. Because of the wide range of technologies that make it possible to catch up on games that occur outside the home market, there is no need to toss out that expensive jersey once bought in a moment of passion just because an athlete got traded to Toronto; the sports world is shrinking due to the glut of information currently available.

Another complication created by information technology is gambling. Gambling has long been an aspect of sports consumption, and it is arguable that without gambling, sports fandom might never have developed to the degree that it has. The games themselves might be exciting, as are competition and rivalries, but having money on the line adds a tangible benefit to caring about the outcome of a game.

Gambling also enhances the social aspect of sports watching by creating an additional playing field, one among the fans watching the athletic spectacle. In the United States two single athletic competitions that are heavily gambled on are the Super Bowl (professional football) and the National Collegiate Athletic Association's (NCAA) "March Madness" basketball championship. Not only are professional football and college basketball among the most popular sports in the United States, they are the two most likely to see office gambling pools created in the form of Super Bowl Squares and March Madness Brackets. These are betting games where everyone throws money into a collective pot and wins money based on the victors of the games, scores of the games, and other characteristics of the outcomes. These examples of social sports gambling are huge. Millions of employees across the United States participate (including those that normally have little interest in the sport in question) in these office pools, with one consulting firm estimating that March Madness alone would cost roughly $3.8 billion in lost work productivity due to employees updating their brackets, checking scores, and watching games online.[102] While the Super Bowl, on the other hand, sees an estimated 112 million Americans betting roughly $8 billion every year through bookies, offshore gambling, and the Internet. All of these bets are in fact illegal—only between $90 and $100 million of those dollars are legally wagered in Nevada.[103] Regardless, this is a considerable amount of money being placed on one sporting event.

Although mostly illegal, the sports betting industry is very large and an integral part of sports. According to the American Gambling Association, $2.58 billion was legally wagered in Nevada's sports books in 2008, with a gross revenue of $136.4 million.[104] This $2.58 billion most likely represents only a tiny fraction of all money wagered on sports in the

United States. The National Gambling Impact Study Commission estimated that illegal wagers could add up to as much as $380 billion annually.[105] Although gambling is morally questionable, it is a large motivating factor for many Americans to pay attention to sports. Without the revenue generated by fans who watch sports because they enjoy competing with their friends and putting a couple bucks on the game, would we recognize the result? Would we see such highly talented and professionalized athletes and such opulent stadiums? What about aspects of most leagues that Americans have come to accept as normal? Would it be cost effective for teams to regularly travel far across the country for games? Would sports information be printed in every major newspaper and would there be special television packages that allow fans to watch every game? Gambling gives many an additional reason to seek out and consume sports information—purchasing a newspaper to read editorial opinions on who is expected to win the big game, using the Internet to look up the latest betting odds, and watching television to view the game in question.

It used to be that the fan's ties to a sports team were limited to what he could glean from attending a game, and the relationships that were formed with other individuals on the basis of attending those games. With each passing decade, relationships between fans and teams grew stronger. Newspapers satisfied a desire for knowing game outcomes and statistics. Radios created opportunities for fans to listen to their matches with groups of friends in the comfort of their own homes, while modern day talk radio sports shows give people the ability to call in and share their opinions on sports matters with the entire city. Television enabled fans to see moving images of their favorite athletes, experience teams playing in distant cities, and even watch entire channels catering specifically to sports and the people interested in them. The Internet has empowered individuals to carry out sports tasks formerly open only to journalists and camera crews. Regular fans can easily make their emotions and opinions heard to thousands of others on message boards and blogs. Fans who dreamed of being athletes and coaches when they were children now have the opportunity to have a stake in the teams they care about through the hugely popular fantasy sports industry. What technology will arise to take the relationship between fan and team to the next level? Which information desires of the fan have yet to be fulfilled? That is still to be seen, but the trend is that sports fans have the desire to be involved with and invested in their teams, so it is likely that whatever comes next will seek to strengthen this bond.

Notes

1. See http://bizoffootball.com/index.php?option=com_content&view=article&id= 334:espn-monday-night-football-audience-up-15-for-year&catid=40:television&Ite mid=57 (accessed April 23, 2009).

2. Irving J. Rein, Philip Kotler, and Ben Shields, *Elusive Fan: Reinventing Sports in a Crowded Marketplace* (New York: McGraw-Hill Professional, 2006), 56.

3. Ibid., 59.

4. Stephanie Busen, "The Addictive World of 'Football Manager,'" CNN.com, March 12, 2009, http://www.cnn.com/2009/SPORT/football/03/12/football.manager .addiction/index.html (accessed April 23, 2009).

5. William Joseph Baker, *Sports in the Western World* (Champaign: University of Illinois Press, 1998), 86.

6. Jennings Bryant and Andrea M. Holt, "A Historical Overview of Sports and Media in the United States," in *Handbook of Sports and Media*, ed. Jennings Bryant and Arthur A. Raney (New York: Routledge, 2006), 21–44, see 22.

7. Ibid.

8. Ibid.

9. Robert W. McChesney, "Media Made Sport: A History of Sports Coverage in the United States," in *Media, Sports, and Society: Foundations for the Communication of Sport*, ed. Lawrence A. Wenner (Thousand Oaks, CA: Sage Publications, 1989), 49–69, see 51.

10. Ibid.

11. Ibid., 52.

12. Ibid.

13. Ibid.

14. Bryant and Holt, "Historical Overview of Sports," 25.

15. Ibid.

16. Ibid., 27–28.

17. McChesney, "Media Made Sport," 52.

18. Ibid., 26.

19. McChesney, "Media Made Sport," 57.

20. Bryant and Holt, "Historical Overview of Sports," 29.

21. Mark Douglas Lowes, *Inside the Sports Pages: Work Routines, Professional Ideologies, and the Manufacture of Sports News* (Toronto: University of Toronto Press, 1999), 100–101.

22. Ibid., 100.

23. ESPN, December 10, 2004.

24. John W. Owens, "The Coverage of Sports on Radio," in *Handbook of Sports and Media*, ed. Jennings Bryant and Arthur A. Raney (New York: Routledge, 2006), 117–130, see 127.

25. McChesney, "Media Made Sport," 59.

26. Bryant and Holt, "Historical Overview of Sports," 32.

27. Ibid., 33.

28. McChesney, "Media Made Sport," 59.

29. Owens, "Coverage of Sports on Radio," 127.

30. Ibid., 128.

31. Bryant and Holt, "Historical Overview of Sports," 31.

32. Ronald A. Smith, *Play-by-Play: Radio, Television, and Big-Time College Sport* (Baltimore: The Johns Hopkins University Press, 2001), 30.

33. Bryant and Holt, "Historical Overview of Sports," 32.

34. Ibid., 31–32.

35. Owens, "Coverage of Sports on Radio," 130.

36. Ibid., 132.

37. Ibid., 131.

38. Ibid., 135.

39. Bryant and Holt, "Historical Overview of Sports," 33.

40. Ibid., 33; Smith, *Play-by-Play*, 7.

41. Bryant and Holt, "Historical Overview of Sports," 33.

42. Ibid., 33–34.

43. McChesney, "Media Made Sport," 63.

44. Bryant and Holt, "Historical Overview of Sports," 33.

45. Ibid., 34.

46. Ibid.

47. Chris Wood and Vince Benigni, "The Coverage of Sports on Cable TV," in *Handbook of Sports and Media*, ed. Jennings Bryant and Arthur A. Raney (New York: Routledge, 2006), 147–170, see 159.

48. Ibid., 160.

49. Ibid., 157.

50. Ibid., 166.

51. Lawrence A. Wenner and Walter Gantz, "The Audience Experience with Sports on Television," in *Media, Sports, and Society: Foundations for the Communication of Sport*, ed. Lawrence A. Wenner (Thousand Oaks, CA: Sage Publications, 1989), 249–269, see 242.

52. Ibid.

53. The Official Site of Major League Baseball, www.mlb.com.

54. Michael Real, "Sports Online: The Newest Player in Mediasport," in *Handbook of Sports and Media*, ed. Jennings Bryant and Arthur A. Raney (New York: Routledge, 2006), 171–184, see 173.

55. Ibid.

56. Ibid., 186.

57. Ibid., 191.

58. Ibid., 192.

59. Ibid., 190.

60. Ibid., 191.

61. James W. Cortada, *Making the Information Society: Experience, Consequences, and Possibilities* (Upper Saddle River, NJ: Prentice-Hall, 2002), 195.

62. Daniel Okrent, "He Does It by the Numbers," *Sports Illustrated*, May 25, 1981, http://vault.sportsillustrated.cnn.com/vault/article/magazine/MAG1124493/4/index.htm (accessed April 23, 2009).

63. Brian K. Smith and Priya Sharma, "Fantasy Sports Games as Cultures for Informal Learning," 2005, unpublished paper, The Pennsylvania State University, http://www.personal.psu.edu/faculty/b/k/bks12/ISE-final.pdf (accessed April 30, 2010), 2.

64. "Fantasy Sports Industry Grows to a $800 Million Industry with 29.9 Million Players," eMediaWire, July 10, 2008, http://www.emediawire.com/releases/2008/7/emw1084994.htm (accessed April 23, 2009).

65. Stephen A. Smith, "To Heck with Fantasy. I'm about What's Real," *ESPN the Magazine* (August 2008), http://sports.espn.go.com/espnmag/story?id=3556641 (accessed April 23, 2009).

66. Ben Klayman, "Technology Spurs Growth of Fantasy Sports in U.S.," Reuters, September 25, 2008, http://www.reuters.com/article/sportsNews/idUSTRE48O02L20080925 (accessed April 23, 2009).

67. Sam Walker, *Fantasyland: A Sportswriter's Obsessive Bid to Win the World's Most Ruthless Fantasy Baseball* (New York: Penguin, 2007), 50.

68. "Compete Inc.'s Latest Research Asks: Are You Ready for Some Fantasy Football?; Yahoo! Dominates, Women Are Joining and Free Sites Satisfy," BusinessWire.com, August 23, 2006, http://www.businesswire.com/portal/site/google/index.jsp?ndmViewId=news_view&newsId=20060823005102&newsLang=en (accessed April 23, 2009).

69. "Fantasy Sports Industry Grows" (note 64).

70. Cortada, *Making the Information Society*, 196.

71. Strat-O-Matic Baseball, http://tabletopbaseball.org/som.html (accessed April 2, 2009).

72. Strat-O-Matic—Wikipedia, http://en.wikipedia.org/wiki/Strat-o-Matic (accessed April 2, 2009).

73. www.whatifsports.com.

74. Ray Vichot, "History of Fantasy Sports and Its Adoption by Sports Journalists," January 2, 2009, http://jag.lcc.gatech.edu/blog/2009/01/history-of-fantasy-sports-and-its-adoption-by-sports-journalists.html (accessed March 15, 2009).

75. Walker, *Fantasyland*, 68.

76. Michael MacCambridge, "Following Baseball in the Abstract and Far Beyond," The Sporting News, July 7, 1997, http://findarticles.com/p/articles/mi_m1208/is_n27_v221/ai_19601089/ (accessed April 30, 2010).

77. Walker, *Fantasyland*, 71.

78. Vichot, "History of Fantasy Sports."

79. For a fascinating description of the intrigue and gamesmanship that can go on during these fantasy drafts, be sure to check out Sam Walker's *Fantasyland*.

80. Walker, *Fantasyland*, 133.

81. Lee K. Farquhar and Robert Meeds, "Types of Fantasy Sports Users and Their Motivations," *Journal of Computer-Mediated Communication* 12 (4) (2007): 1209.

82. "Compete Inc.'s Latest Research Asks" (see note 68).

83. The Official Site of Major League Baseball, www.mlb.com (accessed April 5, 2009).

84. Greg Ambrosius, "Realities of Fantasy: Appeals Heard in Case between MLB, Fantasy Provider," SI.com, June 15, 2007, http://sportsillustrated.cnn.com/2007/fantasy/06/15/fantasy.lawsuit/index.html (accessed April 23, 2009).

85. Bob Van Voris and Jeff St. Onge, "Fantasy Sports Win Right to Player Names, Statistics," Bloomberg.com, October 16, 2007, http://www.bloomberg.com/apps/news?pid=20601079&sid=aVMAY0beLSoA&refer=home (accessed April 23, 2009).

86. Michael Lewis, *Moneyball: The Art of Winning an Unfair Game* (New York: W. W. Norton, 2003), 82.

87. Baseball Prospectus statistics, http://www.baseballprospectus.com/statistics/sortable/ (accessed April 9, 2009).

88. http://www.mlbtraderumors.com/ (accessed April 9, 2009).

89. Smith and Sharma, "Fantasy Sports Games," 8.

90. Evan M. Hochschild, interview by author, Austin, Texas, November 2, 2008, http://www.crawfishboxes.com/ (accessed April 23, 2009).

91. Smith, Brian K., Priya Sharma, and Paula Hooper, "Decision Making in Online Fantasy Sports Communities," *Interactive Technology & Smart Education* 4 (2006): 347–360, see 353.

92. Ibid., 354.

93. Ibid., 355.

94. Chang Wan Woo, Seon-Kyoung An, and Seung Ho Cho, "Sports PR in Message Boards on Major League Baseball Websites," *Public Relations Review* 34, no. 2 (June 2008): 172.

95. Ibid., 173.

96. http://www.sbnation.com/pages/about (accessed April 28, 2009).

97. Stephen L. Higdon, interview by author, Austin, Texas, November 2, 2008, http://www.crawfishboxes.com/ (accessed April 23, 2009).

98. http://sports.espn.go.com/mlb/players/profile?playerId=3115 (accessed April 18, 2009).

99. Alex Rodriguez is expected to make $33 million for the 2009 season. He began the season on the disabled list, however, so it is difficult to calculate his true value as a hitter for 2009.

100. "AP Poll: Fans Says High Cost of Games MLB's No. 1 Problem," Associated Press, March 31, 2009, http://sportsillustrated.cnn.com/2009/baseball/mlb/03/31/ap.poll.baseball.ap/index.html (accessed April 23, 2009).

101. Ben Reiter, "With the Yankee's Current Makeup, Joba Probably Belongs in the 'Pen,'" Si.com, April 24, 2009, http://sportsillustrated.cnn.com/2009/writers/ben _reiter/04/24/joba.role/index.html (accessed April 30, 2010).

102. Jack Shafer, "Productivity Madness," Slate, March 19, 2009, http://www.slate .com/id/2187031/ (accessed April 23, 2009).

103. Al Neuharth, "Super Bowl Betting Spotlights Silly Laws," *USA Today*, February 2, 2007, http://blogs.usatoday.com/oped/2007/02/super_bowl_bett.html (accessed April 23, 2009).

104. American Gaming Association Industry Information, http://www .americangaming.org/Industry/factsheets/issues_detail.cfv?id=16 (accessed April 23, 2009).

105. Ibid.

7 Information in the Hobby of Gourmet Cooking: Four Contexts

Jenna Hartel

Introduction

This chapter aims to characterize the information activities and information resources that underlie the hobby of gourmet cooking in America. Gourmet cooking has roots in French haute cuisine and is a manner of food preparation that entails high quality or exotic ingredients and advanced technical skills (Wilson 2003). It is featured today at many high-end or "white tablecloth" restaurants, associated with cultural icon Julia Child, and has been adopted by millions of Americans as a hobby. Given its complexity to execute, gourmet cooking is information intensive and generates a vast multimedia information universe. Altogether, this hobby is a rich setting in which to study information phenomena in an everyday life and leisure context.

For a decade I was an avid gourmet hobby cook, renowned among family and friends for my decadent chocolate soufflé. When I entered a doctoral program in library and information science, I was unable to continue the hobby, but as an alternative I opted to do a dissertation (Hartel 2007) on its informational features, which had not been the subject of academic inquiry in my field. Taking an ethnographic approach, I conducted fieldwork between 2002 and 2006 by participating in gourmet-cooking social groups, classes, online forums, public lectures, and farmers' markets in greater Los Angeles, California, a culinary nexus. Further, I interviewed twenty gourmet cooks in their homes to learn how they practiced the hobby and used its information resources. The participants lived in Los Angeles and Boston, Massachusetts. My sampling strategy was purposive and opportunistic, and I found subjects through family and friends, or I approached potential informants at culinary-themed events such as cookbook signings. I sought articulate, enthusiastic cooks with substantial experience in the hobby (at least two years), and I did not attempt to examine sociodemographic variables in this exploratory study (Stebbins 2001a).

After the interview, each gourmet cook showed me their personal collections of culinary information in their home (focusing mainly on documentary items like cookbooks and recipes, and secondarily on multimedia and digital resources), which I photographed and diagrammed. I also studied popular and academic writings on gourmet cooking. Altogether, my data set amounted to several hundred pages of transcripts and fieldnotes, almost five hundred photographs, and volumes of literature. I analyzed these materials for patterns and themes using grounded theory (Strauss and Corbin 1998) and NVivo software. The results took the form of a fieldnote-centered ethnography (Emerson, Fretz, and Shaw 1995) that showcased the actual statements of hobbyists (made anonymous to protect their privacy).

The chapter begins with a brief history of gourmet cooking, a vivid example of an actual hands-on episode, and theoretical concepts to specify gourmet cooking in its hobby form. Then, the information experiences and their associated information resources in the hobby are organized and characterized as four contexts— Living a Gourmet Lifestyle, Expressing Culinary Expertise, Staying Informed and Inspired, and Launching a Cooking Episode—presented in chapter sections. Throughout, tables summarize data and list culinary information resources. At the conclusion, there is a survey of Information Age innovations in the hobby.

The Hobby of Gourmet Cooking

The idea of cooking as leisure has only emerged in the last half century in America. Until then, cooking was largely feeding work[1] (DeVault 1991), the required family care and housework performed mostly by female homemakers. Starting in the 1950s, a confluence of technological, economic, and social changes shifted cooking from work to leisure for some populations.

Namely, breakthroughs of the Industrial Revolution changed the nature of domestic food preparation. Home cooks no longer faced arduous tasks such as baking bread and canning perishables, which left time for the more enjoyable finishing touches of assembling and serving meals. Food shipping and storage technologies brought a wider variety of fruit and vegetables, at the peak of ripeness, to consumers year round. Through both immigration and international travel, Americans were exposed to foreign cuisines and ingredients, and eaters became more adventurous. As outlined in *The United States of Arugula: How We Became a Gourmet Nation* (Kamp 2006), charismatic educators such as Julia Child, Craig Claiborne, and James Beard, among others, championed sophisticated, artful ways to cook, eat, and socialize.

While the culinary landscape in America evolved, the amount of free time and disposable income steadily increased. Between 1965 and 2003 men gained six to eight hours per week of free time and women four to eight hours (Aguiar and Hurst 2006). Creative pursuits outside work, like hobbies, became components of identity and markers of a well-rounded life.

By the latter decades of the twentieth century Americans began to take up gourmet cooking as a hobby, and today there are several million participants. A proprietary report by market researchers Mintel Group (2005) calls the gourmet cooking hobbyist "the food enthusiast" or "foodie." Such sorts, "really enjoy cooking" and will "engage in more elaborate cooking from scratch on weekends or for special occasions." The sociodemographics of this community are apparent in the readership profile of *Gourmet* magazine:[2] middle aged (median age forty-nine), mostly female (74 percent), college educated (76 percent), professionally employed (33 percent), and wealthy (median household income around $80,000) (Conde Nast 2008).

Most gourmet hobby cooks have little or no experience in the restaurant industry, which is a form of paid work (Fine 1996). Rather, cooking as a hobby is a type of *serious leisure*, a free-time "activity that participants find so substantial and interesting that, in the typical case, they launch themselves on a leisure career centered on acquiring and expressing its special skills, knowledge, and experience" (Stebbins 1992, 3; 2001b). For most enthusiasts it is a cherished, lifelong pursuit that engenders a deep sense of fulfillment and many personal and social benefits. It is not akin to throwing a couple of hamburgers on the grill when friends drop by, which is *casual leisure*, "the immediately intrinsically rewarding, relatively short-lived pleasurable activity requiring little or no special training to enjoy it" (Stebbins 1997, 18). The hobby of gourmet cooking demands acumen and instead of an immediate and effortless reward, it takes perseverance.

For instance, one summer afternoon while on her weekly trip to the farmer's market in Providence, Rhode Island, gourmet cook Celeste purchased a variety of locally grown, perfectly ripe vegetables. Upon returning home she was pleased to find the latest issue of her favorite serial, *Cook's Illustrated*, had arrived. She flipped through the magazine, lingering on an article about vegetable torta—an elaborate loaf of vegetables baked together with seasonings, eggs, and breadcrumbs. The article outlined the technical challenge of this dish: liquid had to be extracted from each vegetable to concentrate flavors and prevent the torta from being watery and bland. A domestic version of an Italian cheese, asiago, added tang and texture. The recipe piqued her curiosity, was an ideal use of her stash of vegetables, and would be a perfect treat for her son who was arriving home from an overseas trip late the following night. She made up a grocery list and planned to spend the subsequent afternoon on this cooking project.

In her kitchen the next day, Celeste carefully read the article and recipe again, studying its fourteen steps and multiple illustrations. She sliced, salted, pressed, and baked the eggplant. She roasted peppers until they were blistered and brown, and then peeled off the skins. She sliced tomatoes, salted them, and patted away seeds and juices. Because each vegetable required individualized treatment, she recounted, "This went on for hours!" As the vegetables cooked in the oven, Celeste whirled bread in the food processor to form fresh breadcrumbs, which were sautéed in butter, mixed with grated asiago, and pressed into the spring form cake pan to create a thin crust. Next, she made a binder that would hold the torta together: a thin custard of egg, cream, cheese, garlic, and chopped fresh herbs gathered from an herb garden in her back yard.

When all the elements were prepared, they were layered into the pan, one vegetable type at a time, interleaved with custard and more cheese. The torta was baked to an internal temperature of 175 degrees and the result of her effort was a dense golden loaf that, when cut, revealed colorful internal striations. Late the next evening Celeste warmed the torta in the oven, popped it out of the pan, proclaimed it "a thing of great beauty and joy!" and served it up to her hungry, delighted son.

Celeste's project is a snapshot of the unique ethos or body of knowledge, values, and practices native to the hobby of gourmet cooking. As mentioned, it is characterized by food preparation using high quality or exotic ingredients and advanced technical skills. Further, the approach is marked by a refined aesthetic that showcases the intrinsic beauty of food and entails an elegant and orderly cooking process. Though usually executed independently by the gourmet cook, this hobby often takes the form of entertaining to please others. The experience is centered on the home, where the kitchen becomes a culinary workshop.

Gourmet cooking is just one of many hobbies involving food. The hobby of down-home cooking (a near opposite of the gourmet approach) aims to cook with inexpensive and common ingredients using simple or time-saving techniques (Wilson 2003). This cooking type strives for the no-fuss sensibility with which a beloved grandmother might cook. There are also hobby niches that are centered on narrower styles of cooking like baking, barbecue, Asian, or vegetarian. This chapter focuses solely on the hobby of gourmet cooking, which has distinct information phenomena.

Four Information Contexts

In summer 2006, I began descriptively coding interview transcripts and photographs, and naming (coding) all information activity, meaning, any actions related to the acquisition or expression of culinary skill or knowledge. Twenty-four unique codes were produced, such as "watching television," "reading cookbooks," and "eating out." In time, it was apparent that some codes could be grouped together based on similar roles in the hobby; for instance, several pertained to "staying informed and inspired in the hobby" whereas others were relevant to "expressing culinary expertise." By applying obvious associations, the large set of twenty-four was collapsed into six[3] clusters or contexts of information activity; four are featured in this chapter.

In table 7.1 the contexts are numbered, described, and exemplified. They are listed in an order that begins as broadly underlying everyday life (Living a Gourmet Lifestyle), and incrementally lead to the heart of the hobby: hands-on cooking (Launching a Cooking Episode). In the following sections each cluster will be introduced and discussed along with illustrations drawn from the data.

Living a Gourmet Lifestyle

The first cluster of information activities proves that the hobbyists' passion for food and cooking permeates their daily lives. For instance, meals at restaurants or routine food shopping evolve into sessions to study food. Vacations are taken to destinations with interesting cuisines. Friendships become a venue to talk about food or do culinary things together.

Such everyday happenings (eating out, shopping, vacations, and friendships) are typically *not* factored into information research because they are not centered on documents or information systems. One groundbreaking study of everyday-life information seeking (Hektor 2001), coined the term for these *life activities* (activities that physically manipulate the artifacts of everyday life, not information) and placed them outside the research agenda. Taking a different stance, here I propose that some life activities *are* information activities when they set up a platform of knowledge for a

Table 7.1

Four contexts for information in the hobby of gourmet cooking (Hartel 2007).

Context	Description	Examples
1. Living a Gourmet Lifestyle	Everyday-life experiences in which the gourmet cook orients to the culinary dimensions and learns; not centered on documents or information systems	• eating out • experiencing foreign cuisines • adopting regional food cultures • visiting ethnic neighborhoods/markets • having "foodie" friends
2. Expressing Culinary Expertise	The expression of culinary knowledge, mostly to family and friends (in person), but sometimes in an online community or in public	• consulting • teaching • commenting online
3. Staying Informed and Inspired	Any effort to learn about culinary topics and to stay stimulated; centered on published multimedia culinary information and not motivated by an episode at hand	• reading cookbooks • reading works on gastronomy • reading culinary serials • watching television • surfing the Internet • taking cooking classes
4. Launching a Cooking Episode	The actions taken to select and organize a cooking episode	• exploring for ideas • searching home-based collections • searching online recipe databases • Googling for recipes • tapping a "grapevine" • comparing/modifying/amalgamating recipes

hobby. Since these activities blend almost imperceptibly into everyday life and become routines, they are called Living a Gourmet Lifestyle.

Eating Out

Americans love to eat out and gourmet cooks are no exception. However, unlike the average hungry and time-pressed consumer, hobbyists treat visits to upscale restaurants as educational experiences. While eating out, they study menus, food combinations, presentation styles, and flavors to expand their knowledge. Sara reports, "My boyfriend thinks it is irritating because I don't choose what is the most tasty, but what I can learn the most from. It's school, not pleasure!" Similarly, gourmet cook Nancy explains that when she eats out she selects items that are "a little bit unusual" and then proceeds to eat them "very, very slowly" in an effort to identify the ingredients and seasonings.

From the perspective of restaurant wait staff, gourmet cooks may come off as challenging customers who ask a lot of questions. Eric calls restaurants "a living workshop" and says, "I often ask questions . . . the more information you have about it [the meal], the more satisfying it is." Many gourmet cooks relish the chance to go into the kitchen and interrogate the chef directly. Eric continues, "I really want to understand the thinking of the chef in preparing this. I want to understand why they decided to use these specific ingredients to create this kind of flavor." Sara has learned strategies to access the kitchen and points out, "I find that if you are not dining on Friday or Saturday night, chefs are more than willing to talk to you."

Inspired by restaurant foods, gourmet cooks try to duplicate them at home. Hobbyist Katey explains that at a Mexican restaurant she and her boyfriend ate "this incredible chile" stuffed with finely ground beef, ginger, and raisins. It was topped with pumpkin seeds and cilantro and finished with a sauce containing a whiff of vanilla. In her opinion, the dish was "freaking amazing" and she said to her boyfriend, "We are going to figure this out!" Over the next three months Katey gathered, studied, and implemented stuffed chile recipes, while her boyfriend served as taster and provided feedback on their resemblance to the restaurant version. Some cookbooks, culinary Web sites, serials, or weekly newspaper food pages satisfy the gourmet cook's interest in restaurant offerings by publishing restaurant recipes.

Experiencing Foreign Cuisines

Gourmet cooks take overseas vacations and their culinary horizons are expanded. Camilla recounts that in her early days as a gourmet cook, "I started . . . traveling a lot and really relating to the country I was visiting as much through my palate and gullet as through my mind and my eyes." Sara says, "If I travel, the first thing I ask is 'What is done with the food here? What do people eat for lunch?'" In keeping with their penchant for authenticity, gourmet cooks avoid tourist fare and favor highly regarded restaurants or out-of-the-way spots of the locals. Not all countries equally attract these discerning cooks, who generally take the strongest interest in France, the birthplace of the gourmet approach, as well as the Mediterranean region and Asia.

Back at home, these culinary adventures provide ideas and inspiration for months to come. Camilla continues, "It is really nice to experience the cooking of a country you are going to and then come back and incorporate that into your repertoire." For instance, gourmet cook Tom and his wife Alice returned from a trip to France and hosted a dinner party for neighbors with the theme "A Taste of Paris." The event featured a French menu of onion soup, pissaladierre (an appetizer tart with anchovies), escargot (snails), beef bourguignon (a beef stew in red wine), along with French wine and champagne. As the pièce de resistance, a poster of the Eiffel Tower was mounted in their dining room for the night.

In addition to new ideas, gourmet cooks obtain culinary souvenirs on their travels, to stock their personal libraries and kitchens. The most popular keepsake is a cookbook featuring the native cuisine. Memories from the trip are invoked when its recipes are cooked and served. Souvenirs also take the form of ingredients or cooking gadgets. Gourmet cook Nancy takes a keen interest in foreign foodstuffs and explains, "One of the things I do [on holiday] is visit grocery stores, for I love to see what other products are available." She has purchased "pesto bouillon cubes" from Italy that are not available in American supermarkets, spice grinders from Ireland, and an olive-pitting gadget from Italy.

Adopting Regional Food Cultures

As an affluent and professional population, gourmet cooks move around a lot. They embrace and study the regional food cultures where they live temporarily. Rose explains that her husband's U.S. Army career has involved frequent relocations around the world and the United States, which inform her cooking. While in Germany, she was fascinated by the breads and pretzels and began to study baking. When she returned to the United States and lived in Arizona, she pursued Southwestern cookery. During a stint in Tennessee she "got into Southern food such as fried turkey and catfish." Now a resident of Colorado, Rose has learned to adapt to the impact of high-altitude baking. She reports, "Each time we move there is a change in my cooking but overall I have gotten better and more confident." In the same way, gourmet cook Celeste lived in Denmark for eleven years and adopted the cooking and formal entertaining habits of northern Europe. Later, she and her husband moved to the country of Lesotho where she experimented with African cooking and traded recipes with other foreign aid workers from France, Russia, and Germany. She says, "We spent some time on the coast in Kenya and there is an Indian influence; we had access to Indian products and restaurants—that was really fun."

Visiting Ethnic Neighborhoods and Markets

In many cases, gourmet cooks need not travel far from home to step into a stimulating foreign culinary environment. Almost all American cities have ethnic communities, such as Boston's Italian North End or China-town, where markets feature imported products and an array of sights and smells. Katey and her boyfriend regularly host food-themed parties, such as a night of Jamaican or Russian delights. She reports that these events "always involve a couple of shopping expeditions, going to communities where there are authentic shops and immigrants. We'll go before to see what people are buying, and we'll get materials."

And while running workaday errands, cooks keep their eyes open for specialty food shops, to escape into a moment of hobby pleasure. Roland says, "I always look around for bakeries, farms and such." While killing time before a chiropractic appointment, he "discovered this nice vegetable garden and bakery shop, a surprisingly fantastic place." He was impressed by a bunch of fresh, multicolored Swiss chard, which became a featured item for dinner that evening. Even serendipitous visits to such places become a springboard for the hobby.

Having "Foodie" Friends

Over the course of the hobby experience, most gourmet cooks attract friends who share a passion for cooking. Such friends may go to restaurants together, cook for each other, and share recipes. Gourmet cook Celeste explains, "I have one foodie friend in particular and we often will share ideas and try new restaurants. We'll cook for each other." Claire's best friend since childhood also loves to cook and they talk by telephone often about food, and exchange favorite recipes via email.

A benefit of foodie friends is the mutual willingness to honestly evaluate the outcomes of cooking projects. Rose says, "If I made a cheesecake, I would give some to her [a friend]. Then I find out whether or not it [the cooking effort] is worth it. She reciprocates. If she makes a new type of cookie, she will bring one over and ask: 'What do you think?'" Similarly,

Sara reports, "Most of my friends are food-centered enough that they are willing to have a long discussion about: Should we change the flavor? Would it be better with a little this or that?"

Review: Living a Gourmet Lifestyle

The information activities placed under the banner of Living a Gourmet Lifestyle are distinct for being everyday activities that are not usually seen as information-related. Yet, by eating out, traveling to explore new cuisines, embracing local and regional food traditions, visiting ethnic neighborhoods and food markets, and maintaining friendships with other cooks, the hobbyist creates a life rich with culinary knowledge and sets the stage for the hobby.

Expressing Culinary Expertise

A second set of information activities entails what this chapter categorizes as Expressing Culinary Expertise to others. These behaviors are most common when the hobbyist has developed substantial ability. Here, the cook verbalizes his/her knowledge and enthusiasm for all things culinary and acts as an expert. This is a very social activity, since it is based on dialogue and requires an audience. Family and friends participate as listeners, respondents, clients, and students. While expression is the core of this set of activities, in the interplay of words and experiences, cooks acquire skill and knowledge as well.

Consulting

When gourmet cooks provide counsel to others on food-related issues, they are consulting. Usually this process is quick and problem-focused. Katey says, "I get a lot of phone calls—'Kaaateeey, HELP!' I like doing that." She continues, "I've got one friend who is very interested in cooking, she will send me emails or call me with questions, such as, 'Can I just melt the butter to stir it into the cookies?' And I say, 'No, you can't do that! *Please* don't do that!' This has been going on for twenty years. She keeps asking questions and I tell her the answers."

Other consulting projects are larger in scale and draw upon the cook's ability to conceptualize menus and events. Nancy recounts, "People ask me if I have ideas for something. My friend had a fortieth anniversary party and needed a good menu for her Italian family. I put together the menu and gave her the cookbooks." Like any business consultant or even a reference librarian, she has learned techniques for interrogating her "clients":

"I find myself asking questions about the types of guests, the nature of the event, whether they want light or heavy food, that kind of thing. I help to guide them into the foods that they like, presented for this setting. From there I might recommend recipes."

Teaching Cooking

Hobbyists sometimes teach others to cook. In its simplest form, teaching can be a demonstration, without instructional rhetoric. Tom explains, "At a party recently, I made deep fried strawberry and ricotta raviolis. Who would think that would be a good dessert?! It was delicious. I made them at home, brought the fryer to work, and made it in front of everyone."

Hobbyists with young children serve as teachers when they engage them in kitchen tasks; it is a way of blending the hobby with the demands of childcare. Rose's oldest daughter helps her prepare cake batters, and her younger ones shape bread dough. Anne bought a children's cookbook to use with her son and together they followed the recipe for roast chicken.

Going a step further as teachers, some cooks orchestrate cooking lessons for adult family and friends. Celeste designed three cooking classes for her college-aged son and niece. She featured delicious, simple, everyday preparations like lasagna, salad, and a fruit dessert. During the session, Celeste explained culinary principles and techniques, while her "students" received hands-on practice. Similarly, one Thanksgiving Nancy taught her sons and their girlfriends how to make filet mignon and Caesar salad; she explains, "It was such fun to do that together."

Commenting Online

Gourmet cooks can also express knowledge through online channels. For instance, Katey produces a food blog, The Seasonal Cook, where she writes about locally produced foods. Katey's blog enhances her interest in her hobby and, she says, "keeps me from drifting off. I've been doing that since June and found that it keeps me learning things, for I feel that there is always something I have to write and share." Indeed, today the Web is a site for thousands of culinary blogs, and many are authored by avid hobbyists.

A popular online forum is the Web site epicurious.com. Those who have executed an epicurious.com recipe may submit comments, which appear online. This addresses the critical questions tied to any recipe: Is it really good? Does it work? At epicurious.com, after trying a recipe a visitor rates its success using a one-to-four fork rating. The more popular recipes at this Web site have several hundred comments. These dialogues have proven so popular and informative that epicurious.com features the Buzz Box, which is a direct link to a list of "recipes (that) have received the most comments and ratings from our users in the past thirty days." Most of the major online recipe collections feature similar social and interactive capabilities.

Review: Expressing Culinary Expertise

This cluster of information activity entails the expression of culinary knowledge, to family and friends in intimate settings, or to an online public. Cooks willingly serve as on-call troubleshooters or personal culinary advisors in the design of food-based events. As they mature in their hobby, many are motivated to teach. This can be as simple as demonstration cooking or involving a child in a cooking project. Sometimes, more structured lessons are offered to loved ones with a curriculum and hands-on instruction. Often hobbyists use online forums to register their opinions on a recipe. For the gourmet cook, these information activities are fun and natural expressions of their culinary passion.

Staying Informed and Inspired

Gourmet cooks stay informed and inspired through regular contact with the culinary literature and media. These behaviors increase basic knowledge, keep the cook abreast of culinary trends, and supply a constant stream of cooking ideas. Such activities are *not* done for the immediate purpose of steering hands-on cooking and rarely occur in the kitchen. Instead, regular immersion in culinary information is a routine and is the cherished part of a hobbyist's everyday life. For instance, cooks read cookbooks in bed before falling asleep, enjoy a newspaper's weekly food section while relaxing at a coffee shop, or tune in regularly to a favorite television cooking show. Such behaviors differ from Living a Gourmet Lifestyle by being explicitly information based, and per Hektor qualify as unfolding, "continually directed attention towards an information system and the symbolic display it offers, for instance by looking and listening, and thereby taking part in a content" (2001, 170).

The extent to which cooks do this set of activities varies. Some make an effort to stay informed on a daily basis; others are less riveted to culinary media and may gain knowledge and inspiration simply by Living a Gourmet Lifestyle. This constellation of information activities is more prevalent, and pursued with greater zeal, by newcomers who face the steep learning curve of the hobby. Intensity can lessen as skill and knowledge accumulate. One mature cook reveals, "I don't really watch cooking shows much anymore. Every once in a while I get a new book and follow it through. I suppose I have my ideas at this point!"

Often, Staying Informed and Inspired focuses on a specific culinary subject. A topic such as homemade pasta or braising can capture the cook's imagination and temporarily focus their reading and learning. Sometimes hobbyists will follow a self-developed curriculum to thoroughly study a culinary subject area. To restate, these information activities are *not* immediately directed at hands-on cooking, which is a different context and set of information activities discussed next as Launching a Cooking Episode.

Because these behaviors demand focused attention on an information resource they are usually done individually. In fact, cooks may covet moments away from family and friends to savor culinary learning. At the same time, Staying Informed and Inspired has social implications because in these private moments cooks transfer information across the hobby community. Such exchanges are mediated via published or public artifacts like books, magazines, Web sites, and the like. Non-hobbyists (engaged in feeding work) and dabblers (those who infrequently partake in a hobby) may have an occasional interest in these materials as well.

The universe of information artifacts and resources within the hobby of gourmet cooking is vast; it is one of the largest popular information domains. Resources come in print, digital, and multimedia forms through various public channels. In 2007, 13.9 million books were sold in the food and entertaining category, according to Nielsen BookScan, which tracks about 70 percent of sales (Rich 2008). On a monthly basis, culinary magazines with nationwide circulation are mailed to several million households and also sold at newsstands and grocery checkouts. Most newspapers feature a weekly food column or section focused on the local culinary scene. Online, there are numerous gourmet cooking Web sites that serve as reference sources and portals to databases with upward of twenty-five thousand recipes. The Food Network cable channel supplies twenty-four-hour-a-day culinary programming to ninety-six million American households (Television Food Network 2009), while many other television channels feature cooking-themed content. Further, there are newsletters, newsgroups and radio broadcasts on a wide range of culinary topics.

Reading Cookbooks
Above all, cooks get informed and inspired by reading cookbooks. A hobbyist states, "Certain things like Fanny Farmer or Escoffier are just good reads, you don't necessarily do any cooking from them." Some cooks have a dedicated time for reading and a favorite location in the home. Nancy says, "Saturday morning is my reading time. I sit right where you are sitting!" (a comfortable chair in her dining room). Gourmet cook Celeste

settles into a loveseat in her den during the evening and reads while the
television plays in the background.

Cookbooks are valued for offering the most extensive treatment on
culinary topics. Sara explains, "I find that cookbooks go into greater detail
than any other source. I have never been able to get someone to have an
hour-long conversation with me about the 'emulsifying process' . . . but
there are books that do this. Books give me the broadest and deepest information." For gourmet cook Ken, the contemporary cookbook offers much
more than recipes: "I use them for background knowledge about cuisines,
types of ingredients, menu planning, so it is not just to go and look at a
recipe. I use them much more diversely."

The cookbooks used by hobbyists today have evolved over the centuries.[4] A millennium ago, the very first cookbooks were simple lists of
foods and menus, often of the favorites served at court. Settlers to America
carried European cookbooks across the Atlantic, and for decades these
texts were used with modifications to accommodate the pioneer culture.
The first cookbook *by* an American *for* the American kitchen was Amelia
Simmons's 1796 *American Cookery*, which featured local ingredients such
as cornmeal and recipes for native foods like Indian pudding and slap-
jacks. Throughout the 1800s there was a cookbook boom, spearheaded
by reform-minded women whose texts championed American themes of
economy and frugality, management and organization, diet and health,
and temperance. At the same time, specialty cookbooks emerged, documenting the foodways of American regions like New England or the
South. In the twentieth century, continued immigration and globaliza-
tion contributed to cookbooks featuring cuisines from around the world.

The genre is still evolving and nowadays many cookbooks include personal, social, cultural, and historical discussions of food. Nancy says, "A
lot of cookbooks are more like social studies . . . I like this trend; there is
more background and depth. So I like reading cookbooks." Contextual
information enriches a cook's understanding and motivates future cooking.
During her interview, Nancy picks up a French cookbook, opens it to a
chapter on olives and exclaims, "There are all these awesome recipes for
olives, and it is not just about making it, but background on where it comes
from. . . . Look at all this stuff about olives and olive oils! I love to read
about this, like about home-cured olives. Then I will look for a recipe
because I get excited about it."

The cookbook is far from obsolete in this digital age; it was the favored
resource of cooks in this study because of exceptional content, ease of use,
familiarity, and portability. A cook explains, "I like looking at books, Web-

crawling is not quite the same." Sara elaborates, "[Cookbooks] are a very convenient source. If I'm online, at the library, or at work, I can use them. I can use them while I'm watching a movie or whatever. I can carry them around with me so they are where I am. Also, I am a reader; it is what I'm most familiar with."

Reading Gastronomy

In addition to cookbooks, hobbyists read materials on gastronomy, a literary genre developed in France in the 1800s that addresses proper eating and cooking habits, culinary history and myth, and the "nostalgic evocation of memorable meals" (Mennell 1986, 270–271). Recently in the United States, gastronomy has been popularized through Random House's The Modern Library Food Series, edited by Ruth Reichl, cookbook author and former editor of *Gourmet* magazine. This collection features classic gastronomic works in translation, life stories of renowned cooks, and nonacademic culinary social history. Another popular type of gastronomy falls under the Library of Congress subject heading Cookery—Fiction. These novels present a cooking-themed story interspersed with recipes.[5] For instance, Laura Esquivel's 1995 bestseller, *Like Water for Chocolate: A Novel in Monthly Installments with Recipes, Romances, and Home Remedies,* features a love story that unfolds around the kitchen hearth in one Mexican family.

Some cooks in this study report being touched by gastronomy. Celeste says, "I read Laurie Colwin and got inspired. I wished I had her life! Laurie . . . was a food writer and wrote essays that were originally published in *The New Yorker.* She had an interest in culinary history and old cookbooks. She would write an essay about, say, gingerbread and then have a recipe. . . . I've always liked her and would love to do that kind of journalism but never did." Celeste went on to create her own culinary diaries in the same style as Colwin.

One work of gastronomy repeatedly mentioned by hobbyists was *Larousse Gastronomique.* This huge—1,350 pages, eight pounds!—culinary encyclopedia was written by French chef Prosper Montagné and first published in 1938 (and translated into English in 1961). It documents the history of the culinary arts in France through alphabetically listed articles on foods, techniques, and equipment, as well as significant chefs, restaurants, and cuisines; recipes are interspersed throughout. *Larousse Gastronomique* is known for its unsettling accounts of culinary practices now out of fashion, such as the preparation of animelles (testicles) and pig's head. Still, it is considered *the* authoritative source on gourmet cooking; hobbyist Dorene remarks with a wry smile, "Whenever we want to know something *for sure*, my husband is the first to run in and get *Larousse Gastronomique*, to check that I'm not giving him the business."

Reading Culinary Serials

All the cooks in this study have subscribed to culinary magazines. Some actively had three or more arriving per month. Cooks eagerly await new issues, and one states that she "reads them from cover to cover every month." Serials contain articles about foods with recipes and menus; reviews of equipment and restaurants; and reader exchanges or recipes and tips, among other things. They arrive just in time to supply cooking ideas as foods come into season or as holiday meals loom. Roland explains, "I love the magazine *Bon Appétit*. I don't subscribe right now, but I'll pick it up, especially at the holidays." Monthly magazines also keep the gourmet cook abreast of fashions in the gourmet cooking social world, which is constantly championing new foods, techniques, chefs, and restaurants. Mandy attests "magazines are more innovative." As they arrive month after month, serials accumulate in the homes of gourmet cooks, who are hesitant to throw them away. Usually the most recent issues are kept at hand on a coffee table or bedside and later moved to a permanent storage shelf in an office, den, or garage, to function as reference materials.

Culinary serials have roots in the earliest American women's magazines. Founded in 1866, *Harper's Bazaar* had a primary emphasis on fashion, but devoted pages to recipes and household management. *Good Housekeeping* was launched in 1885 with a sharper focus on domestic life. In 1900, the magazine established a research unit that would become The Good Housekeeping Institute, in a move to test the performance of household devices and the purity of foodstuffs. Over the course of the century, the magazine and its institute were a force to police the home products industries and champion improvements to the American diet. *Good Housekeeping* had mainstream appeal (closer to the ideals of down-home cooking), and in 1941 *Gourmet* magazine was launched with a distinctly more continental, artful, and indulgent orientation to food and cooking. Today there are dozens of food-themed magazines with nationwide circulation, and hundreds of topical or regional publications of smaller scale that, for the most part, perpetuate the elements of the original *Gourmet*. Those favored by participants in this study are profiled in table 7.2.

One culinary magazine mentioned with enthusiasm by interviewees was *Cook's Illustrated*, a relatively slim, black-and white monthly publication with no advertising or glossy photographs. Feature articles in this magazine focus on a food concept (such as apple pie) and discuss the outcomes of

Table 7.2

Survey of popular gourmet cooking magazines, drawn from an article by Wilson 2003

Magazine (launch)	Paid circulation	Frequency	Points of distinction
Gourmet (1941–2009)	983,836	Monthly; no longer published	Placed a dual focus on food and travel, with attention to ethnic, regional, and seasonal foods. Most recipes required access to specialty ingredients.
Bon Appétit (1956)	1,371,495	Monthly	Known as "America's Food and Entertaining Magazine," most articles feature menus and entertaining ideas; geared to a range of culinary skill levels.
Food & Wine (1978)	958,348	Monthly	Puts equal emphasis on food and wine; recipes utilize predominantly upscale ingredients and are designed for more advanced cooks.
Cook's Illustrated (1992)	600,000	Bimonthly	Contains no advertisements or photography, and features a strongly educational and scientific tone; emphasizes cooking technique through master recipes.
Fine Cooking (1994)	240,000	Bimonthly	Focuses on holiday and seasonal preparations and is geared to all skill levels. An emphasis on hands-on cooking is conveyed through concise directions, photos, and tips for the amateur home cook.
Saveur (1994)	381,585	Nine times per year	Celebrates food and travel, and introduces world cuisines to the amateur cook. Known for striking photography and artful design.

different versions of recipes, explicating the impact of variables such as ingredients, techniques, temperature, and cooking times. To illustrate, an article entitled "Chocolate Mousse Perfected" from the May 2006 issue described how different proportions and handling of four basic ingredients (chocolate, eggs, sugar, and fat) influenced the results. These articles lead to a recommendation for an ideal or master recipe. *Cook's Illustrated* has popularized a scientific approach to cooking and sparked a trend among publications to focus on culinary technique and principles, instead of lifestyle, and this serial is now widely imitated.

Watching Television

Television has become a popular medium for food and cooking information, and emerged from a radio model.[6] The first multimedia culinary sensation in America, Betty Crocker, was invented by flour purveyor Washburn Crosby Company in 1921. At first, Betty (actually, a team of home economists) responded by letter to homemakers with baking quandaries. In 1924 she became the star of a radio program about cooking. American gastronome James Beard is credited as the first TV cooking show host. No footage remains of his fifteen-minute program, *I Love to Eat*, which aired from 1946 to 1947 on WNBT-TV in New York, but in one surviving audio recording Beard discusses a luncheon for skiers. Another pioneer was Dione Lucas, who was the first female graduate of Le Cordon Bleu cooking academy and a renowned culinary educator; she presented *The Queen's Taste* and long-running *The Dione Lucas Cooking Show,* both broadcast in the United States by CBS. In February of 1963 public television station WGBH in Boston debuted *The French Chef*, featuring the tall (six feet, two inches), no-nonsense, and very good-humored Julia Child. Her program was a breakthrough that demystified French cuisine for regular home cooks and attracted a huge new audience to gourmet cooking.

On November 22, 1993, the landscape of culinary television changed forever, when Food Network was launched by E. W. Scripps Company with twenty-four-hour-a-day programming about cooking (Meister 2001; Ketchum 2005). At first, shows on Food Network followed the conventions set by Lucas and Child. Later in the 1990s, the concept was broadened to "all things food" such as leisure and travel, as well as reality-show formats. Food Network has contributed to the rise of chefs in popular culture. Multiple Food Network hosts with larger-than-life personalities have become celebrities with their own brand empire, such as Emeril Lagasse and Bobby Flay. Today Food Network is one of the most successful lifestyle cable channels, though recently its programs have been losing audience share to imitators as food-themed broadcasting goes mainstream.

Most of the gourmet hobby cooks in the study were enthusiastic about culinary television, in particular Food Network and its stars. Gourmet cook Roland claimed to be in love with Rachel Ray, the latest Food Network sensation. For Mandy, a turning point in her hobby career was when her parents got cable and she had access to Food Network. She says "I became a Food Network addict. I watched so much of the Food Network and I learned a lot. Not just about technical things but about combining ingredients. It was inspiring more than anything." Ken values programming on Food Network because "watching a good chef cook, watching how they use heat in a pan, *that* is what I'm looking for when I'm watching television—it is the techniques, not a recipe." At the same time, some found the Food Network shows too far removed from reality; one cook says, "I can't relate to it in my everyday cooking. It is interesting and entertaining, like Emeril, but I would not go cook their ideas." Another concurs, "I used to like Emeril but I don't like him so much anymore, he is too glitzy for me now."

Surfing the Internet

The Internet provides access to a plethora of food- and cooking-themed Web sites and online communities, which many of the cooks in this study use for inspiration and to stay informed. The heaviest user, Dorene, was the oldest at age seventy-three. She keeps an elaborate hierarchy of bookmarks in her AOL Web browser that link to her "favorite places" online. She has bookmarked collections of general culinary Web sites, newspapers with food sections, companies offering gourmet products, Web sites affiliated with culinary serials, and topical areas such as breads, chefs, pressure cookers, and restaurants. She surfs the Internet for one or two hours every day, following culinary news and gathering recipes that spark her interest. Katey participates now and then in Chowhound, an online discussion forum designed "For Those Who Love to Eat"; she reads their restaurant reviews and posts comments. Gourmet cooks in this study were more likely to use the Internet when on a mission to find a recipe, a different type of information activity discussed shortly as Launching a Cooking Episode.

Taking Cooking Classes

Cooking classes are offered in most American cities by community education programs, culinary arts institutes, and gourmet retailers. During cooking classes, a topic or recipe is introduced and then students execute it for hands-on practice; the resulting foodstuff is sampled and discussed. Often, students don't get a complete cooking experience because the educators eliminate time-consuming prep work. Roland reports, "The topic was weekend breakfasts . . . they gave us ten recipes to try. Each person would get a recipe. We'd go back into the kitchen area and each had a workstation. Everything was all prepared and chopped up."

Some gourmet cooks in this study attended a few cooking classes, though not to the point of it being a routine. This may be because cooking classes cater to novices and this can bore the advanced gourmet cook. For instance, Nancy recounts, "The class was nice, good chef, but it was too basic, very broad for the newbie. . . . Our recipe was so easy I finished it, and then went to help others and was giving them advice and tips."

Review: Staying Informed and Inspired

To summarize, gourmet cooks stay informed and inspired through steady consumption of culinary-themed literature and media. They perform a constellation of information-intensive activities that are not tied to any particular hands-on cooking project. Rather, the experience of learning via the culinary literature and media functions as a routine and cherished part of everyday life. The extent of such activity varies across cooks and is most prevalent during the early stages of the hobby career. Above all, gourmet cooks read cookbooks and the occasional work of gastronomy and culinary fiction. Culinary serials are valued for their timely coverage of seasonal foods, holidays, and trends, while weekly newspaper columns cover local culinary scenes. Food Network provides twenty-four-hour-a-day food-themed instruction and entertainment; and the Web is yet another culinary resource. The most information-hungry gourmet cooks orchestrate all these mediums to stay well informed and inspired, providing a springboard for hands-on cooking, discussed next.

Launching a Cooking Episode

Thus far, I have presented how hobbyists develop, share, and sustain culinary knowledge. Now, I will focus on the role of information as gourmet cooks begin the hands-on cooking process.[7] Many life factors can influence the decision to enter the kitchen in hobby mode. A project may be tied to a holiday or social gathering, or inspired by a season and its produce. It could fill a window of available time such as a weekend, vacation, or snowstorm; or it could be sparked by serendipitously encountering a recipe.

Hobbyists with an itch to cook often seek *inspiration* first, not information. This may take the form of daydreaming about cooking possibilities, or browsing cookbooks, Web sites, magazines, or personal recipe files. At the start, the cook may harbor a vague sense of a food category, technique, or cuisine of interest, but there is no commitment or focus. A hobbyist recounts this moment as, "So, I flipped through my various magazines and cookbooks . . . *what did I want to make?*" In this step, cookbooks and recipes are engaged superficially (sometimes just skimming the text or looking at pictures). The cook is in a mode of fantasy and an upbeat sense of opportunity. Nondocumentary forms of exploring are to visit markets or restaurants; ideas may come from a display of produce or a beguiling flavor encountered during a meal out. At some point the hobbyist's interest is piqued, and crystallized, resulting in a more precise vision for a cooking project such as duck a l'orange or pad thai.

Once hobbyists have committed to a food concept, they begin planning. Since gourmet cooking involves advanced culinary techniques, a necessary step is to obtain instructions—in other words, a recipe. Even for seemingly straightforward preparations, most hobbyists will consult at least one recipe. One cook explains, "I'll usually look up three or so recipes and see the common ways, then decide what to do." This sets up the moment of traditional interest in information science: the search.

A "Berrypicking" Recipe Search

Overall, the search for recipes resembles Bates's model of information "berrypicking" (1989). In this model, searching is an iterative process in which a variety of search techniques into different information repositories are used sequentially. In the same way, cooks attempt a variety of search strategies leading into different information collections. If one path fails to deliver, or if more information is desired, another is attempted. As in berrypicking, potential recipes are gathered "a bit-at-a-time" from each repository along the way. Bates asserts that during berrypicking the search query changes with each iteration. Likewise, in gourmet cooking the search experience may cause the cook to modify their original vision. For instance, a hunt for lasagna might end up with a recipe for the Italian casserole cannelloni, if along the search route a compelling reason arises to change.

Searching Home-Based Collections

A first stop for recipes typically is within arm's reach. Over the course of their lives, gourmet cooks build a culinary collection within their households. It contains cookbooks, serials, recipes, and other sorts of culinary ephemera gathered from many sources. In this study, these personal libraries varied in size from small (twenty texts) to large (one thousand-plus texts). Such a collection has advantages as the natural first stop. It is nearby and usually based in a room adjacent to the kitchen. It is relatively small and does not induce information overload. Materials have been hand picked and organized into a schema based on the cook's own logic. The collection is familiar, since the cook has read many of the items while staying informed and inspired. Often these resources are marked with annotations, Post-it notes, bookmarks, or folded pages. Unlike a public library or bookstore, the collection is steeped in sentiment, sometimes containing cherished documentary artifacts passed through generations, such as a grandmother's recipe box.

As they go berrypicking for recipes in their personal collections, cooks apply a variety of search heuristics and techniques. They employ standard bibliographic metrics such as genre, subject, and author to isolate the most promising items to search. When they look into individual artifacts, they use tables of contents, indexes, and culinary-themed ordering principles to hone in on recipes. (An example of a culinary-themed ordering principle is the sequence of a meal—from appetizer to dessert—a common linear format for many cookbooks; another ordering principle is the seasons). A cook explains, "What happens is I'll look in the back of books, if there is a particular ingredient. I'll go through all the indexes, or look in my own recipe collection." With lasagna as one possible goal, various berrypicking strategies are:

• search within classic reference cookbooks (often this establishes a standard or benchmark), such as *Joy of Cooking*;
• search cookbooks of the appropriate subject area, such as all Italian cookbooks;
• search cookbooks of an author who is a subject authority, such as those by Italian cook Marcella Hazan;
• search personal recipe collections (on cards or pages) in fitting subject file, such as Italian, pasta or casseroles;
• search the months of serials that are appropriate (not effective for the all-season dish of lasagna, but for some seasonal foods);
• search in a personal file containing recipes that have been earmarked as promising, to see if there is a lasagna recipe.

Searching Online Recipe Databases

For some hobbyists, the Internet is the preferred resource for recipes and the starting point of a search. Millions of Web sites supply recipes posted by other hobby cooks, restaurants, product manufacturers, or ingredient marketers, among others; and most food-themed newsletters and magazines put their recipes online. Some repositories require membership fees, though most are free. Hobbyists in this study favor Web sites with recipes that embody the refined style of gourmet cuisine; a sample appears in table 7.3.

Table 7.3
A small sample of sources for gourmet-style recipes

---

chow.com
cookinglight.com
cooksillustrated.com
epicurious.com
foodandwine.com
foodtv.com
ichef.com
theworldwidegourmet.com
epicurean.com
marthstewartliving.com
saveur.com

By far the most popular online destination for recipes among cooks in this study is epicurious.com. It is the digital archive and index to the twenty-five thousand-plus recipes that have appeared in *Gourmet* and *Bon Appétit*. Increasingly, the site serves as a clearinghouse for recipes from other magazines and cookbooks, as well as for more than seventy-five thousand recipes submitted by users. A hobbyist attests, "You can get anything online by going to epicurious.com. I do that often."

There are good reasons why epicurious.com is a favorite. Most of the recipes are rigorously pretested and employ consistent terms for ingredients, measurements, and techniques. There are striking color photographs for many items, a boon in a hobby that celebrates elegant food presentation. Frequently, recipes are presented in the context of a complete menu, geared to the gourmet cook's penchant for entertaining. Further, as already mentioned, the community of users ranks and comments upon recipes, which provides an additional level of information.

The search capabilities at epicurious.com are also state of the art, and they offer multiple access points to recipes. Searching can be done using keywords or an advanced template that makes it easy to tick off key facets such as dietary consideration, cuisine, meal/course, type of dish, season/occasion, preparation method, or main ingredient. Users can also browse through all the recipes for popular foods, or view a recipe slideshow of the best recipes in a food category.

Online resources such as epicurious.com have a number of benefits over home-based paper recipe or cookbook collections. They are free, quick, easy, and effective to search. A cook explains, "I go to the Internet. The problem with having so many cookbooks is it's hard to remember where things are and look them up . . . [the other day] I needed a pumpkin soup recipe and just went online. I probably do that more often than not." Online recipes can be custom formatted (as cards or full pages) and then printed to use in the kitchen; these documents are treated as disposable and can be marked up, splattered on, and thrown away since it is easy to reprint next time. Conveniently, the Web sites can be accessed while away from home or at work. Sara says, "I use the computer more [for recipes] because it is in front of me at work all day. I'll do it on my lunch break or while waiting for a fax."

Googling for Recipes

Some cooks in this study use the Google search engine to locate recipes. Google functions more as an Internet portal, leading to other online recipe collections or to recipes directly. Sara says, "If I am cooking something unusual, I'll just enter it into Alta Vista or Google, and I can get so many hits, hundreds and hundreds of pages." Since these searches can return a mixed bag of hits, Tom prefers a Google image search. He explains, "I wanted to make cantaloupe soup. I went to Google, and if you do an 'image' search on Google, you get pictures and this ferrets out a lot of [irrelevant, low-quality] things. So the recipe is not mixed with other stuff. And clicking on the image takes you right to the site."

Tapping a "Grapevine"

In a minority of cases, cooks turn to friends and family to acquire recipes. Gourmet cook Claire calls this her "grapevine," which serves as a first stop during recipe searching. "I go to them for recipes. I love to eat their food. I know that if they recommend it, it will be yummy." When Roland sought to make a traditional French-Canadian tortilla (pork pie), he reached out to friends: "We collected recipes from French friends. I asked people to pass me their tortilla recipes. All the recipes were different, some with pork, hamburger, potato, and different spices. We gathered about a hundred different recipes, counting those from the Internet, too."

After the Search

Gourmet cooks gather several recipes through their searching and often compare them. Dorene states, "I like to check recipe against recipe and I'll know just what it is that I want." Recipes are not considered indelible and some cooks will change them in various ways, but this tinkering is likely performed more often by advanced hobbyists. Sometimes, two or more recipes can be amalgamated to bring together their best features. When making a Thanksgiving dessert Rose recounts, "One recipe called for pumpkin cheesecake, plain. The other was chocolate pumpkin cheesecake with a pecan topping. I didn't want the chocolate. I thought, 'Why don't I follow the pumpkin cheesecake recipe and take the idea of the topping from the other recipe?' And I did that." Once committed to a recipe (or recipes, if a more elaborate project is planned), hobbyists go into the kitchen and continue to *use* culinary information in various ways—a distinct, complex, and fascinating practice that is outside the scope of this chapter.

Review: Launching a Cooking Episode

To recap, one of the most information intensive moments in the hobby of gourmet cooking occurs when launching a cooking episode. Fueled by a desire to cook and to have a life rich with cooking experiences, hobbyists explore the culinary literature and environments for hands-on opportunities. When a food concept captures their interest, they conduct a recipe search. They use a berrypicking strategy that entails multiple probes into different information repositories. Home-based collections are a typical first stop and the Internet is another favorite recipe trove; in a berrypicking mode, many cooks search *both*. Recipes that meet the cook's needs are gathered, compared, and sometimes amended before being implemented in the kitchen.

Information Age Innovations

Over the past decade, Information Age innovations have diversified the resources and tools available to cooks, and in turn they have generated new information-gathering practices. Already, this chapter has mentioned how gourmet cooks tap the Internet for recipes, and how the home computer is employed to follow culinary trends and communicate with other cooks. Practices such as these have been well assimilated into the hobby of gourmet cooking.

This exploratory ethnography did not generate any definitive statistics on the use of such contemporary approaches versus the "traditional" forms of paper recipes and cookbooks. In the sample of twenty informants, a few favored online resources almost exclusively, while an equal number displayed an attachment to paper sources. *Most* cooks in this study were strategic users of *both* mediums. Hobbyists tend to reach for the resource that suits the moment. To illustrate, if they are on a break at work, they may hop onto epicurious.com; on a lazy afternoon at home, they will curl up with a favorite cookbook. One thing is certain: the universe of culinary information is constantly changing. In this final section a few of the latest innovations are profiled.

A new culinary genre has emerged: the "cook-through" blog (Gomes 2008). In this writing style, a blogger picks a seminal or noteworthy cookbook and implements every recipe sequentially and reports the adventure online. Readers follow the trials and tribulations in detail through both narrative and images. The first of this kind was The Julie/Julia Project, when in 2002 hobbyist Julie Powell cooked all the recipes from *Mastering the Art of French Cooking* by Julia Child and collaborators, and reported the experience online. Powell's blog gained a following, was adapted to a bestselling book, and later became a movie starring Meryl Streep as Julia Child. Today, there are hundreds of cook-through blogs and the Web site Cooking the Books (http://cookthroughblogroll.blogspot.com/) serves as a clearinghouse.

In some circles, Twitter, the social networking tool that broadcasts short messages about daily happenings, is being used to disseminate recipes (Downes 2009). The catch is that "tweets" (Twitter postings) must not contain more than 140 characters—far fewer than the typical recipe. The hobby cook who launched this trend, Maureen Evans, relishes the challenge of reducing the complex ingredients, actions, quantities, times, and temperatures of a recipe into the essentials, such as: ⟨*Strudel Pastry: cut2Tbutter/1cflour/mash tater. Knead w 2t yeast/2T h2O; rise 1h. On flour cloth gently pull 17x25"; trim-1"/butter well*⟩. In their reduced form, recipes become a puzzle to decode, adding a playful challenge to the cooking process. Those of Evans's six thousand fans who actually implement the "tiny recipes" attest that they really do work.

The cell phone has become, for some, the ultimate cooking technology (Moskin 2009). Cooks use it to keep grocery lists, search for recipes, photograph handiwork, convert ingredients, keep time, and conduct research. Fans of the approach say it is handy, familiar, and a quick kitchen tool, and ever more software products and Web sites are advancing the trend. *Grocery iQ* and *Jot* are applications for making shopping lists and *Eggy* is a cute cell phone widget for producing perfect hard-boiled eggs. BigOven. com, a Web site with 167,000 recipes, has a free iPhone application that has been downloaded more than a million times. The most sophisticated phone technology comes from food manufacturer Kraft (not a favored supplier to the gourmet set); their *iFood Assistant* features a GPS that directs users to stores carrying desired ingredients.

New Web-based products are breaking down the traditional barrier between professional culinary knowledge and the home cook (Chernova 2006). *Chefs Line* is a service that links trained chefs with paying customers to troubleshoot kitchen problems in real time, take cooking lessons, and design menus. On a smaller scale, the Web site Chef.com answers culinary email questions for free. Some restaurants invite Web-based communication with their chef as a means to build loyalty and encourage the love of food and cooking. Interestingly, the consulting role performed by professional chefs in these services is performed by hobby cooks to less skilled family and friends.

Finally, fantastical projects have been underway in the basement of MIT's media lab, by the Counter Intelligence Research Group (http://www .media.mit.edu/ci/research-index.html). This technologically oriented culinary think tank aims to produce the next-generation "smart" kitchen. The group's prototype CounterActive (Ju et al. 2001) is an interactive information system that projects recipes onto a countertop. The glowing recipes contain hyperlinks that lead to background details in an effort to "enhance the experience of cooking." The inventors claim, "This is the first computerized recipe system we have seen that not only expands on the conventional cookbook by incorporating pictures, audio, and video, but also deals better with being covered with spilled milk" (Bell and Kaye 2002, 55). Another innovative culinary technology (still in development) is the Living Cookbook (Terrenghi, Hilliges, and Butz 2007). In this system, cooks narrate and explain their cooking experience in the form of a "kitchen story" that is videotaped live and made available to others. The goal of Living Cookbook is to capture the social nature of cooking and to transfer culinary knowledge from one generation to the next. These futuristic technologies from creative academics and inventors may have uncertain outcomes, but they surely stir the imagination.

Conclusion

The hobby of gourmet cooking entails the preparation of food using high-quality or exotic ingredients and advanced culinary techniques. As a form of serious leisure, it is pursued by millions of Americans in free time and for pleasure. Gourmet cooks favor an orderly cooking process and strive for delicious, artful results that are often shared with family and friends. A vast multimedia information universe underlies this hobby. In an effort to make sense of the information phenomena therein, this chapter draws upon ethnographic fieldwork with twenty gourmet hobby cooks. Four contexts of information activities and their associated information resources have been outlined.

To review, through the context Living a Gourmet Lifestyle, routine events of the hobbyist's life are oriented to culinary matters. Trips to restaurants, vacations in exotic places, food shopping, and relationships with like-minded friends are used as opportunities to learn about food and cooking. In the Expressing Culinary Expertise context, hobbyists share culinary acumen with family, friends, and the public. Cooks enjoy rescuing others from kitchen debacles and play a consulting role in the design of meals, menus, and parties. Hobbyists sometimes teach family and friends to cook through simple demonstrations or informal classes. Online communities and blogs are yet another means to opine about culinary matters. The Staying Informed and Inspired context involves frequent engagement with the literature and media of cooking. These activities are not preparation for a cooking event, but rather are self-directed learning for its own sake. The array of strategies is broad; gourmet cooks read cookbooks, culinary magazines, and works on gastronomy; they watch food-themed television and surf the Internet. Finally, a most information-intensive moment in the hobby occurs in the Launching a Culinary Episode context—the tasks that precede hands-on cooking. At this time, cooks conduct a berrypicking search in personal collections and online resources to gather recipes, formulate menus, and obtain specifications on techniques or ingredients. Though isolated here for analytical purposes, in reality, these four contexts overlap with one another, and blend into the hobby and everyday life.

A different perspective is to view information phenomena in this hobby longitudinally across the lives of individual cooks. All hobbyists experience a continuous leisure career (Stebbins 2001b, 9–10) with turning points and increasing ability. At the beginning is a steep learning curve, lasting a few years, in which cooks strive to master the fundamentals of ingredients and techniques. This often entails an experimental sensibility in the kitchen, repetitive hands-on practice, and voracious reading (elsewhere called Staying Informed and Inspired). Once hobbyists have mastered the basics, they enter a long-running period of deeper study of different cuisines, techniques, or ingredients, one at a time. For instance, cooks in this project described focusing on "Japanese cookery," "baking," or "chocolate" for a time, before moving onto another subject. These topical pursuits keep the hobby fresh and engage a constant stream of new information resources. During the later stages of the hobby career, when cooks are confident and accomplished, information consumption often declines, and attention shifts to sharing expertise with others.

Given these patterns, the hobby of gourmet cooking is an interesting case study into everyday-life information phenomena, and the nature of culinary knowledge, specifically. This project creates a vantage point to reflect on the informational issues tied to food, cooking, and eating outside the leisure context, a matter with huge cultural, social, and economic implications. The past half-century has seen a steady decline in home cooking, and increasing consumption of less nutritious processed, take-away, or restaurant foods. As a result, upcoming generations have not witnessed cooking firsthand, and may experience "culinary illiteracy" and its associated health problems. It is timely to ask: What is culinary knowledge? Is it being lost? How can it be captured and disseminated? While this chapter did not focus upon these sweeping changes and important questions, it establishes useful concepts and a point of comparison to one population that savors culinary information and knowledge to its fullest.

Acknowledgments

This chapter is drawn from dissertation research and was made possible by my committee: Greg Leazer (chair), Mary Niles Maack, Leah Lievrouw, Sanna Talja, and the late Melvin Pollner. I am also indebted to Robert A. Stebbins for support and inspiration over the years. Thanks also to the editors of this collection for feedback and guidance, and to Ashleigh Burnet for help in preparing the manuscript.

Notes

1. Marjorie DeVault provides an introduction to the work women perform to feed their families in *Feeding the Family: The Social Organization of Caring as Gendered Work*. Several other texts offer sociologically oriented insights into cooking in its non-hobby form: namely, Short 2006, Symons 2000, Haber 2002, and Levenstein 2003.

2. Conde Nast stopped publication of Gourmet magazine in 2009, but continued the brand on the Web site http://www.gourmet.com/.

3. In addition to the four contexts discussed in this paper, my dissertation (Hartel 2007) described "Using Information During a Cooking Episode" (Hartel 2006) and "Managing a Personal Culinary Library" (publication forthcoming).

4. The history of the cookbook in America is a fascinating story. Starting points for this subject are Neuhaus 2002 and the online collection of Feeding America: The Historic American Cookbook Project, http://digital.lib.msu.edu/projects/cookbooks/index.html, accessed May 1, 2009, a digital archive of seventy-six of the most important and influential American cookbooks from the late eighteenth century to early twentieth century.

5. The use of recipes in novels is the topic of an essay by Sceats 2003 and is included in a collection of academic writings on recipes available in *The Recipe Reader: Narratives, Contexts, Traditions* (2003).

6. This section draws from the detailed history of culinary television shows available in Collins 2009.

7. In Hartel 2006, I model the cooking process as a nine-step episode and describe the use of information throughout; this chapter focuses on the first two steps of the episode.

References

Aguiar, Mark, and Erik Hurst. 2006. Measuring Trends in Leisure: The Allocation of Time over Five Decades. Working paper. The Federal Reserve Bank of Boston. http://www.bos.frb.org/economic/wp/wp2006/wp0602.htm.

Bates, Marcia J. 1989. The Design of Browsing and Berrypicking Techniques for the Online Search Interface. *Online Review* 13 (5): 407–424.

Bell, Genevieve, and Joseph Kaye. 2002. Designing Technology for Domestic Spaces: A Kitchen Manifesto. *Gastronomica—The Journal of Food and Culture* 2 (2): 46–62.

Chernova, Yuliya. 2006. Soggy Stuffing, Dry Turkey? Now You Can IM a Chef. *The Wall Street Journal*, November 22: D1.

Collins, Kathleen. 2009. *Watching What We Eat: The Evolution of Television Cooking Shows*. New York: Continuum.

Conde Nast Publications, Inc. 2008. Gourmet Demographic Profile: Adult Readership, MRI (Spring) 2008 Survey (a report cited by *Gourmet*'s owner, Conde Nast, that included information about *Gourmet* magazine).

DeVault, Marjorie L. 1991. *Feeding the Family: The Social Organization of Caring as Gendered Work*. Chicago: University of Chicago Press.

Downes, Lawrence. 2009. Take 1 Recipe, Mince, Reduce, Serve. *The New York Times*, April 22. http://www.nytimes.com/2009/04/22/dining/22twit.html.

Emerson, Robert M., Rachel I. Fretz, and Linda L. Shaw. 1995. *Writing Ethnographic Fieldnotes*. Chicago: University of Chicago.

Fine, Gary Alan. 1996. *Kitchens: The Culture of Restaurant Work*. Berkeley: University of California.

Gomes, Lee. 2008. Latest Web Bloggers Give Cooking the Books a Whole New Meaning. *The Wall Street Journal*, May 28:B6.

Haber, Barbara. 2002. *From Hardtack to Home Fries: An Uncommon History of American Cooks and Meals*. New York: Free Press.

Hartel, Jenna. 2006. Information Activities and Resources in an Episode of Gourmet Cooking. *Information Research* 12 (1): paper 281. http://InformationR.net/ir/12-1/paper282.html.

Hartel, Jenna. 2007. Information Activities, Resources, and Spaces in the Hobby of Gourmet Cooking. PhD diss., University of California.

Hektor, Anders. 2001. *What's the Use: Internet and Information Behavior in Everyday Life*. Linköping: Linkoping University.

Ju, Wendy, Rebecca Hurwitz, Tilke Judd, and Bonny Lee. 2001. CounterActive: An Interactive Cookbook for the Kitchen Counter. Abstract. In *Proceedings of CHI '01: Human Factors in Computing Systems Conference*, Seattle, WA, 269–270.

Kamp, David. 2006. *The United States of Arugula: How We Became a Gourmet Nation*. New York: Broadway Books.

Ketchum, Cheri. 2005. The Essence of Cooking Shows: How the Food Network Constructs Consumer Fantasies. *Journal of Communication Inquiry* 29 (3): 217–234.

Levenstein, Harvey A. 2003. *Paradox of Plenty: A Social History of Eating in Modern America*. Berkeley: University of California.

Meister, Mark. 2001. Cultural Feeding, Good Life Science and the TV Food Network. *Mass Communication & Society* 4: 165–182.

Mennell, Stephen. 1986. *All Manners of Food: Eating and Taste in England and France from the Middle Ages to the Present*. Oxford: Basil Blackwell, Inc.

Mintel Group. 2005. Report Cover: Cooking—US. http://reports.mintel.com/sinatra/reports/display/id=150074#about (accessed July 2005).

Moskin, Julia. 2009. Top Kitchen Toy? The Cell Phone. *The New York Times*, January 20. http://www.nytimes.com/2009/01/21/dining/21tele.html.

Neuhaus, Jessamyn. 2002. *Manly Meals and Mom's Home Cooking: Cookbooks and Gender in Modern America*. Baltimore, MD: Johns Hopkins.

Rich, Motoko. 2008. A Plan to Sell Cookbooks: Give Away Recipes Online. *The New York Times*, November 1. http://www.nytimes.com/2008/11/01/books/01cook.html.

Sceats, Sarah. 2003. Regulation and Creativity: The Use of Recipes in Contemporary Fiction. *The Recipe Reader: Narratives, Contexts, Traditions*, ed. J. Floyd and Laurel Forster, 169–186. Surrey, UK: Ashgate.

Short, Frances. 2006. *Kitchen Secrets: The Meaning of Cooking in Everyday Life*. Oxford, UK: Berg.

Stebbins, Robert A. 1992. *Amateurs, Professionals, and Serious Leisure*. Montreal and Kingston, Canada: McGill-Queen's University Press.

Stebbins, Robert A. 1997. Casual Leisure: A Conceptual Statement. *Leisure Studies* 1 (16): 17–25.

Stebbins, Robert A. 2001a. *Exploratory Research in the Social Sciences*. Thousand Oaks, CA: SAGE Publications.

Stebbins, Robert A. 2001b. *New Directions in the Theory and Research of Serious Leisure.* New York: Edwin Mellen Press.

Strauss, Anselm M., and Juliet M. Corbin. 1998. *Basics of Qualitative Research: Techniques and Procedures for Developing Grounded Theory.* 2nd ed. Thousand Oaks, CA: SAGE Publications.

Symons, Michael. 2000. *A History of Cooks and Cooking.* Champaign: University of Illinois.

Television Food Network. 2009. About Foodnetwork.com. http://www.foodnetwork .com/home/about-foodnetworkcom/index.html (accessed June 1).

Terrenghi, Lucia, Otmar Hilliges, and Andreas Butz. 2007. Kitchen Stories: Sharing Recipes with the Living Cookbook. *Personal and Ubiquitous Computing* 11 (5): 409–414.

Wilson, Terrie L. 2003. Tasty Selections: An Evaluation of Gourmet Food Magazines. *Journal of Agricultural & Food Information* 5 (2): 49–66.

8 The Transformation of Public Information in the United States

Gary Chapman and Angela Newell

Introduction

The United States has always been a country in which public access to government information has been both a widely shared value as well as a source of political conflict. What government information the public should be able to access, and what should not be shared, falls into the category of what political scientists call "essentially contested concepts," such as the character of democracy or the appropriate use of force. Struggles over sensitive information held by the government have long been part of the tumultuous political history of the United States, while at the same time public access to government information has dramatically expanded over the course of the nation's history. The United States is a world leader in freedom of information and has been the source of major revolutions in this concept, particularly in the era of the Internet, which was invented in the United States and originally launched with federal government funding. But this leading position in freedom of access to government information has also had the paradoxical effect of highlighting, by contrast, what the government withholds from public access, so that these cases tend to become more contentious and significant than they might be in a country with less open access.

This conflict, or constant friction, between open access to government information and cases of sensitive information withheld from public access, is entering a new and uncertain phase because of the effects of globally networked information on the Internet. President Barack Obama signaled the significance of this issue with his first official acts as president, when he signed two presidential memoranda on freedom of information and open government, on Inauguration Day, January 21, 2009. Since then, the Obama administration has launched several initiatives on the World Wide Web that represent new ways for the public to see and query what

the U.S. federal government is doing. The Web sites recovery.gov and data.
gov, run by the White House Office of Management and Budget (OMB),
are two examples of innovation in government information accessibility—
the former tracks the stimulus money authorized by Congress to help the
nation recover from the economic meltdown of 2008; the latter is meant
to be a repository of government databases that can be used by independent programmers and Web developers as the basis of new, online views
of government activity.

However, the Obama administration, despite its intentions of reversing
some of the Bush administration's policies that kept sensitive government
information from public view, has also found itself embroiled in new
controversies about sensitive information, predictably about national security issues, such as whether to release information about prisoner abuse,
warrantless wiretapping, or even who has visited the White House
(Nicholas 2009). The Obama White House has continued the Bush administration's policies of exempting the White House Office of Administration
and the National Security Council from Freedom of Information Act
requests (Rood 2009).

Thus as government transparency and access to information have
expanded, exceptions have become more pronounced and in some cases
more controversial and acrimonious. The growth of the Internet has contributed to this and introduced new complications. Information revealed
by the government, sensitive or not, can be distributed instantaneously,
at no cost, and in ways that government officials cannot control. There
are now more "eyes" watching government, via the Internet, which makes
it harder for the government to manage its information resources or to
cover up mistakes or misdeeds. And perhaps the most profound new phenomenon is that the Internet is blurring the lines between the government
and the citizenry, and between government levels and agencies, which
makes the concept of "official" government information difficult to define.
For approximately two hundred years, the concept of public access to
information meant that the public could request information generated
by government officials or activity, and this information was either passed
to the public or kept from the public. This model was not unlike a visitor
requesting something of a clerk at a window, or someone on the other side
of a door, whereupon a decision would be made about whether to comply
with this request or not. If the decision was to comply, the item requested
was passed to the person who requested it.

The Internet has changed this concept dramatically. Now information
exchange is much more likely to be between multiple parties, there is no

longer a door or a window, and increasingly there need be no "request"—
the information is already available, in digital formats that can be searched,
shared, copied, distributed, and processed for essentially zero cost. There
are immense benefits to both the public and governments because of this
technological revolution. But there are also complexities and vexations for
both government officials and citizens. Government agencies are no longer
simply producers and custodians of information as they were during an era
of government documents and databases. Today government officials are
becoming participants in information-sharing activities that in themselves
create new sources of information that are not clearly controlled by anyone.
Information can pass nearly instantaneously through government computers, Google server farms, personal laptops, and mobile phones and then
back again, picking up new attributes along the way. FBI agents can search
Google while working on cases, thanks to the Patriot Act. U.S. military
commanders overseas are confronted with the immense challenge of managing troops who have digital cameras and access to Facebook, YouTube,
and other nonmilitary Web sites. Sensors that monitor air pollution from
oil refineries in Texas have provided real-time data to commodities traders
all over the world, who use the sensor data to nudge the minute-by-minute
price of oil, a price that is itself monitored by government officials. Increasingly, government agencies are exploring "user-generated content" online,
which means that government Web sites will be collecting and displaying
information produced by nonofficial users.

All of these trends are new and not yet well understood. Government
authorities and public-sector information technology managers are just
beginning to explore how the public uses government information when
it is provided in these new ways, under new rules. The Internet is clearly
having a disruptive impact on many industries whose basic performance
and processes were taken for granted for a long time—newspapers are
perhaps the most dramatic example of a threatened business model. No
one, at present, knows whether newspapers will survive or, if so, how.
Similar challenges are looming for government. In particular, online services such as Facebook, Twitter, YouTube, and text messaging—all popular
with young users—strongly suggest that the era of documents is nearing
an end. The bureaucratic model of vertically integrated organizations with
highly defined boundaries of authority is also fading. Most modern governments were built using these two conceptual models, documents and
bureaucracy, and still continue to use these models for their basic architecture. How well governments will adapt to the historic transformation
all organizations are experiencing now is presently uncertain.

The Informed Citizen

The very idea that citizens should know what their government is doing is relatively recent. For most of human history, governments either conducted their business without regard to informing citizens or else "managed" the news that citizens were allowed to learn. Apart from the two significant exceptions of Athenian democracy and the Roman Republic, most arrangements of public affairs were developed to reinforce the status hierarchy of a society. Leaders and elites had the authority to run public affairs as they saw fit, and everyone else was supposed to know their place. Illiteracy among the lower classes was the rule, and it was not uncommon for the ruling classes to believe that lower-class illiteracy should be enforced for the public good.

This long-standing arrangement began to break down with the Protestant Reformation in the sixteenth century. Protestant groups were encouraged to read the Bible so that they could experience God's word personally. Gutenberg had published his Bible in the mid-fifteenth century, the first book in the West produced with moveable type; by the time the Protestant Reformation ended a hundred years later, there were commercially printed Bibles for people to read. The Reformation not only had a large impact on literacy, which became more and more common among Protestant congregants, but it also encouraged people to regard long-frozen social arrangements as mutable and subject to rational criticism.

The Protestant colonialists who first settled the Eastern coast of North America carried with them a high regard for literacy, for primarily religious purposes. But the colonialists of New England did not challenge the older forms of social hierarchy that gave the privileges of public affairs to a small elite and made everyone else their obedient subjects. There was no routine public access to information about what colonial governments were doing, except for the occasional decree from authorities about what citizens were required to do or not do. Although a printing press arrived in Boston in 1638, its output was dedicated exclusively to printing material aimed at the elite, including government documents, religious works, and jobs for Harvard College (Brown 2000, 40).

It was the flammable combination of the Enlightenment and the revolutionary zeal of the late eighteenth century that introduced the idea of the informed citizen. The philosophy of the Enlightenment that swept over Europe and the American colonies in the latter half of the eighteenth century championed human reason's ability to both comprehend and possibly remake the world. The growing potential for revolutionary upheaval made the mass opinion of the citizenry a potent political weapon. The increasing proliferation of printing presses, and of printers looking for jobs, set up the conditions for an explosion of political pamphlets, small newspapers, and general readers for populations increasingly hungry for learning, news, and ideas. During the American Revolution, the popularity of Thomas Paine's *Common Sense* was a good measure of this phenomenon. Historian Richard D. Brown notes, "Before 1776 was over, Thomas Paine's pamphlet was published in nineteen American editions and achieved a circulation exceeding 100,000 copies in a country that possessed no more than 500,000 households" (Ibid., 45–46). Paine was the right man at the right time, using a direct and colloquial style in his argument that appealed to the middle-class colonialists, and taking advantage of a new network of commercial printers that had only just been established in America.

The Enlightenment had extended to Scandinavia in the eighteenth century, and it was Sweden that passed the first freedom of information act in 1766 (Burkert 2007, 126), part of a freedom of the press law. The motivation for these measures was partisan; Sweden's government was driven by factions, as would become common for most other modern governments since then. When one faction took over from the other, it passed a freedom of information law meant to prevent rule through secrecy.

In the new United States, the prospect of factions was famously hailed as a public good by James Madison, who regarded the preservation of legitimate factions as the most reliable protection against tyranny. As the Swedes had already figured out, factional balance requires transparency when one faction is in charge. Thomas Jefferson took this a step further, by arguing repeatedly that the general public was the most effective protection against tyranny, and thus people should be informed and self-educated to protect democracy. There are numerous Jefferson quotes that express this theme; for example, he wrote, "An informed citizenry is the only true repository of the public will. . . . The People cannot be safe without information. When the press is free, and every man is able to read, all is safe." Madison's thoughts are also well represented in the famous quote from his 1822 letter to W. T. Barry: "A popular Government, without popular information, or the means of acquiring it, is but a prologue to a Farce or a Tragedy; or perhaps both" (Madison 1910, 103).

At this time the advocates of popular rule by an informed citizenry—particularly Madison, Jefferson, and Benjamin Rush—believed that important government information would be provided to the public by a free press, and that citizens would become informed because of their desire for information from an unfettered press. Thus in the earliest days of our modern concepts of freedom of information and the informed citizen, freedom of the press is a right that implies access to public information.

Unfortunately, it would be a long time before people were granted direct, legal access to government information itself. The government was still free to impose a wide variety of forms of censorship, secrecy, prior restraint, and various other methods of preventing the press from revealing information that government officials preferred to keep from the public. The Swedish government, which passed a freedom of information law two hundred years before the U.S. Congress passed one, routinely invoked secrecy protections to prevent the release of government information to the public (Burkert 2007, 134). The U.S. government has repeatedly invoked similar reasons for secrecy, primarily relying on the concept of "state secrets" for national security. As Herbert Burkert notes, "So far, at least anecdotal evidence—from Watergate to the aftermath of 9/11—suggests that societies undergo pendulum swings in their appreciation of secrecy and transparency" (Ibid., 134–135).

The Freedom of Information Act

Americans' access to public information evolved very slowly, although the United States was always one of the world's leading nations in innovations in information access. The U.S. Postal Service was inaugurated by Benjamin Franklin in 1775, and importantly the U.S. Congress subsidized postal delivery of newspapers and magazines (John 2000, 60–61). The Library of Congress was established in 1800, although it did not acquire the status it has today until the end of the nineteenth century. Throughout most of the nineteenth century, Americans' access to government information was haphazard and irregular, focused mostly on commercial information such as patents and contracts, agricultural information, or war news for common citizens. The U.S. Postal Service continued to play an important role in the dissemination of information, and it was supplemented by a national railroad network, a national telegraph system, and, in the early years of the twentieth century, a national telephone network.

But for a long time each individual agency of the federal government was responsible for its own record keeping and maintenance, which varied widely among agencies and which led to a chaotic form of nonpolicy for government information. The National Archives and Records Administration, or NARA, was not created until 1934 (National Archives and Records Administration n.d.). It was not until 1897 that the Library of Congress created its Geography & Map Division, even though maps of the continent were among the first documents collected by the Library of Congress (1995). The Declaration of Independence and the U.S. Constitution did not become part of the Library until 1923 (Coyle n.d.). The *Federal Register*, the official compendium of regulations and rules filed by the U.S. federal government, did not appear until March 16, 1936.

This long evolution of routinized and formalized access to government information culminated, one might say, when President Lyndon B. Johnson signed the Freedom of Information Act on July 4, 1966, the 190th birthday of the United States. But Johnson signed this bill reluctantly, without ceremony, and with a presidential "signing statement" that stressed the need for government secrecy. FOIA, as it has come to be known, was eleven years in gestation in the House of Representatives, six years in the Senate, and President Johnson did everything he could to stall the legislation. Bill Moyers, Johnson's press secretary, recalled that Johnson "hated the very idea of the Freedom of Information Act; hated the thought of journalists rummaging in government closets and opening government files; hated them challenging the official view of reality" (National Security Archive 2006). Johnson had to sign the bill that had been the pet project of Rep. John Moss, a Democrat from California, because of the pressure exerted by newspaper publishers all over the country.

The Freedom of Information Act established the rules of accessing federal government information, and it has been modified and strengthened since its inception in 1966. Any person can file a FOIA request for government information, and the government must either comply or show reason why it should not comply. FOIA requesters can go to court to attempt to force the government to release information. This changed the status quo ante, in which government officials could decide what information to release to the public and what to keep out of public circulation. After Watergate in the 1970s, FOIA was fortified considerably, and in 1996 it was extended to electronic records. Importantly, FOIA became the standard for most state and local governments in the United States and similar laws were adopted by many other countries. Complying with open records requests has, over the years, become a public sector profession.

How people use FOIA and state and local open records is nearly impossible to characterize, for two reasons. First, governments are prevented by law from asking the purpose of the open records request. Second, anecdotal evidence about what people do with public information is so diverse and idiosyncratic that it is impossible to describe except in the broadest terms. A 2006 report on FOIA use by the nonprofit group the

Coalition of Journalists for Open Government noted that individual requests included "a movie producer doing research for *The Road to Guantanamo*, a divorcee searching for hidden assets and UFO enthusiasts seeking evidence of other worldly visitations." The report went on: "There were also requests from a local police department mining for information on federal grants, a whistleblower trying to shore up a claim of government wrongdoing, historians digging into original source material, a cryptologist trying to recover a Navy intelligence report he had worked on years earlier, and a lawyer in the Texas Attorney General's office trying to locate parents overdue on child support payments" (Coalition of Journalists for Open Government 2007).

The federal government does collect information on the types of requestors of public information. The Coalition's 2006 report noted that, in a survey of 6,439 requests made in September 2005, about two-thirds came from businesses—which are billed differently for FOIA compliance costs—and a third came from "other," which are likely individuals. However, the report authors noted that the U.S. Army routinely classifies all its FOIA requests as "other," which results in an underreporting of business requests. The Coalition report noted that in its survey, only 6 percent of FOIA requests in September 2005 came from journalists, about half the figure that came from law firms. Nonprofit organizations were the source of only 3 percent of the requests. A similar survey of FOIA users done by the Heritage Foundation in 2003 found that 40 percent of requests came from corporations and a quarter came from law firms (Coalition of Journalists for Open Government 2007).

In fiscal year 2007, in data reported in 2008, the U.S. Department of Justice revealed that the entire U.S. government received 21,758,651 FOIA requests, an increase of 2 percent over the previous year (Department of Justice 2008). This figure illustrates one of the chief criticisms of FOIA, that it has overwhelmed federal agencies and created a massive backlog of requests, long delays in compliance, and large expenses for the government. In fiscal year 2006, the total cost to the government for complying with FOIA requests was $398,500,000, a significant amount of money even for the U.S. federal government. The government spent $20 million litigating FOIA cases in fiscal year 2006 (Department of Justice 2008).

These kinds of problems with FOIA appeared shortly after the Act was implemented in 1967, and have only grown over time. In 2006 President George W. Bush issued an executive order (13,392) imposing reforms meant to clear FOIA backlogs, and in 2008 Congress passed legislation reforming and streamlining the FOIA process in order to shorten compliance times. Nevertheless, the nonprofit Sunshine in Government Initiative reported in 2009 that the CIA, the Department of Defense, and the National Archives each hold FOIA requests dating from 1992—sixteen years old—and the Department of Homeland Security has FOIA requests three years older than the agency itself (Weitzel 2009, 2). An analysis of FOIA requests from 2001 to 2008 revealed that compliance costs for the government went up, but personnel assigned to complying with FOIA requests went down (Ibid.).

After the events of September 11, 2009, the sovereignty of FOIA, in the domain of the public's right to know, was damaged even further. Lotte E. Feinberg, of the Jon Jay College of Criminal Justice, wrote in *Government Information Quarterly*, "The most striking point in assessing 'federal information policy' in the aftermath of the September 11 attacks on the U.S. is the almost complete absence of any integrated or coherent government policy that responds to the changing needs to balance access, privacy, and secrecy. The Freedom of Information Act (FOIA), the centerpiece of access policy since 1966, no longer holds this position" (Feinberg 2004, 439). Feinberg warned that "In many ways access policy is entering uncharted territory. After being part of the landscape for thirty-five years, and very much taken for granted as a 'right,' there is growing concern that this right is being winnowed away, keeping citizens less informed and, at times, leading to abuses by government that secrecy not-infrequently produces" (Ibid., 454).

Concerns about the Bush administration's classification policies, its relocation of federal oversight of FOIA from the National Archives to the Department of Justice, and many instances of the invocation of secrecy for national security purposes revived a constituency of right-to-know activists in organizations such as the American Civil Liberties Union, OMB Watch, and the newly formed Sunlight Foundation. Barack Obama was a good fit for this constituency as a presidential candidate, because Obama, as a senator, had helped craft and pass the Federal Transparency Act of 2006, which, among other things, created a searchable online database of federal government contracts and spending data (as of fiscal year 2007 and later).

Thus when Obama was elected president in 2008, he was already familiar with many of the concerns of the U.S. right-to-know community. It was this background that led President Obama to address FOIA and open government in his first two official acts as president, on Inauguration Day (White House 2009a, 2009b). This illustration of the new president's priorities, and his many statements during the campaign about accountability and open government, have created expectations that the Obama administration will be more proactive and innovative in providing access to government information than previous administrations. Obama hired several prominent people from the field of new technologies use to expand access to government information in new ways, particularly Vivek Kundra, who stepped into a new White House position as federal chief information officer; and Beth Simone Noveck, a law professor from New York University who is the author of a book titled *Wiki Government*, and who joined the White House Office of Science and Technology Policy (Noveck 2009).

Frustrations with FOIA, the emergence of new technologies of information access, and a new president apparently committed to both open access to government information and Internet-based innovation, have all set up the conditions for a new phase of public access to government information—a new phase of both hope and possible disappointment.

Post-FOIA America?

Ellen Miller, cofounder and executive director of the Sunlight Foundation and a longtime activist on campaign finance, wrote a speculative "obituary" for the Freedom of Information Act for a compendium of articles titled *Rebooting America* (Miller 2008). In 2009 (i.e., the future), Miller said, "Congress passed the Government for All Act that became the gold standard for government transparency of personal and financial relationships. The law required that all public reports be filed electronically and shared within twenty-four hours of their filing. It forced the Senate to follow the House's lead and make Personal Financial Disclosure reports available online. It also required Senators to file, and make public within twenty-four hours, campaign finance reports (Ibid., 62). She added, "Ultimately, FOIA's demise was necessary to allow transparency and information to flow freely" (Ibid. 2008, 63).

Some advocates of expanded access to public information now describe a potential "post-FOIA" era in which government information is made available online in useful and accessible ways as the default and legally required means of public disclosure. This "dream" of transparent access to government information online is a counterargument to the FOIA model of any person requesting government information or records; instead, literally any person with online access would be able to search, find, and use government information in a timely fashion, in some cases in close to real-time.

This discussion among some open government activists has now moved on to how government agencies have used the World Wide Web to actually cloud government transparency—using Web pages as promotional "brochureware" instead of as useful repositories of information for accountability—and how access to databases could allow citizen-programmers to actually process and assess government performance in ways that government officials don't pursue themselves. A team of Princeton University scholars published an article in the *Yale Journal of Law and Technology* in 2009 titled "Government Data and the Invisible Hand," in which they say:

If President Barack Obama's new administration really wants to embrace the potential of Internet-enabled government transparency, it should follow a counter-intuitive but ultimately compelling strategy: *reduce* the federal role in presenting important government information to citizens. Today, government bodies consider their own Web sites to be a higher priority than technical infrastructures that open up their data for others to use. We argue that this understanding is a mistake. It would be preferable for government to understand providing reusable data, rather than providing Web sites, as the core of its online publishing responsibility. (Robinson et. al. 2009)

The Princeton team asserted "Government must provide data, but we argue that Web sites that provide interactive access for the public can best be built by private parties" (Robinson et al. 2009, 161). Funding such independent projects of Web sites built on public data has become the main activity of the Sunlight Foundation, based in Washington, DC, which has helped build sites such as OpenCongress.org, the Library of Unified Information Services (Louis), and the Open Senate Project, among other activities. Access to both the public data and the necessary application programming interfaces (APIs) that programmers need to use the data is one of the "gold standard" features of the Web site USAspending.gov (http://www.usaspending.gov/), run by the White House Office of Management and Budget. Using open-standard data sources and public APIs, independent programmers are able to build their own Web sites that analyze government data and performance, such as the nonprofit Web site Fedspending.org (http://www.fedspending.org/), developed by the nonprofit watchdog organization OMB Watch. Extending this model of open transparency of government data has been announced as one of the goals of the Obama White House, which launched a new Web site, data.gov, in May 2009. The purpose of data.gov, according to the White House, is "to increase public access to high value, machine readable datasets generated by the Executive Branch of the Federal Government."

Governmental agencies face significant challenges in adopting this new model of information access, however. In addition to the normal obstacles of secrecy, privacy, and bureaucratic defensiveness, there is the simple fact that much of the data produced by the government is neither readily open to independent programmers, nor easily adapted to common functions on the Internet, such as searchability, interoperability, or transportability. Government agencies use a wide array of different approaches and electronic devices to create, store, and archive data—including ancient batch-processing machines—and most government agencies have only recently made the transition to electronic documents, but in formats (such as portable document format or PDF) that are not easily processed by other programs. Public-sector information technology managers, a huge population, are professionally trained to support the people who work for their agencies, and public access to information is not yet a high priority. Public records managers responsible for complying with open records requests have been trained in the FOIA model, and a new, post-FOIA model of open data access is not yet part of their professional canon (DiMaio 2009). And of course there are only a few examples of legislation that require entire databases to be available to the public in ways that could lead the way to a post-FOIA era.

At the same time that public officials are confronted with these novel developments in public access to information, the nature of the information is also changing dramatically. The most significant new element is the "blurring" of boundaries between public institutions and between the government and the public at large, between authorities and experts, and between "providers" and "users."

Knocking Down Silos: The Blurring of Boundaries

It is commonplace now in the field of e-government to identify three different phases of e-government evolution: the "informational" phase of purely informational, noninteractive Web sites; the "transactional" phase of online transactions such as renewing licenses or paying taxes; and the "transformational" phase, not yet clearly defined but generally described as the phase in which government activity and organizations are transformed by the Internet, the Web, and other networked information technologies (such as mobile devices).

The most evident phenomenon of the transformational phase of e-government is also prominent in the private sector: the blurring of organizational boundaries as the Internet allows the interpenetration of both information and business processes, the two fundamental components of any organization. Amazon.com, for example, is a "platform" for sellers who are blended together into one interface for interested buyers. Over the past decade, most governments adapting to the Web have launched single-entry "portals" that allow citizens looking for information to begin at one page, instead of needing to know in advance which agency is the appropriate custodian of the information. This "one-stop shopping" model has contributed to interagency collaborations in other ways, leading to more public sector transformation.

One of the original government data portals was the Geospatial One-Stop geographical and mapping data portal (ESRI 2003). The portal was created under Executive Order 12906, "Coordinating Geographic Data Acquisition and Access: The National Spatial Data Infrastructure," which was signed on April 11, 1994, by President Bill Clinton (Clinton 1994). The goal of the executive order was to standardize the documentation of geospatial data under standards set forth by the Federal Geographic Data Committee. The purpose of standardization was to allow government employee users of the data to pull geographic data from many sources so that they could create detailed, even multilayered maps. Initially, the portal contained consolidated geographic data in tabular form from all agencies collecting data with any geospatial parameters, like a street or city name, a geographic descriptor (lake, river, desert), or any locality.

The value of the portal was derived from the blurring of the boundaries between agency geospatial databases, which allowed the government to cull data points from multiple sources of data and combine these data to produce a more complete map or picture of a given physical location. In the opening statement of the executive order, Clinton defined the importance of geographic data for government purposes: "Geographic information is critical to promote economic development, improve our stewardship of natural resources, and protect the environment. Modern technology now permits improved acquisition, distribution, and utilization of geographic (or geospatial) data and mapping." The data was published on a Web page accessible to government employees. And while Clinton recognized a future need to open the portal to citizens, the order was limited to internal governmental use of geographic data and did not set forth a plan for citizen use of the data.

In 2002, the Bush administration added to the purpose of the geospatial data site, incorporating the idea of data for decision making. President Bush recognized the purpose of Geospatial One-Stop as being twofold. The first purpose of the site was to support the business of government. Bush noted that "Almost every aspect of government, including, but not limited to, disaster management, recreation, planning, homeland security, public health, and environmental protection, has a geographic component and requires geospatial data and tools to appropriately manage it." He supported the continued cooperation of agencies to provide data for the portal and acknowledged the importance of cross-agency access to geospatial data as a means to conducting government business.

However, he also recognized geospatial data as an aid to citizens. Bush added a second purpose to the Geospatial One-Stop, to support decision making. He noted that "issues occur in places (e.g., floods, events, crimes) and decisions addressing one issue often have broader implications, sometimes affecting entire communities. Geospatial information allows decisions to be viewed in a community context and can facilitate cross-agency coordination." By cross-agency cooperation, he not only meant cross-agency cooperation within the federal government, but also cooperation across all levels of government and with business, academic, and not-for-profit organizations. He also noted that access to data by citizens and local government could aid in local decision making.

The Geospatial One-Stop evolved into geodata.gov (http://gos2.geodata .gov/wps/portal/gos). In its current evolution, the site provides a portal for geospatial data collected from every major federal agency, state and local governments, businesses, not-for profit organizations, academic institutions, and from citizens who access the site, upload, and maintain local, state, and national geodata. Not only do users access and deposit data at the site, they use the site to coordinate specialized "community" action. Users of the site can participate in "communities" specific to their geodata specifications, interests, data uses, and needs. Each community has different attributes that allow users to search and filter geospatial data, create multilayered maps, and connect and collaborate with multiple agencies and people to develop interactive tools for the community and the general public.

Some of the communities produce maps associated with current events, such as wild fires, hurricanes, and tornadoes, in order to contribute to local disaster management systems. Other mapping communities document the progress of strategic goals like charting energy use over time by location, documenting property acquisition and value changes in given areas, mapping unemployment rates across varied spaces, following disease patterns and rates, identifying service provision gaps by geographic area, determining traffic patterns and rates, and many others. Many of the communities that begin on the geodata site are exported to other sites to facilitate action and provide trusted data to contribute to localized or specified community information and decision-making systems and processes.

Portals such as geodata.gov represent one of the first steps in the ongoing blurring of boundaries between former "silos" of public authority and information. Breaking down the old silos has become something of a mantra for information managers in the public sector. Two critical features of online information, search and filter, have decoupled public information from any particular location, which has huge implications for the boundaries of authority and control of government agencies. New interactive Web-based tools extend the decoupling of public information and blur the boundaries of authority by allowing users to personalize or localize data in a way that matches the needs of a given "community" of users to support localized needs, decisions, and processes.

The public sector is still adapting to these changes that have emerged as part of the new information age characterized by ubiquitous Internet access. But of course the Internet is still evolving faster than most government agencies can adapt. "Web 2.0" is the latest challenge, even though that term, and the basic outlines of the concept, have been around now for several years (O'Reilly 2005). The essence of the Web 2.0 phenomenon is that because large numbers of people are using the Internet today, with high-speed connections, the information that these Internet users generate online is a new, significant source of knowledge as well as a vast new universe of information itself. Wikipedia is probably the best-known example and clearest illustration—an encyclopedia of human knowledge built on the contributions of its users and offered online for free. YouTube, the online video-sharing site owned by Google, is only about four years old but is already so embedded in American culture that it seems like it has been around a lot longer. In January 2009 alone, YouTube served up a staggering 14.8 billion videos, and "147 million U.S. internet users watched an average of 101 videos per person" in that month (YouTube 2009). The White House itself launched a YouTube channel for its own videos in January 2009. YouTube users can not only watch videos produced by the White House, but comment on them as well.

In fact, the White House has generated a completely interactive Web platform to which citizens can connect via social networking sites like Facebook and Twitter and through which citizens can seek out information and comment on current issues and policy decisions. WhiteHouse.gov (http://www.whitehouse.gov/) has moved from a one-way search engine of press releases, radio addresses, photos, and Web pages to a forum for citizen interaction and engagement.

Users of WhiteHouse.gov can comment on the president's blog, comment and provide video feedback on videos produced by the White House, listen to and leave audio exchanges via podcasts and iTunes, send the president an email message, write on the "wall" of the White House Facebook page, follow the real-time White House updates via Twitter, and even download widgets to connect an external Web site to WhiteHouse.gov to simultaneously update these Web sites with the latest information on pressing issues, such as a potential pandemic flu. The White House regularly solicits public commentary and offers regular policy updates that often reflect the input garnered via the interactive Web tools. The expectation is that citizens will view the information and participate in the process of sharing and generating new information.

This new domain of "user-generated content" is a tidal wave washing over the Internet. Facebook, the most successful of the social-networking Web sites, reported over four hundred million active users in September 2009 (Facebook 2010). Every minute of every day, YouTube users upload twenty-four hours of video (YouTube 2009). In the summer of 2008, social networking sites passed pornography sites as the most accessed online (Goldsmith 2008). The microblogging service Twitter is the latest rage, experiencing a 1,400 percent growth rate in 2009 (McCarthy 2009). These high-profile and hugely popular online services are supplemented by hundreds of thousands of Web sites where people contribute thoughts, comments, ideas, expertise, photos, computer programs, ad infinitum.

The other significant way that these developments blur boundaries is that many of the more popular services and applications on the Internet can be "pulled into" other Web sites, proliferating themselves by their interoperability. YouTube videos can be embedded into multiple university course and government agency Web pages; Google Maps have become a de facto way to display geographic information; items found on blogs can be shared with Facebook friends with a simple button click. Then all of these things can be searched globally, usually using Google, the dominant search engine service. The result is that the Internet is a vast but integrated information resource that is increasingly taken for granted as the principal way people get information today. The very rapid rise of this global information universe stands in sharp contrast to the way that governments have handled information for centuries. The impact that the Internet is having on the public sector is beginning to change the very shape and practice of government, although, for some, not nearly fast enough.

Public Information: From Bricks and Mortar to Sandbox

Until very recently, government data has been accessible to citizens as a means to an end of the information-gathering process. In the process of gathering data—regardless of whether through a FOIA request, the accessing and mining of an individual agency databank, or through a multiagency portal—information was sought and provided within a limited purpose of information exchange. The primary driver of electronic information delivery was that of lowering the transaction cost of information access. With the maturation of government portals and the simultaneous maturation of Web 2.0 information tools, however, a shift in the purpose of information delivery began to take shape. The purpose of information accessible to any person began to transform to information interactively engaged with any person.

This idea that citizens can engage and interact with government data online in support of government service provision and in support of local "community"-based decision making and action was reflected in legislation and in line with business documents or documentation justifying continued tax-dollar allocation associated with the various Web portals and government-run Web sites. In the Federal Funding Accountability and Transparency Act of 2006, legislation sponsors Senators Barack Obama and Tom Coburn explicitly mandated the use of the Internet and interactive tools for citizen interaction with federal spending data (Obama and Coburn 2006). The Department of Defense developed contracting requirements for its contractors to post contract data and information on a public Web server displaying real-time data. The United States Geological Survey (USGS) built a metadata dictionary and guide into their line of business documentation so that citizens could more easily post and use data on the USGS site. Providing information around which citizens could act became an important function of government.

For the geodata.gov portal, this prompted a move from a consolidated data warehouse serving insulated, authorized communities to a consolidated warehouse connecting organizations with interactive services through which individuals could access government data. An example is the Environmental Protection Agency's (EPA) use of the geodata portal data to inform, educate, and facilitate citizen action for protection of the environment. The EPA is a member of several geodata communities and provides geographic, biological, environmental, and ecological data to the geodata portal. EPA combines the data the agency uses with data provided by other agencies and by individual users to generate maps of

environmental hazards or maps of environments in need of civic attention. These maps can be regionalized so that users can see the status of
the environment in their local areas.

The maps have become part of both the geodata site and the EPA Web
site (http://www.epa.gov/). They are searchable under specific topics
(watersheds, chemical spills, extremely endangered or contaminated areas)
as well as by locality (county, state, region). Users can identify the major
environmental hazards and problems in their area and learn steps that may
be taken to allay these hazards and problems. Upon learning about the
issues, through the EPA Web site, users can then join a community on the
geodata site as well as take individual action through or join a community
associated with "Pick 5 for the Environment." (Environmental Protection
Agency 2005). If an individual becomes a user member of the geodata site,
he or she can post new data, talk with fellow citizens and specialists
working to improve the environment, and monitor the progress of any
action.

If individuals become users of the Pick 5 for the Environment site, they
can chart, monitor, share, and promote community activism by choosing
and promoting five steps to remedy environmental damage and hazards.
They can make videos of their progress and post to Flickr and Facebook.
They can also upload any current statuses or events through Twitter and
Facebook. Users can interact with their local community to develop best
practices and exchange ideas for future action. EPA has connected with
social networking tools such as Facebook and Twitter in a way that allows
for simultaneous updates by individuals to appear on the individual and
community sites along with the social networking site of the EPA. The
agency also offers an alerting system for when new data or information is
produced so that users have access to the latest data and information.

In addition to individual users having the capacity to use and take
action with government-provided data and information, users can also
load widgets onto their individual or organization Web sites that will
connect them to the newest information and tools available from the
agency and allow them to embed that information and those tools into
their own Web site. This gives users the capacity to share data and tools
with other users through multiple Web sites. The government agency and
the underlying portal become a sort of backbone provider of data and
information to the user and the user's community.

The Web 2.0 phenomenon has transformed government information
resources from data bounded by the "bricks and mortar" of the respective
government institution to an open "sandbox" of collaboration, mashups,

experimentation, and the profound mixing of conventional and authoritative information with new user-generated content. This is posing new challenges to government managers who are confronted with issues about data reliability, compliance with the law, privacy, archiving, and so on. There are benefits to this transformation, but costs as well. At the moment, it is not clear how well governments will manage the transition to these new forms of information dissemination and accessibility.

Government Innovation: Transparency, Collaboration, Participation

In his book *Web 2.0 Heroes: Interviews with 20 Web 2.0 Influencers*, Bradley Jones describes the interactive Web. "Web 2.0 is not about mass marketing. It's about understanding the masses. And it's not about controlling the message. It's about engaging the audience and actually hearing what they have to say. It's about enabling creativity, realizing a culture of contribution, and putting the user in control" (Jones 2008). The Obama administration views users of government information in much the same way. Upon entering office, President Obama challenged federal agencies and the general public to look at government data in a new way. He suggested three primary purposes for data: to provide transparency, to promote participation, and to enhance collaboration.

The Obama administration has taken a novel approach to the promotion of transparency, participation, and collaboration with government data. White House officials have asked agency managers to post machine-readable data and to engage the public in a discussion of effective ways to use the data.

The approach is similar to that going on in the business world. Companies like IBM, Intel, Amazon.com, Dell, and Apple have used interactive Web 2.0 tools to engage their customer bases in new ways to determine new individual user-specific and user-oriented products and services. Dell, most famous for IdeaStorm (http://www.ideastorm.com), created an interactive site where users could view, post, vote, and see suggestions in action. As of September 2009, nearly 13,000 people had contributed more than 85,000 ideas and promoted nearly 700,000 product renditions. Dell implemented 366 of the product chain suggestions. The ideas range from the development of computers with no operating systems and laptops that have color options to chip and hardware designs. To offer another example, Amazon.com uses interactive tools to create communities of users that connect similar individual interests with related products using miniblogs, videos, and user chats.

With the understanding that a primary government product is information, President Obama called on government agencies to engage citizens in a discussion of the optimal means and methods for accessing information and about the online tools that would best facilitate ongoing interaction with government. Agency leaders are taking up this challenge to work with the public to determine user-oriented, interactive government information services. The White House Office of Management and Budget's data.gov site is meant to be a repository of high-value data sets in machine-readable formats for public use. Accompanying the data are a compendium of tools and tutorials for reading, visualizing, and using the data and a request for users to suggest potential new data sets and new tools for using and interacting with the government-provided data.

The open government approach is not without its critics. At a recent conference on open government and innovation, a participant quoted Henry Ford saying "If I had asked people what they wanted, they would have said 'a faster horse'" (GovLoop 2009). The conference participant was referring to the experience that people often don't know what they want until it's actually built and they have a chance to use it. People are just beginning to interact with the data on data.gov, and with other government data repositories. It remains to be seen whether the availability of public, machine-readable data from government sources will have a significant impact on citizen use of public information.

Some observers fear that the trend of opening up government data sources and expecting citizens to use the data are at a "peak of inflated expectations," similar to other patterns of a technology "hype cycle," and that constraints of cost, personnel time, and other limitations—including a potential realization that the user community is very small—will deflate the "bubble" of enthusiasm for government transparency through data accessibility. There is also the looming and constantly growing problem of privacy, not only for citizens whose private information might be revealed by the expansion of public information online, but also for government employees, who are under increasing pressure to reveal their activities.

Citizens and government officials hold regular discussions in which they seek to find the balance between privacy and transparency. One of the most vivid examples of an issue that requires privacy but depends on certain levels of transparency is health care. In response to Obama's double mission of increasing access to government information and working with citizens to design interactive information tools and to reform health care, the Department of Health and Human Services created HealthReform.gov (http://www.healthreform.gov/). HealthReform

.gov allows individual users to take part in the health care reform process. The site hosts Web casts of Health Care Stakeholder Discussions and White House forums that were held across America. On the site, users are encouraged to join important discussions about health care reform by monitoring the crafting of the legislation and sharing their thoughts directly with the Obama administration to develop the legislation.

Users are able to access current data and information related to health care and to health care reform. They are able to post their personal stories and to comment on legislation and activities associated with reforms. They are also able to participate in interactive health care quizzes, search quick facts related to individual states, watch videos, and read and respond to posts made by fellow citizens and the administration. They can also sign up for real-time alerts and notices so that they can take action in real-time on the Web site via chat and forum functions. Users are able to interact with and act on government data and information in real time.

Sharing data related to health is risky for any individual for personal and professional reasons. However, understanding the health care needs and tools that can assist in health care delivery is as important as health information sharing is risky. In the first three months of operation, more than ten thousand people shared their personal stories about the need for health care reform. More than fourteen thousand users signed a statement in support of enacting comprehensive health reform. More than twenty-one thousand people joined a health care reform distribution list that provides updates about new reports, forums, and features on HealthReform.gov. Tens of thousands of Americans watched and commented on events associated with the reform movement including the White House Forum on Health Reform; the Regional White House Forums on Health Reform, which featured a bipartisan group of governors and top White House officials; and several Health Care Stakeholder Discussions at the White House.

Not only did the development of the government information sandbox lead to new and interesting ideas for the use of government information, it also led to new and interesting government partnerships. As noted earlier in the chapter, in 2006 the Federal Funding Accountability and Transparency Act, sponsored by Obama and Coburn, passed through Congress. The Act served to make transparent data produced and maintained by the Office of Management and Budget and requires that all entities and organizations receiving funding from the federal government disclose such funding on an OMB Web site. The Act prompted an unprecedented partnership between OMB Watch and the OMB to produce the

Web site USAspending.gov. The USAspending.gov site was built collab-
oratively using the same technology as the OMB Watch site and offers
many of the same capacities and services.

Users of the FedSpending and USASpending sites can sort and filter
budget data, access data updates, build widgets to connect data from the
sites to an individual Web site, and build and use application programming
interfaces that allow users to define data comparisons and characteristics
so that users may build data projects that fit their interests and needs.

Increasingly users are able to observe and apply government data in
a manner defined by them. It is this user-defined use of government
information that underpins the vast majority of the innovative open
government projects today.

Conclusion

The history and phenomena we have described are affecting governments
and citizens around the world, and in epochal ways. Facebook, Twitter,
and the dissemination of government data to Web sites far and wide are
all examples that point to the end of the document era, which has lasted
longer than anyone could have reasonably imagined. Given the disrup-
tions affecting nearly every major sector of the economy, it is no surprise
that immensely important values attached to modern government,
such as transparency and accountability, are taking on new forms, and
consequently presenting government officials with new and vexing
challenges.

The Freedom of Information Act, for example, has come from a time
when there was a bright line of distinction between official information
and unofficial information, a line that is now being dimmed or blurred.
The founders of the United States clearly imagined a nation that was set
up to inform its citizens, but the founders obviously could not have imagined the staggering volume of information that comes out of government
or that comes into government, nor the modern ways in which these data
are being intermingled in complex ways. Privacy activists, to take one
example of civic activism on government information policy, used to argue
over whether the government or the private sector was the bigger threat
to privacy rights. Now the distinction makes little sense—information
passes between these sectors so routinely that the old argument has lost
its significance.

At present we are on the leading edge of an important transformation in how government information will be managed. The Freedom

of Information Act and the role of the press—the hand-in-glove model of transparency and accountability that is a legacy from history—are still the predominant means for informing citizens, but both are weakening. FOIA requests are in slow decline, in part because of the availability of public data and the utility of tools like Google. And newspapers and other journalistic outlets are rapidly fading, especially among young people. The Obama administration has enthusiastically adopted the tools of Web 2.0, which helped elect President Obama in 2008. But it is too early to tell whether these very large-scale trends—the crisis in journalism and the new commitment to government transparency online—will support or harm the goal of the early founders of the United States: an informed citizenry.

A concern that has emerged in tandem with these new trends is that civic engagement and the concept of the informed citizen may become limited to a small class of digitally savvy, highly educated, and interested citizens, who will coexist with a huge mass of ill-informed or even misinformed citizens susceptible to demagoguery. The political polarization of the United States, the explosion of partisan sources of information such as blogs and cable and radio talk shows, and the decline of news readership and viewership have all fueled this worry over the modern civic culture of the United States. A 2009 report on the Internet and civic engagement by the Pew Internet & American Life Project "shows that the Internet is not changing the fundamental socio-economic character of civic engagement in America. When it comes to online activities such as contributing money, contacting a government official or signing an online petition, the wealthy and well-educated continue to lead the way" (Pew Internet & American Life Project 2009). Matthew Hindman reached a similar conclusion in his 2008 book *The Myth of Digital Democracy* (Hindman 2008).

Skeptics about this concern might respond that civic engagement has always been limited to a minority and the American public is not generally known for being highly informed about public affairs—nevertheless, democracy and the functions of government continue to muddle along. The Pew report does note that "There are hints that the new forms of civic engagement anchored in blogs and social networking sites could alter long-standing patterns. Some 19% of Internet users have posted material online about political or social issues or used a social networking site for some form of civic or political engagement. And this group of activists is disproportionately young (Pew Internet & American Life Project 2009).

Cuillier and Piotrowski found through their surveys of 2004 to 2006 that public support of access to government information was positively correlated with Internet use, meaning that the more people used the Internet, the more they were inclined to support access to government information. However, they noted:

A potential negative implication of our findings is the potential for a widening political participation gap, or knowledge gap, in society. We found that those who are more educated and active in seeking information online are more supportive of access to public records. If those people actually acquire government documents and other information helpful to their lives then they will benefit. Those who do not have internet access may be less likely to see the value of government records and not access information that could help their lives, and they will suffer. This could result in a widening digital divide separating the haves from the have nots. (Cuillier and Piotrowski 2009)

This is a possible new dimension of the "digital divide," the phrase used during the Clinton years as shorthand for the gap between online users and everyone else. This will be subject matter for researchers in the future: whether the expansion of public information online exacerbates a "digital divide" in civic participation, specifically the gap between the well-informed and the uninformed.

To conclude on a more optimistic note, however, we believe that the Internet, the Web, and related new tools give government the means to inform citizens like never before. And there are encouraging signs in the United States and in other countries that government officials have absorbed this idea and are working hard on new forms of transparency and accountability, in response to citizen demands. This is likely to be a jerky, "two-steps-forward, one-step back" process, as the government and the public sort out the proper relationships and authority, and figure out what is useful and what is not. Government is going to be swept up in the huge changes that the Internet is imposing on the entire world and every institution in it. It may take another generation for the results to become clear. Fortunately, the generation that will assume responsibility for democracy appears committed to an informed citizenry and a transparent public sector.

References

Brown, Richard D. 2000. Early American Origins of the Information Age. In *A Nation Transformed by Information: How Information Has Shaped the United States from Colonial Times to the Present*, ed. Alfred D. Chandler, Jr., and James W. Cortada, 39–53. Oxford: Oxford University Press.

Burkert, Herbert. 2007. Freedom of Information and Electronic Government. In *Governance and Information Technology: From Electronic Government to Information*

Government, ed. Victor Mayer-Schönberger and David Lazer, 125–141. Cambridge, MA: MIT Press.

Clinton, William J. 1994. The White House. Executive Order 12906: Coordinating Geographic Data Access. April 11. http://govinfo.library.unt.edu/npr/library/direct/orders/20fa.html.

Coalition of Journalists for Open Government. 2007. http://www. cjog.net/ (accessed August 29, 2009).

Coyle, John Y. n.d. Amassing American "Stuff": The Library of Congress and the Federal Arts Projects of the 1930s. Federal Theater Project collection. http://memory.loc.gov/ammem/fedtp/ftcole01.html (accessed April 30, 2010).

Cuillier, David, and Suzanne J. Piotrowski. 2009. Internet Information-Seeking and Its Relation to Support for Access to Government Records. *Government Information Quarterly* 26: 441–449.

Department of Justice. 2008. FOIA Post: Summary of Annual FOIA Reports for FY 2007. http://www.usdoj.gov/oip/foiapost/2008foiapost23.htm.

DiMaio, Andrea. 2009. Open Data and Application Contests: Government 2.0 at the Peak of Inflated Expectations. Gartner blog network, September 22. http://blogs.gartner.com/andrea_dimaio/2009/09/22/open-data-and-application-contests-government-2-0-at-the-peak-of-inflated-expectations/.

Environmental Protection Agency (EPA). 2009. Pick 5 for the Environment. April. http://www.epa.gov/pick5/.

ESRI. 2003. Geospatial One-Stop Portal Is Key to President's E-Government Strategy. Federal GIS Connections (Fall). http://www.esri.com/library/reprints/pdfs/gos-portal.pdf.

Facebook. 2010. Facebook Statistics. http://www.facebook.com/press/info.php?statistics.

Feinberg, Lotte. 2004. FOIA, Federal Information Policy, and Information Availability in a Post-9/11 World. *Government Information Quarterly* 21: 439–460.

Goldsmith, Belinda. 2008. Porn Passed over as Web Users Become Social: Author. Reuters, September 16. http://www.reuters.com/article/internetNews/idUSSP31943720080916.

GovLoop Conference Participants. 2009. OGI Conference TweetBook. Open Government and Innovations Conference, July 21–22. http://1105govinfoevents.com/OGITweetBook_FINALrev1.pdf.

Hindman, Matthew. 2008. *The Myth of Digital Democracy*. Princeton: Princeton University Press.

John, Richard R. 2000. Recasting the Information Infrastructure for the Industrial Age. In *A Nation Transformed by Information: How Information Has Shaped the United States from Colonial Times to the Present*, ed. Alfred D. Chandler, Jr., and James W. Cortada, 55–105. Oxford: Oxford University Press.

Jones, Bradley L. 2008. Web 2.0 Heroes: Interviews with 20 Web 2.0 Influencers. Indianapolis, IN: Wiley Publishing, Inc. http://www.wiley.com/WileyCDA/ WileyTitle/productCd-0470241993.html.

Library of Congress. 1995. History and Background. Geography & Map Reading Room, Geography & Map Division, Library of Congress. http://www.loc.gov/rr/ geogmap/gmhist.html (accessed April 30, 2010).

Madison, James. 1910. Letter to W. T. Barry, August 4, 1822. In *Writings of James Madison*, vol. 9, ed. Gaillard Hunt, 103. New York: G. P. Putnam.

McCarthy, Caroline. 2009. Nielsen: Twitter's Growing Really, Really, Really, Really Fast. CNet News.com, March 19. http://news.cnet.com/8301-13577_3-10200161-36 .html.

Miller, Ellen. 2008. The Merciful Death of the Freedom of Information Act and the Birth of True Government Transparency: A Short History. In *Rebooting America: Ideas for Redesigning American Democracy for the Internet Age*, ed. Allison Fine, Micah L. Sifry, Andrew Rasiej, and Josh Levy, 59–63. http://rebooting.personaldemocracy .com/files/Rebooting_America.pdf.

National Archives and Records Administration. n.d. History. http://www.archives .gov/about/history/ (accessed April 30, 2010).

National Security Archive. 2006. Freedom of Information at 40, July 4. http://www .gwu.edu/~nsarchiv/NSAEBB/NSAEBB194/index.htm.

Nicholas, Peter. 2009. White House Declines to Disclose Visits by Health Industry Executives. *The Los Angeles Times*, July 22. http://www.latimes.com/news/ nationworld/nation/la-na-healthcare-talks22-2009jul22,0,7434392.story.

Noveck, Beth Simone. 2009. *Wiki Government: How Technology Can Make Government Better, Democracy Stronger, and Citizens More Powerful*. Washington, DC: Brookings Institution.

O'Reilly, Tim. 2005. What is Web 2.0?, September 30. http://oreilly.com/web2/ archive/what-is-web-20.html (accessed August 28, 2009).

Obama, Barack, and Tom Coburn (cosponsors). 2006. Federal Funding Account-ability and Transparency Act of 2006. Public Law No: 109–282. April 6. http:// thomas.loc.gov/cgi-bin/bdquery/z?d109:s.2590.

Pew Internet & American Life Project. 2009. The Internet and Civic Engagement, September 1. http://www.pewinternet.org/Reports/2009/15--The-Internet-and-Civic-Engagement.aspx.

Robinson, David, Harlan Yu, William P. Zeller, and Edward W. Felten. 2009. Government Data and the Invisible Hand. *Yale Journal of Law and Technology* 11 (Fall): 160–167. http://www.yjolt.org/files/robinson-11-YJOLT-.pdf (accessed April 30, 2010).

Rood, Justin. 2009. Like Bush, Obama White House Chooses Secrecy for Key Office. *ABC News*, May 15. http://abcnews.go.com/Blotter/story?id=7589622&page=1.

Weitzel, Pete. 2009. Fewer Requests, Fewer Responses, More Denials: An Analysis of Federal Agencies Performance in Responding to Freedom of Information Act Requests in 2008. Sunshine in Government Initiative. http://www.sunshineingovernment.org/stats/highlights.pdf.

White House. 2009a. White House Open Government Innovation Gallery. http://www.whitehouse.gov/open/innovations.

White House. 2009b. Freedom of Information Act, January 21. http://www.whitehouse.gov/the_press_office/Freedom_of_Information_Act/.

YouTube. 2009. YouTube Report: YouTube Statistics. http://youtubereport2009.com/category/youtube-statistics/.

9 Active Readership: The Case of the American Comics Reader

George Royer, Beth Nettels, and William Aspray

Introduction

This chapter explores information issues related to the practice of reading. More specifically, the focus is on comics, detailing the transition of the comics reader from the passive consumer to an active participant in shaping both the future of the medium and a participatory reading culture.

The patterns of readership of comics in America since the end of the nineteenth century form a richly textured tapestry. In his book *Comic Book Nation: The Transformation of Youth Culture in America*, Brad Wright explains: "Few enduring expressions of American popular culture are so instantly recognizable and still so poorly understood as comic books" and "Just as each generation writes its own history, each reads its own comic books" (Wright 2001 xiii). Underneath the fascinating and often sensational history of the comics medium is the complicated history of the comics reader. As technology, social values, and cultural norms have exhibited radical shifts over America's history, so too have the identities and practices of the comics reader. Over their history, comics, in both strip and book forms, have been viewed as everything from ephemera to high art. We seek to understand how people have interacted with comics, and what they took from these interactions, as well as what the readers have contributed. To make sense of seemingly volatile patterns of consumption by the general public and to unearth the tale of the comics reader's transition from passive consumer to active participant, one must understand how the medium and the reader developed together—each influencing the other.

While increasing amounts of scholarly attention are being directed toward the study of comics, most studies focus on cultural studies or aesthetic histories, rather than the role of the comics reader. This chapter is not intended to serve as either an aesthetic or comprehensive history of comics in the United States, but instead uses history to identify the readers and illustrate their practices. Details regarding creators and publishers are included insofar as they illustrate important points surrounding reader-ship. By identifying various types of comics, analyzing patterns in their respective readerships, scrutinizing the methods by which comics have been obtained and consumed, and examining the increasingly active roles taken by readers, this chapter provides an overview of the readership of the comics medium and alludes to broader issues that intersect with both the study of comics and the practice of participatory reading. These are examples of a type of information-seeking behavior experienced by a large portion of the American population as part of their everyday lives.

The Birth of a New Medium: Comic Strips and the Early Readers

The story of the readership of comics in America begins in 1895 with a battle for readers between publishing giants Joseph Pulitzer and Randolph Hearst. Both Pulitzer and Hearst were interested in increasing the circula-tion of their newspapers within the urban immigrant population—and both saw the comic strip as a means by which they could sell papers to a demographic that was perceived as being boorish, dubiously literate, and hungry for accessible materials that reflected their everyday experiences. In 1895, Pulitzer acquired a printing press capable of mass producing full-color reproductions on newsprint. He used this special press to introduce a Sunday supplement, a four-page full-color insert full of scintillating mate-rial, and ran it weekly in his *New York World*. One item published in this Sunday supplement was artist Richard Felton Outcault's *Hogan's Alley*, a series of sequential illustrations set in the gutters of Manhattan's Lower East Side. *Hogan's Alley* starred a grinning, barefoot, yellow-shirted carica-ture of New York's immigrant poor. This Yellow Kid exhibited many of the qualities ascribed to the immigrant poor: he was violent, ugly, and spoke in garbled English. The rest of the cast of *Hogan's Alley* formed a chorus line of racial and ethnic stereotypes. The inhabitants of *Hogan's Alley* delighted in subjecting adults, authority figures, and proper society to mockery and physical abuse.

The success of the Yellow Kid character and *Hogan's Alley* did not escape Hearst's notice. He purchased his own color press and hired away many of Pulitzer's staff—including *Hogan's Alley*'s creator, Outcault. In 1896 Hearst's *New York Journal* published what is considered to be the first true comic strip—Outcault's *The Yellow Kid*, employing the conventions of frames and dialogue balloons (Inge 1990). Outcault's comic strip was a major success with Hearst's target audience (Sabin 1993, 134). The popularity of the Yellow Kid character quickly led to licensing deals—he appeared on everything from buttons to cigarette packs and even in a Broadway play. The addition of comic strips increased newspaper circulation substantially and comics quickly became common in newspapers. By today's standards, the first comics appear crude—but possess an undeniable vitality. *The Yellow Kid* strongly resonated with immigrant populations; as author David Hajdu explains (2008, 11), "The funnies were *theirs*, made for and about them. Unlike movements in the fine arts that crossed class lines to evoke the lives of working people, newspaper comics were proletarian in a contained, inclusive way." Another scholar comments, "Underneath the horseplay and absurdity we find a world of anguish and pain. *The Yellow Kid* is not really an innocent entertainment. . . . It expresses, though perhaps not consciously, a sense of malaise, a feeling that the old, rural, natural, America is being destroyed" (Berger 1973, 31). A medium new to America had been introduced, and with it, a new kind of reader.

There were many reasons *The Yellow Kid* appealed to readers—and not just among its main target audience of the urban poor. The marriage of the visual with the temporal narrative that is fundamental to comics enabled Outcault to create a vibrant world and make the product eye catching, accessible, and memorable. The wretched slum in which the Yellow Kid resides interested readers with its topical setting and bizarre adult-child characters. One author explains that the weird inhabitants of Hogan's Alley "wear derbies, smoke cigars and cigarettes, and others are bearded. They are ragamuffins, children of the poor, and as such, are allowed all manner of exotic actions" (Berger 1973, 25). Rather than attempting to escape from the gutters in search of a better life, the children of Hogan's Alley content themselves with base pleasures, joyously inhabiting their caricatured slums. Implicit in this construct is that the poor remain poor because they do not wish otherwise—they do not desire the same things as the rest of society. By romanticizing the lives of the poor in his comic strip, Outcault created a place for urban immigrant readers to identify with even as he allowed New York's more fortunate citizens to relieve themselves of the responsibility for social action (Ibid.). Instead of pitying the poor, other social classes were afforded a regularly delivered laugh at the expense of their less fortunate neighbors.

Newspaper publishers were quick to embrace comic strips and the increased circulation they afforded. Many comic strips appeared within a few years of the publication of *The Yellow Kid*. The strips during this period largely abandoned the merciless, satirical qualities of *The Yellow Kid* and gravitated more toward lighthearted romps. Prominent examples include Outcault's *Buster Brown* (1902), *The Katzenjammer Kids* (1897), and *Mutt and Jeff* (1908). By 1903 newspapers across the country featured comic strip supplements in the Sunday editions. Syndication and the consolidation of control in the newspaper industry exposed increasingly large numbers of Americans to comic strips. For example, the comics appearing in Hearst's papers appeared in at least seventeen newspapers across the country (Gordon 1998, 38). Syndication removed comic strips from their earlier localized nature and encouraged creators to appeal to a broader audience of Americans. Increasingly accessible to the public, comic strips during this period were enjoyed by mixed-age audiences, and their place in the Sunday edition ensured deployment to living rooms across the nation (Sabin 1993, 133). Vibrant and large, the comics supplement to the Sunday editions of newspapers was perfect for spreading out on the floor, inviting individuals to enjoy the entertainment of the reading experience together.

Comic strips' commercial success created a complete entertainment industry around them (Inge 1990, 17). The increases in circulation afforded by the addition of comic strips encouraged publishers to scramble to add more strips to their papers. By 1908, "syndicates had brought comic strips to the nation and established methods of distributing their product that continue to the present. Between 1903 and 1908 newspapers in twenty-four locations added comic strips while only two dropped them. Almost 75 percent of Sunday newspapers had strips" (Gordon 1998, 41). As distribution expanded to include rural regions, so too did the standardization of the medium.

During the 1920s and the 1930s, the comic strip continued to rise in popularity. Sabin cites two major forces chiefly responsible for the strip's increasing popularity: "the gradual standardization of humorous strips into situational comedies, and the emergence of a new genre—adventure." Publishers realized that minimizing the localized nature of the earlier strips, which were primarily concerned with urban living, would increase their appeal in rural areas, where much of the American populace then lived. As a result, comic strips "began to speak much less to the fringes of American society, and more to 'middle America'" (Sabin 1993, 137). Some strips from this period, such as *Gasoline Alley* (1918), *The Gumps* (1917) and *Little Orphan Annie (1924)*, deal more with what was perceived to be the common American experience. These strips "emphasized the pathos of lower middle-class life and the impact of industrialism and technology on the ordinary family" (Ibid.), but did so with a consistent sense of humor. As the urbanites of New York had seen their lives reflected in the late-nineteenth-century strips, Middle America found a pleasing reflection in the syndicated strips of the 1920s. These strips were so popular that the stock market once suspended operations to see if a character in *The Gumps* followed through with matrimony (Inge 1990, 5). Others, such as *Buck Rogers* (1929) and *Terry and the Pirates* (1934), concern themselves with distant, fantastical settings full of escapism and thrills. Escapist fare proved quite popular during the Depression. The net result was that the working-class-oriented urban comic strips gave way to a much less challenging and much more inclusive brand of entertainment that appealed to a broad spectrum of readers. A mid-1920s survey of children in Kansas revealed that reading comic strips was the most frequently engaged-in activity, regardless of race or sex, during all seasons of the year (86). George Gallup's 1930s surveys revealed that the most popular and widely read section of the newspaper was the comics page, and a person was more likely to read his favorite comic strip than the lead front-page story (85). By the 1930s, the strips had become an entrenched feature of daily American life (81). Having been familiarized with the comics medium through exposure to strips, the American readership was next introduced to the comic book.

The Golden Age of Comic Books and the Genesis of Participatory Readership

In 1933, five years before *Action Comics #1* would hit the newsstands, a group of employees at the Eastern Color Printing Company conducted a printing experiment that resulted in *Funnies on Parade # 1*, the first comic book (Benton 1989, 15). It comprised reprints of Sunday strips including *Mutt and Jeff*, *Skippy*, and *Joe Palooka*. Max C. Gaines, a salesman at Eastern Color, sold ten thousand copies of this new comic book to Procter and Gamble, which gave it away to customers as a premium with soap product purchases.

The idea of collecting reprints of newspaper comic strips was not new. As early as 1897, publishers began to release hardcover collections of cartoons that had appeared in humor magazines such as *Puck* and *Judge*. It did not take long before similar collections of the Sunday comics began to appear. Collections of *Mutt and Jeff* were highly successful—enough so to merit five separate editions (in 1910, 1911, 1912, 1915, and 1916). Emcee Publishing began the first serialized publication of comics collections in 1922, but printed only twelve issues. Each monthly issue contained a collection of a different strip. It was a book of comics, but not a comic book as we know it today. In 1929, the Dell Publishing Company brought the first full-color collections of comics to the newsstands in a magazine called *The Funnies*. *The Funnies* was printed in a newspaper format. Its "unwieldy size and the fact that it looked like an incomplete portion of a regular Sunday newspaper, spelled its demise" (Benton 1989, 15). Dell's failure, however, was not insignificant. Eastern Color had printed *The Funnies* in 1929, and the idea of a monthly comics magazine had made an impression on some of the employees. In 1933, the company's printers discovered a way to reduce the size of Sunday comic strips and print two pages side by side on one sheet of tabloid-sized paper. Folded in half and bound together, the result was a seven-and-a-half-by-eleven-inch full-color book of comics, the comic book as we know it.

Gaines knew he had something special when the entire 10,000-copy print run of *Funnies on Parade* was gone in a matter of weeks. He found other manufacturers, including Canada Dry, Kinney Shoes, and Milk-O-Malt, that wished to use the new comic books as premiums. *Skippy's Own Book of Comics*, printed in 1934, was given away by Phillips Toothpaste to listeners of the new *Skippy* radio show based on the comic strip adventures of fifth-grader Skippy Skinner. The target audience for these premiums was, of course, children, whom manufacturers were hoping would convince their parents to buy the products being advertised (Gordon 1998, 129–130). Eastern Color wanted a greater share of the profits from its comics printing venture and decided to print comics for direct sale to the public. Eastern Color staff members approached George Delacorte at Dell Publishing, for whom they had printed *The Funnies* five years earlier, and found him willing to take a risk on a print run of 35,000 copies. *Famous Funnies: Series 1* was sold through chain department stores rather than through standard magazine distribution outlets and became the standard in terms of price (ten cents) and size for the rest of the decade (Benton 1989, 17).

For the next three years, the business of publishing comic books grew steadily. In 1936, the syndicates that provided comic strip material to newspapers entered the field, printing reprints of their own material. At the end of 1936, the comic book industry consisted of six publishers and ten titles (Benton 1989, 20). The last years of the decade, however, saw explosive growth, doubling practically every year of the late 1930s. With help from characters such as Superman and other costumed superheroes, comics containing original content surpassed reprint comics in sales figures by the end of 1940. At the beginning of 1942, over 143 distinct titles were available, read by over fifty million readers every month (35). The early 1940s saw the introduction of a host of superheroes. Detective Comics (now known as DC Comics) had found the winning formula, with characters such as Superman and Batman; other publishers quickly followed.

At this point, children still constituted the majority of comic books' readership. David Hajdu has characterized these early comics as "erratic, inelegant, often clumsy, but boundlessly energetic and wide-eyed" (2008, 36)—in a sense, much like children themselves. Children could afford to pay ten cents for an issue of their favorite comic book, and once purchased, the comic had longitudinal value in its ability to be passed on to friends. In the hands of a child, the comic book became a precious treasure offering substantial social capital in the currency of bragging rights and swap value. A child may have owned only a few comics, but through the practice of group readings and pass-alongs, may have read hundreds. Easily hidden from parents and conveniently sized so as to be readable just about any-where (e.g., school, home, a treehouse), comics were the one medium over which children had exclusive control (93).

The possession of even one comic book imbued a child with sufficient social capital to allow access to scores of additional comics material, through communal reading and sharing. Without commercial guidance, children developed informal information-sharing networks to increase their comic book consumption. Clubhouses, schoolyards, bedrooms, street corners, and any other place in which children may have gathered served as potential hotspots for the unsupervised distribution of comics. A 1943 article examining the comics reading practices of children notes that they extended their purchasing power by employing a communal exchange system in which each child purchased a different comic book and traded with other children. One mother reported that in a single day no less than sixteen children came to the household seeking to trade comic books with her child (Strang 1943, 336). These comics were passed along until just tatters remained. Young readers were not yet trained to value the physical elements of the artifact. The practices exhibited by children in the 1930s and 1940s contrast with the later fetishizing of the comic book as an intrinsically valuable commodity but foreshadow the creation of spaces devoted solely to the practice of reading, trading, and collecting comics. Specialty stores, discussed later in this chapter, often found themselves plagued by broad characterizations of juvenile culture and exclusivity.

The war years were a boon to comics publishers. Although paper alloca-tions forced publishers to limit the number of titles, issues, and pages in individual issues they printed, sales were never better, doubling to twenty million copies sold between 1941 and 1944 (Gordon 1998, 139). Citing studies conducted by the National Market Research Company, Zorbaugh paints a picture of the broad appeal of comic books to readers of all ages, genders, and backgrounds in 1944. Boys and girls considered regular readers consumed an average of twelve to thirteen comics each month, young men and women read seven to eight each month, and adults read six a month. Among children aged 6 to 11 years, 95 percent of boys and 91 percent of girls were regular readers. Among adolescents (12–17 years old), 87 percent of boys and 81 percent of girls were regular readers. 41 percent of men and 28 percent of women between the ages of 18 and 30 read comics regularly, as did 16 percent of men and 12 percent of women 31 years and over. Except among adults aged 18–30, the difference between the genders in each age bracket is slim.

Several scholars have attributed this particular gender gap, as well as the overall growth in comic book sales, to the reading habits of American servicemen.[1] As of 1944, of the 189 periodicals approved for distribution to U.S. soldiers fighting in World War II, from a list drawn up on the basis of soldier preferences, nearly 50 were comic books (Zorbaugh 1944, 199). Comic books owed their popularity among soldiers to the fact that they were inexpensive, portable, and easy reading. At home and abroad, soldiers were reading comic books. One survey revealed that 44 percent of men in training camps read comics regularly and 13 percent were occasional readers (198). Comics outsold popular magazines such as *The Saturday Evening Post*, *Life*, and *Reader's Digest* ten to one in domestic post exchanges. The demand for comics overseas was so great that the Navy made the decision to ship comic books to soldiers on Midway Atoll in the North Pacific. They were passed from man to man until they were too worn to read. In all probability, many soldiers who read comics were the same young males who read them before the war (Gordon 1998, 140). The childhood distribution model of comics shared among friends easily transferred to the barracks, where entertainment and icons from home were precious.

The close of World War II did not spell the end for the comic book boom. With paper restrictions lifted, the industry expanded even more, culminating at the end of 1953 in an average monthly circulation of sixty-eight million copies. In ten years, comic books had quadrupled their sales. By the late 1940s, however, publishers noticed a drop off in the sale of superhero comics. No superhero introduced after 1944 was successful, and many publishers canceled superhero titles as a result of poor sales (Wright 2001, 57). The decline of superhero comics forced publishers to find new stories to sell, including genres of romance, crime and true crime, horror, and westerns. Romance and crime comics in particular proved to be immensely popular. Surveys had shown that women read almost as many comics as men, but there were as of yet very few titles specifically targeting young girls and women.[2] Archie Comics (formerly MLJ Publications) was the first company to find a solid footing among preteen female readers (72). Other publishing companies had attempted teen comics in the vein of the Archie Andrews stories even before the end of World War II, but none were as successful. So it was that in 1947, Jack Kirby and Joe Simon, veteran superhero creators, were seeking alternatives to superhero comics (Benton 1989, 43). What they developed for Feature Publications was *Young Romance*, an instant bestseller that spawned a romance comics genre that was over one hundred titles strong by 1949 (142). These comics specifically courted adult women, and *Young Romance* even advertised itself as being "For the More Adult Readers of Comics!" (Wright 2001, 128). Such comics mostly reinforced gender norms of the Cold War era rather than challenging them. There were no female superheroes in their pages, no Wonder Woman, Miss Fury, or Mary Marvel. Gone also were the covers depicting women tied up in poses suggestive of bondage that might alienate female readers. Instead, these comics were "cautionary morality tales told from the perspective of a female protagonist (though generally written by men) [that] illustrated the perils of female independence and celebrated the virtues of domesticity" (Ibid.). Everett Raymond Kinstler, an artist who worked on several romance comics from this time remarks that "Romance comics dealt with a range of emotions, some of them quite subtle and sophisticated. . . . The stories had heart and passion, and they were sexy. They weren't kid stuff" (Hajdu 2008, 163). Romance comics sold well enough to cut into the market share of the true- confessions magazines from which they had taken their formula (Wright 2001).[3] The plan had worked: adult women were buying and reading comics. Readership of comics was even more diverse than it had been before, and women were lining up alongside adolescent males to buy their comics at newsstands and drugstores around the country. A study conducted in 1950 in Dayton, Ohio, found that among both children and adults, males made up 52 percent of the readership, and females 48 percent. Of the entire population, nearly 35 percent had read at least one comic book within the past month.

The participatory fan culture, which would eventually evolve into one of the comic book industry's most vital assets, began to develop during this period. Publishers invited readers to join "fan clubs" devoted to their favorite comic books, often incentivizing membership with the promise of exclusive collectibles. For example, Captain America fans could, for the mere price of a dime payable to Timely Comics, receive a membership card and official badge identifying them as "Sentinels of Liberty." The publisher-sponsored comic book fan clubs do not appear to have been forums for the development of community, but rather merely instruments designed to extract more dimes from children. The first fan-published magazine, or "fanzine" devoted to comics, *Comic Collector's News*, was published in October 1947. It offered its readers a variety of information resources, including sections devoted to classified advertisements and a forum through which communication with geographically distant fans became possible (Pustz 1999).

Two other new genres, crime and horror comics, also sold well. The first crime comic book, *Crime Does Not Pay*, was published in 1942 and sold well enough to survive in a market overwhelmingly dominated by superheroes and funny books. It achieved circulation of about two hundred thousand copies per month in 1946. Between 1947 and 1951 almost every comic book publisher introduced a crime comic. In 1948, *Crime Does Not Pay* and *Crime and Punishment* each sold over one and a half million copies per month (Benton 1989). Along with their rise in popularity, crime comics quickly drew criticism. While accusations of promoting juvenile delinquency had been leveled at comics since the first decade of the century, by 1950 the medium's critics had gathered enough momentum to lead to a Senate hearing on possible connections between comics and juvenile delinquency. The Special Committee to Investigate Crime in Interstate Commerce, chaired by Democratic Senator Estes Kefauver, had been formed to investigate racketeering, but felt its work would be incomplete if it did not also look into the "frequently heard charges that juvenile delinquency has increased considerably during the past five years and that this increase has been stimulated by the publication of the so-called crime comic books" (Hajdu 2008, 172). A survey conducted in the late 1940s by the publisher of *Crime Does Not Pay* found that 57 percent of the readers were over twenty-one years of age. Horror comics were also born in the mid-1940s, but their peak in popularity was 1951 to 1954 (Benton 1989). In 1952, horror titles accounted for almost 30 percent of all comics published.

The legendary horror publisher EC Comics, which would later bear the brunt of anti-comic book sentiment, appears to have been the first publisher to capitalize on the potential inherent in interacting with an adult fan community. EC Comics employed a number of forward-thinking practices including the interspersing of reader comments with creators' comments on the letters pages of its magazines, the publication of letters from readers inviting other readers to join their mailing lists and fan clubs, and the inclusion of a subscription to the official *E.C. Fan-Addict Club Bulletin* with fan club membership. EC's fan-centered publication simultaneously fostered a sense of community within its readers and actively engaged them in the creative process with its offer of forums for the discussion of ideas for new comics (Pustz 1999). Most important, the *E.C. Fan-Addict Club Bulletin* created a direct bond between readers and the creators. This type of participatory, inclusive model of readership would later be refined by personalities such as Marvel Comics' Stan Lee into what would ultimately save the comic book from extinction.

The Crucible: The Anti-Comics Movement and Creation of the Comics Code

The meteoric rise in popularity of genre comics, including horror and crime comics, was met with vehement opposition by self-styled public guardians. Watchdogs grew increasingly convinced of a direct correlation between youth readership of comic books and juvenile delinquency (Hajdu 2008; Wright 2001). Many believed that reading crime-related comics was causally linked to a propensity for indulging in deviant acts. Civic organizations, concerned about the deleterious effect of comic book consumption on the part of America's youth, organized efforts to prevent the distribution and sale of comic books. Parent-teacher associations, women's clubs, and the National Education Association took various actions, including national publicity campaigns against comic books, calls for anti-comics legislation, and public mass burning of comic books (Wright 2001; Nyberg 1998). By October 1948, fifty American cities had enacted laws designed to curb youth access to objectionable comics (Wright 2001). Penalties for selling objectionable comic books, namely those depicting criminal acts such as burglary or violent acts such as murder, included substantial fines and possible imprisonment. Anti-comic book regulations were no joke— the city of Cleveland's police department dedicated two officers to "the comic-book beat" (Hajdu 2008, 145). The early stirrings of large-scale organized resistance to youth readership of comic books foreshadowed events that would nearly destroy the entire industry.

Fearing backlash over youth readership of inappropriate materials, comic book publishers banded together in an attempt to allay public concern. In July 1948, twelve comic book publishers collectively formed the Association of Comics Magazine Publishers (ACMP) and agreed to adopt their own code of self-imposed standards (Wright 2001). The ACMP's standards included restrictions on the presentation of crime in an effort to ensure that unlawful activities would only be presented in a negative light; authority figures including policemen, judges, and other government officials would not be portrayed in any kind of disrespectful fashion; and depiction of torture, sexy images, and profanity would be prohibited. If an issue from a participating publisher met these standards and was given the thumbs-up by the ACMP's staff of reviewers, it would be allowed to place a seal of approval on its cover (Nyberg 1998). The ACMP was "doomed from the start" due to the fact that only twelve out of the comic book industry's thirty-four publishers became members. Comic books sales covered by the ACMP's code accounted for only fifteen million out of fifty million comic books sold per month. Many of the larger publishers, including Marvel, DC, and Dell, claimed their comics were wholesome beyond reproach and declined to continue participating in the ACMP. By 1950, the initial attempt at organized self-regulation on the part of the comic book publishers was "effectively defunct," (Wright 2001). Four years later, only three publishers even remained members of the ACMP (Nyberg 1998).

Some publishers did their best to sidestep the gathering storm of public outrage while others either failed to check the weather or weren't afraid to get wet. Lev Gleason, publisher of *Crime Does Not Pay*, pioneer of the crime comic book genre, and target of substantial criticism, argued that his comics contained an overt and unmistakable anticrime message. In an impassioned letter to *The New York Times*, he argued that comics' detractors sought to, "set up an intellectual dictatorship over the reading habits of the American people," and that inasmuch as many of them were adults, his readers had a right to like their comics any way they pleased (Wright 2001, 102–103). In spite of his indignant stance toward censorship, Gleason was one of the few major publishers who joined and did not leave the ACMP (104). By 1950 *Crime Does Not Pay* had shed its bloody imagery and featured a banner across the top of the cover promoting the book as "a force for good in the community" (Hajdu 2008, 154). Indeed, readers need not take Gleason's good intentions on his word alone: covers now included a photo of a stern policewoman. Mary Sullivan, former chief of the NYPD Women's Bureau now acting as editorial consultant to *Crime Does Not Pay*, graced the corner of each issue with the reassuring statement, "I approve of this magazine as a good moral influence on our youth. I recommend it to parents as a powerful lesson for good behavior" (155).

While some of the publishers working in the crime genre were partly tamed by public outcry, the publishers of the horror genre displayed little interest in reform. EC Comics, under the leadership of William Gaines, was the oft-imitated leader in the horror comics genre. Interest in horror comics skyrocketed in the early 1950s: from 1950 to 1954 twenty-eight publishers introduced nearly one hundred new horror titles (Wright 2001). As Wright explains (156), "Brutality, sadism, death, and a grisly variety of 'living deaths' became lucrative commodities bought and sold in youth culture." The imagery offered by EC's horror titles, which included *The Vault of Horror*, *The Crypt of Terror*, and *The Haunt of Fear*, were "some of the most feverishly intense visuals yet seen in comics." EC's horror titles were turning substantial profits and inspired other publishers to produce hackwork imitations. Unable to reproduce the levels of artistry and clever writing that defined EC's products, other publishers instead attempted to appeal to readers by upping the shock value and graphic display of gore and violence (Sabin 1996, 67). Upping the gross-out factor to attract readers soon proved to have drastic consequences not only for horror comics, but also for the entire comic book industry.

Frederick Wertham, a German-born New York psychiatrist and longtime foe of the comic book industry, published his magnum opus, *The Seduction of the Innocent*, in 1954. His book was, "pompous, polemical, and sensational, it aimed to impress a popular audience with professional expertise and moral outrage" (Wright 2001, 158). Wertham characterized readers of comic books exclusively as children and ignored the adolescent and young adult segments of readership (Hajdu 2008). His attack claimed that comic books were responsible for violence, deviant acts, low self-esteem, and even sexual maladjustment in children. Even traditionally all-American super-heroes failed to escape Wertham's scrutiny, where he argued that allowing children to read *Superman* created an experience that mirrored the "blunting of sensibilities in the direction of cruelty that has characterized a whole generation of central European youth fed on the Nietzsche-Nazi myth" (Wright 2001, 159, citing *Seduction of the Innocent*). Wertham had words for the crime-fighting detective Batman as well, stating that his relationship with Robin was a "wish dream of two homosexuals living together" (159). All of these observations were rendered by Wertham in his professional capacity as a medical doctor. His evangelical zeal and sensationalism "were enough to inspire widespread moral panic" (Sabin 1996, 68). When Senate subcommittee hearings were called in 1954, Wertham insisted on testifying.

The initial Senate inquiries into comic books in 1950 had aroused Senator Estes Kefauver's suspicions that comic books were linked to juvenile delinquency. In early 1953, a Senate subcommittee on juvenile delinquency was formed, known as the "Kefauver committee." Under his leadership, the subcommittee investigated the comic book industry in the spring of 1954. Over a three-day session, twenty-two witnesses testified and thirty-three exhibits were accepted into evidence. Wertham informed the senators that comic books were a serious contributing factor in juvenile delinquency and that his study of comic books was the only existing large-scale study. However, it was Kefauver's interaction with EC Comics publisher William Gaines that garnered the most publicity. Gaines stated that the only limits to what he would or would not put in a comic book, including those that might be seen by children, were those imposed upon him by his concept of good taste. Seizing the moment, Kefauver presented Gaines with an issue of EC's *Crime SuspenStories* featuring a cover illustration of a murderer holding the severed head of a woman in one hand and a bloody axe in the other. The woman's lifeless, bloody body lay in the background. Kefauver asked Gaines whether he considered this image to be in good taste, to which Gaines replied, "Yes sir, I do, for the cover of a horror comic" (Nyberg 1998, 61–63). Their exchange was reported on the front page of *The New York Times* and proved unflattering for the entire comic book industry. Gaines's feeble defense demonstrated to the public that the publishers of comic books were apathetic to the welfare of young readers. Another senator, Robert Hendrickson, stated that "not even the Communist conspiracy could devise a more effective way to demoralize, disrupt, confuse, and destroy our future citizens than apathy on the part of adult Americans to the scourge known as juvenile delinquency" (Wright 2001, 172). By the time the subcommittee adjourned, the senators had made it clear that it was their opinion that crime and horror comic books needed to be brought under control—or stamped out entirely (Ibid.).

Following the Senate subcommittee hearings, the comic book publishers knew they had little choice but to collectively take action before action was taken against them (Wright 2001). In so doing, they tacitly admitted that comics were harmful for children. They agreed to adopt and abide by strict rules governing what could and could not appear in a comic book. To accomplish this goal, publishers formed the Comics Magazine Association of America (CMAA)—an independent entity with the purpose to police and sanitize inappropriate content in comic books. It was agreed that the CMAA would establish and enforce a code governing the industry. To indicate compliance with the CMAA's code, comic books would display a seal of approval on their covers. The code adopted by the CMAA was rigidly designed to avoid public protest. The new code, known as the "Comics Code," required comic books to, among other things, emphasize the sanctity of marriage, respect authority, depict crime in a negative light, refrain from displaying excess violence, ensure that good always triumphed over evil, and prohibit content that depicted horror (Nyberg 1998). The Comics Code was so intent on stamping out horror comics that it expressly forbid the use the words "horror" or "terror" on the cover of a comic book (Wright 2001). Eager to turn public sentiment more positive, spokesmen for the Comics Code boasted that it, "imposed restrictions far greater than any in force in competing media" (173).

The strictness of the code and the zeal with which it was enforced effectively flattened the previously rich landscape of American comic books. "Reflecting a bland consensus vision of America, comic books now championed without criticism American institutions, authority figures, and middle-class mores. The alternatives simply disappeared" (Wright 2001, 176). The creative frenzy that surrounded the comic book industry in the years leading up to the code instilled in readers "a disregard for the niceties of proper society, a passion for wild ideas and fast action, a cynicism toward authority of all sorts, and a tolerance, if not an appetite for images of prurience and violence" (Hajdu 2008, 330). These concepts, which would become prominent aspects of popular American media in the 1960s, must have resonated with the readers of the mid-1950s. When the code crippled the comic book industry's ability to create stories readers cared about, readership dried up. In the "financial disaster" that followed the introduction of the Comics Code, numerous publishers folded, genres died, and the number of comic book titles published in America dropped from roughly 630 in 1952 to 250 in 1956 (Sabin 1993). The American comic book would never again enjoy such widespread readership.

After the Comics Code, the trend toward the inclusion of readers in the mainstream publications and the emergence of independent, reader-created material continued. In 1958, DC Comics editors introduced pages with correspondence from readers in the *Superman* comics. These letters pages proved to be a popular addition and soon became a standard practice. They not only demonstrated that publishers were asking comics readers what they wanted to read but also, due to the frequent publication of the letter writer's full address, allowed readers to locate and contact one another. While overall readership numbers were dropping, hardcore fans demonstrated increasingly dedicated and sophisticated community-building practices. Fanzines, such as *Alter-Ego* produced by Jerry Bails and future industry leader Roy Thomas, emerged during this period and commonly featured indexes and reviews of comics, fan-created amateur comics, personal information about the fans, details of fan meetings, and information on collecting. It was through these fanzines that communities were able to organize and rally themselves around a foundering industry. The first comic book convention, held in New York City in 1964, was attended by prominent fans, dealers, and industry personalities, and facilitated through fan-organized activities (Pustz 1999). This first convention, a humble antecedent to today's sprawling multinational expos, was organized primarily by a sixteen-year-old and some of his friends (Serchay 1998). Eventually, some of these devoted readers transitioned from fan to professional—further blurring the line between readers and creators.

The Comic Book Comeback: Marvel Comics and the New Readership

The industry looked bleak in the aftermath of the Comics Code, when readership declined steadily and publishers continued to exit the business. In the early 1960s, two publishers, Marvel Comics and DC Comics, dominated the much-diminished market (Sabin 1996). The industry revived itself by harkening back to the conventions of superheroics and capitalizing on inclusive interactions with readers. By the late 1960s, reading comic books meant more than picking up a magazine—it meant belonging to an inclusive, participatory, and vibrant subculture.

Stan Lee, the charismatic editor of Marvel Comics, struggled to reach new readers in the face of floundering sales figures (Wright 2001). As the creator of iconic characters such as Hulk, The Fantastic Four, and Spider-Man, Lee instilled his superpowered creations with human flaws and tied Marvel Comics together into an integrated universe. Marvel's characters behaved in ways fundamentally different from the superheroes of previous eras; for example, "Spider-Man seemed to spend most of his time agonizing over his personal life, something that Batman would have never done" (Pustz 1999, 52–53). The Marvel heroes from this period were almost uniformly antiestablishment and yet refrained from taking overt political positions. These creations appealed to adolescent and young adult audiences because they could relate to them, yet the figures remained sufficiently fantastic. Spider-Man's adventures took place in real cities as opposed to Superman's Metropolis or Batman's Gotham (Wright 2001). By creating unrealistic characters with realistic problems, Lee had found a way to write within the confines of the Comics Code while making his creations attractive to readers.

The reader response was nothing short of phenomenal. Marvel's comic book sales figures doubled from 1962 to 1967 while the sales figures of its competitors declined or remained constant. Although comic books would never again hit pre-television levels of popularity, they gained a loyal new readership. Lee's careful orchestration of an integrated universe, in which characters native to one book would make guest appearances in another, was a clever marketing plan designed to simultaneously encourage fans to purchase multiple titles and instill a sense of community in readers (Wright 2001). Understanding exactly why the Hulk was flinging boulders at Spider-Man in this month's issue of *The Amazing Spider-Man* might require a reader to have purchased an issue of *The Incredible Hulk* several months before. Thus, familiarity with back issues enriched the reader's experience with current issues. Soliciting and responding to fan mail was an important element of Marvel's strategy to engage readers. The editorial sections at the end of each issue published and responded to fan mail in a warm and humorous fashion. Readers understood that, unlike the outcast Hulk or grief-driven Spider-Man, they were not in this alone. As Wright explains (2001, 218), the editorials and fan mail sections were "all designed to impart that there was more to the Marvel experience than just reading a comic book and throwing it away." Another Marvel strategy was to praise the audience, to tell the readers "how smart and hip they were" (Pustz 1999, 53). Marvel built a new model for readership, and added value to the experience of readership by acknowledging and including its fans. Marvel wasn't just selling comic books—they were selling a "participatory world for readers" (Pustz 1999). Marvel streamlined and enhanced the patterns of interaction with readers that had previously been employed by EC Comics; by embracing fan culture, Marvel created unprecedented brand loyalty and helped to transform comic book reading from an individual to a communal experience.

Marvel's creations proved extremely popular with college students. The publisher's office received substantial fan mail from more than 225 colleges on a daily basis. Some fifty thousand individuals paid one dollar apiece to join Marvel's fan club, the Merry Marvel Marching Society. Lee himself became a celebrity, which included speaking invitations at many colleges and universities, including Columbia and New York University. One of Lee's lectures, at Bard College, drew a larger audience than former President Eisenhower (Wright 2001). Marvel's readers came to view Lee's suffering superheroes as iconic counterculture figures. A 1965 poll of colleges conducted by *Esquire* reported that student radicals "ranked Spider-Man and the Hulk alongside the likes of Bob Dylan and Che Guevara as their favorite revolutionary icons" (cited in Wright 2001, 223). Lee's success lay in his finding a new way to take advantage of the familiar medium of comic books. The convention of monthly publication lent itself to gripping cliffhanger endings, perpetuation of community-building practices through fan mail and news bulletins, and tantalizing sales pitches for the following month's issue. All of this worked due to an inclusive humanization of the fantastic—adolescent and young adult readers saw their own lives mirrored in the plights of the Marvel superheroes, plus the personal recognition of their own letters in print. These elements, together with fanzines, facilitated the rise of geographically scattered reader communities and heralded the creation of specialized venues for the consumption, discussion, and adoration of the comic book.

The Rise of the Comic Book Shop and the New Home of the Comic Book

Because Marvel and its competitors adopted the practice of delivering a larger story arc over multiple issues and multiple titles, familiarity with back issues became an important part of reading comics. The proliferation of issues and titles complicated the reading experience. "By the 1970s Marvel was publishing comics set in all three of these periods [past, present, and future], requiring fans have knowledge of hundreds of years of events to completely appreciate Marvel universe and its continuity based realism" (Wright 2001, 223). This need to acquire previous issues evidenced a new kind of information-seeking behavior on the part of the comic book reader. Newsstands would not typically carry issues from three months ago—much less three years ago. Fans looking for a specific back issue were limited to scavenging bins at flea markets or second-hand bookstores (Wright 2001, 6). The artifacts themselves transitioned from being disposable items to valued collectables. Businesses specializing in the collection and sale of current and back issues of comic books capitalized on this change in the publisher's strategy. A number of specialized stores catering to the needs of comic book readers appeared during the 1970s, and by the end of the 1980s they numbered approximately four thousand in the United States (Sabin 1996). As they have increased in number, comic book shops have steadily become a fixture in the landscape of American culture. They are easily recognizable by the brightly colored posters of costumed men and partially costumed women prominently displayed in shop windows. By creating a place for comic books, these specialty shops have impacted the popular perceptions of comic book readers, potential audiences for comic books, ideas about the monetary value of the comic book, and the distribution practices of comic book publishers. The advent of the comic shop has resulted in the tying of the newsstand-format comic book to unique retailer—effectively locking the physical artifact in a distinctive locale.

By the late 1970s, a new distribution practice known as direct market distribution began to replace the traditional newsstand and mail subscription practices. Under the traditional newsstand distribution model, comic books were distributed by companies that handle a variety of periodicals, including traditional magazines (Nyberg 1998). Unsold copies of comic books were sent back to their source; and "as many as seven of every ten copies of a comic book were returned to the publisher." This distribution model, which as of 1998 still accounted for roughly one-quarter of comic book distribution, was overtaken by the direct market model. The direct market model allows specialty shops to solicit orders for upcoming issues in advance—but they are not able to return unsold copies. Under this model, comic book shops are able to "develop an inventory of back issues that then may be sold to fans and collectors at a later date" (144). The model eliminates some of the risk involved on the part of the publisher—under a solicitation system, a company has advance notice as to how many issues to print; and it also allows the specialty shop to accumulate a large store of back issues in a more economical fashion. The direct market model is now so pervasive that specialty stores are now known as "the direct market" inside the comics industry (Wolk 2003). The shift in distribution models of comic books indicates a fundamental difference in how businesses had come to view the comic book: they were no longer disposable items but rather were assets to be stockpiled for future gain.

For many patrons, the comic book store is a beloved place, rich in community and culture. Pustz describes the veteran fan's experience of the comic shop as "gold mines, places to find buried treasure, catch up with old friends, and make new acquaintances with like-minded souls" (1999).[4] However, for the rest of society, comic book stores have developed a reputation for being rabidly devoted to a specific subculture while alienating outsiders. Having witnessed the success of the Marvel line of comics, which focused almost entirely on superheroes, most comic book publishers wasted little time in devoting their periodicals to the superhero genre. The "shops sold an endless array of superhero comics, which to an outsider seemed like pointless riffing on the same theme, but which to a fan were continually fascinating, and hence limited their clientele to males between the ages of about twelve and twenty-five" (Sabin 1996, 157). The narrow scope of subject matter, in distinct contrast to the broad appeal enjoyed by the comics of previous eras, proved alienating to potential female readers: "Women were certainly not welcome: the comics were in the traditional mould of power fantasies about macho muscle men, and commercially speaking there was very little incentive for publishers to think about any other kind of subject matter" (Ibid.). This comic shop environment did little to invite female readers to stay and "hang out" as it did their male counterparts. Even in one comic book store regarded to be more female-friendly than most, female readers tend to "come in to get what they want and leave quickly" while male readers are "talking with employees, gossiping with friends about the new artist on *Wonder Woman*, or just silently counting money in their wallets" (Pustz 1999, 5). Female patrons "commonly feel uncomfortable or feel unwelcome as a result of the gazes of male patrons who are surprised to see women in that setting or by posters that frequently objectify women"; female readers are "turned off by the fact that there do not seem to be any comics interesting to female readers" (8). Author Scott McCloud refers to this kind of environment as "the stereotype of the dimly-lit, cultish 'no gurlz-allowed' breed of the comics store" but is optimistic about efforts on the part of some retailers to create environments that fight to reinvent the image of the comic shop (McCloud 2000, 84). He explains that progressive retailers are seeking to expand their reader base by offering a broader selection of types of comic books and working to create an atmosphere that "welcomes new readers rather than driving them away" (Ibid.). For all his optimism, McCloud notes that "99.9 percent of the reading public would never dream of entering a comics store" (85).

Gender issues aside, the experience of entering a comic book store often proves to be overwhelming experience for the uninitiated. Information-seeking behavior on the part of the inexperienced visitor is frustrated by what Pustz terms "the bowels of the shop," which feature "boxes of hundreds of thousands of carefully sorted old comics and with racks of new issues of hundreds of titles" (1999, 5). While a seasoned comic shop visitor would have little difficulty navigating the store's unique environment to unearth a particular item, a new patron would likely be overwhelmed by the fact that there are typically no indexes. Visitors are left with little recourse but to root around in alphabetized boxes for a particular item. Of course, finding a particular item in this arrangement requires that the patron possess previous knowledge of its existence. With the exception of the new issues displayed on stands, a comic book store's holdings are uniformly sealed in plastic bags with acid-free backboards. The message implicit in the practice of bagging, sealing, and filing away comic books leaves visitors with a markedly different impression regarding the proper treatment of comic books, contrary to the days in which children would haphazardly pass comic books around until they were worn to the point of falling apart.

In recent years, comics have increasingly escaped the confines of the specialty shop and are increasingly appearing in more traditional venues such as bookstores. These venues offer readers the opportunity to enjoy comics without feeling as though they are entering a place where they don't belong. Critical and public attention given to celebrated graphic novels has helped to catapult the comics medium from the realm of juvenilia into the ranks of respectable literature. The comic books offered in these environments are typically graphic novels and trade paperbacks, as opposed to the single-issue format that now resides almost exclusively in the specialty shop. In the bookstore environment, the graphic novel is appealing to readers. Sales of graphic novels have been "zooming" upward in both traditional bookstores and specialty shops (Wolk 2003). Publisher Paul Levitz attributes the success of comics in the traditional bookstore to "book buyers and media critics who grew up reading groundbreaking comic works like *Watchmen* and *Maus*" in the 1980s (MacDonald 2005). Sales figures from 2005 of graphics novels showed a 35 percent growth rate, with $140 million through bookstores and $67 million through comic shops (Reid 2005). These kinds of sales figures indicate that comics clearly have the capability to appeal to a reader base beyond the young adult male demographic that frequents the specialty shop. By placing the comics in a different context, the books become accessible to a wider reader base.

The Digital Age: Readers, Creators, and Readers as Creators on the Web

The Internet has enhanced the well-established practice of reader interaction with creative forces, offering many fan-created Web sites (or e-zines), personal Web sites and blogs of writers and artists, official Web sites of publishers, and everything in between. Readers and comics industry professionals have realized substantial gains from interactions through Web-based communications such as listservs.

The collaborative nature of fan-based communities has flourished in the free-form nature of digital composition. Media scholar Henry Jenkins describes fan organizations as "expansive self-organizing groups" that are, among other things, "focused around the collective production, debate, and circulation of meanings, interpretations, and fantasies in response to various artifacts of contemporary culture" (2006, 137). In many circumstances, to be a comic book reader today demands that an individual engage not only with an expansive history but also with an expansive community. Jenkins cites Nancy Baym's statement with regard to television soap fans, that "the fan community pools its knowledge because no single fan can know everything necessary to fully appreciate the series" (Baym 1998, 115–116). Baym's observation is equally applicable to the information-pooling practices demonstrated by comic book fans because, for example, no one fan may reasonably be expected to know the forty-plus year history of Spider-Man. Through collaboration, fans act as creators of valuable and labor-intensive information resources in the construction of sites such as http://www.spiderfan.org, which features a vast array of fan-created articles about the character and his foes, original webcomics, and discussion forums. The content offered by fan-generated Web material satiates the same needs that urged the creation of fan magazines, mailing lists, and early conventions. Through these kinds of collaborative processes, readers are now actively employing the affordances of digital technologies in creating persistent, codified hubs that not only represent a collection of information regarding their favorite comic books but also stand on their own as cultural artifacts.

Comic book scholar Pustz predicts that "only established comic readers will find the websites and newsgroups that will be a necessary part of participating in future comics culture, thereby maintaining the borders between comic book fans and the rest of the American popular culture audience" (1999, 198). One topic neglected by Pustz is webcomics. They represent the culmination of reader-generated material—hyperfocused interest groups are, for the first time in large numbers, generating, distributing, and reading exactly the kinds of comics material they want.

Webcomics are comics material made specifically for first publication on the Web, involving no previous print publication. Online versions of print comics are not considered here as webcomics. Early antecedents of webcomics include ANSI art such as the comic *Inspector Dangerfuck* done by an artist who styled himself Eerie.[5] In April 1992, Hans Bjordahl posted his comic strip *Where the Buffalo Roam* to the Internet for the first time using GIF and PostScript formats and made it available to readers through the USENET newsgroup alt.comics.buffalo-roam. His audience at the time was bounded by the limitations of the Internet.

The story of just how Bjordahl came to post his comic is particularly relevant in the context of this chapter. He began drawing his strip in 1987 for the *Colorado Daily* while he was a student at the University of Colorado.[6] He was approached by Herb Morreale, an IT employee of XOR Network Engineering, who urged him to post his comic to the Internet. Morreale remembers thinking "Clarinet is now offering the UPI wire in USENET format, so some of the 'morning paper' is already available (though at a price). So what's missing? Not just *Dear Abby*, but the comics!" (Campbell

2006, under "From Out of the Desert . . .). In Morreale's vision, comics were not acquiring a revolutionary new space. Rather, it was their familiar setting, the newspaper, now being roughly translated to the Internet, which sparked his desire.

The first regularly updated comic on the World Wide Web, *Doctor Fun*, appeared in September 1993 (Allen 2007). As the Web expanded with graphic capabilities, the number of webcomics grew. The Web became a place where artists could publish their work and circumvent the syndicates. The creators of webcomics of the mid-1990s, however, did not yet constitute either a community or a subculture. Webcomics have grown tremendously in popularity over the last several years.

Penny Arcade and *PvP* (an acronym for Player vs. Player), two video game-themed strips, are the examples cited most frequently by the media as representatives of hugely successful webcomics. Both were formed in the 1990s when the World Wide Web was still young, and both now reach millions of readers every month (Grossman 2007). They illustrate the subcultural appeal held by webcomics that transcends the blanket conception of comics. Other popular strips cater to hackers, emo kids, Ph.D. students, and cancer patients. Webcomics do not necessarily rely on the subculture of comics readers, but instead move among, contain, reinforce, and serve the needs of any of a diverse range of communities that exist completely independent of comics.

Already there are numerous works discussing the economic advantages and style flexibility offered to creators by webcomics (Allen 2007; Campbell 2006). Communities of webcomics readers, while still emerging, have grown sufficiently over the past ten years to take on a shape distinct from both the comics community and the general audience for popular culture of all kinds (Fenty, Houp, and Taylor 2005).

In order to be a webcomics reader one has to have access to the Internet and have some navigation and search skills. This suggests that the digital divide might be at play in availability to webcomics because there still are differences across the American population in network access and skills. As with other parts of the Internet, webcomics are available only to those with the means to discover them. Another important question is how readers go about discovering webcomics. There are portals such as Modern Tales and Webcomics Nation, which host webcomics for creators, but making use of them presupposes knowledge of them. There is a chance for a user to become a webcomics reader after serendipitously stumbling upon it. Word of mouth among friends and acquaintances may play an important role in how readers find webcomics. It is a simple matter for the average Internet user to follow a link provided to them in an email from a friend, or to go to a site by entering a URL into a browser. However, so far there is little research available on how individuals become webcomics readers and what the profile of the typical webcomics reader is (typical in terms of both population demographics and degree of technical savvy).

The absence of physical space probably affects webcomics readership. By not being tied to the physical space of the comic book shop, with its racks of material and rows of longboxes, the webcomic may provide a safer "space" for female readers. The comic book shop has a stigma from which the Internet is free. Another way in which webcomics are perhaps "safer" for readers than comic books and comic book shops is the way in which the Internet lends itself to the archiving of old material. As mentioned earlier in this chapter, the uninitiated reader of comics is often unaware of large portions of relevant storyline. To become familiar with the continuity of the average Marvel comic book means acquiring back issues in some form, through a trade paperback or single issues. The Internet allows a reader to browse at home, at the office, on a cell phone, or anywhere else an individual has online access.

There are other open questions about the Internet and comics and webcomic reading. Is online shopping killing the comic book shop? Through sites such as Amazon.com it is easy to buy comics from the comfort of home. The comic book shop is a locus for a community; further investigation into the role of webcomics sites as an online locus of a comics subculture is needed.

Conclusion

In little over a century since the inception of comics, the patterns and practices of American comics readers have undergone major changes. While each major change in the comics industry was accompanied by either a gain or loss in readers, the trend toward an increasingly active reader base continued unabated. As society's views of comics shifted, the industry itself repeatedly demonstrated a propensity for adaptation that allowed it to accommodate broad economic shifts and reader preferences, and survive the vicissitudes of public opinion. So too did the reader adapt and evolve—eventually reaching a point where the practice of reading comics has become nearly synonymous with membership in a distinct subculture. The subculture of comics readers has its own information needs, which readers have taken it upon themselves to satisfy by employing sophisticated strategies of information pooling and sharing. From the neighborhood swap sessions of the children from the 1940s, to the self-published, community organized resources of fanzines and collaborative Web sites, to the community coming together at specialty comic book shops, comics readers have worked together not only to support one another but also to advance the experience of reading comics.

A number of exogenous forces have changed the nature of comics and comics reading, and these changes have changed the information needs and information-seeking behavior of comics readers. The exogenous forces include the rise of immigration and immigrants' settlement in lower-class communities within large cities, the development of the modern newspaper, war, the increases in youth violence and the perceived causes of it, government intervention, and the wide spread of the Internet. There have also been endogenous forces that have shaped the comics reading experience and its information needs and seeking behavior. These endogenous changes include the evolution of new forms of comics (strips, books, webcomics), new forms of distribution (in newspapers, on newsstands, in specialty bookstores, direct marketing, online), and new ways of organizing the industry (trade associations, fan clubs).

Many assume that readership of comics will never again reach the pre-Comics Code heights, but the advent of the World Wide Web and the webcomic may ultimately reverse the medium's fortune. The creation of the Comics Code and the resultant loss of readership drove the comics industry into a serious crisis. The fashion in which comics adapted to reach readers, and invite them into a participatory environment, in the inhospitable environment of the postcode world helped to set comics on the path that would lead them to the specialty comic book shop. Inhabiting the confines of the comic book shop, comic books and their readers would grow slowly, in isolation. While the specialty shop arguably kept the comic book alive, it is not entirely unfair to liken it to an iron lung. Critical attention given to groundbreaking titles in the 1980s sparked a renewed interest in comics. This spark of interest slowly grew, resulting in alternative venues for readers to access comics. Sold in bookstores and circulated by libraries, comics appeared to gradually remove the shackles of the specialty shop. So too, have the information pooling and seeking practices of readers become unhinged from physical space: the reader-created material on the Web guides, enhances, and increasingly enables the interactive practices of the participatory culture of comic readers.

Notes

1. Gordon and Benton are two such scholars. See also Sabin 1993.

2. Some titles that did: *Patsy Walker, Millie the Model,* and *Nellie the Nurse* (Marvel Comics).

3. Also see *Time* magazine (August 22, 1949) in which Macfadden Publications employee Dwight Yellen admits, "No doubt about it—the confession comics have hurtourfieldalot."Seehttp://www.time.com/time/magazine/article/0,9171,800653,00 .html (accessed April 28, 2010).

4. Pustz further explains, "comic book fans who might otherwise be afraid to talk about their hobby for fear of ridicule can go to a comic book shop and find themselves reaffirmed. At the comic shop, reading comics is normal: it is what everyone there does" (1999, 6).

5. Very little information is available about this strip in printed or Web literature. Art using either the ANSI or ASCII character set, however, is familiar to users of the Internet in its early days, and is produced by typing characters on a page in such a way as to create the shape of whatever the artist desires to depict. See, for example, http://en.wikipedia.org/wiki/ANSI_art.

6. See Bjordahl 1987.

Bibliography

Allen, Todd. 2007. *The Economics of Web Comics, 2nd Edition: A Study in Converting Content into Revenue.* Chicago: Indignant Media.

Baym, Nancy. 1998. Talking about Soaps: Communication Practices in a Computer Mediated-Culture. In *Theorizing Fandom: Fans, Subculture, and Identity,* ed. Cheryl Harris and Alison Alexander, 115–116. New York: Hampton Press.

Benton, Mike. 1989. *The Comic Book in America: An Illustrated History.* Dallas, TX: Taylor Publishing.

Berger, Arthur Asa. 1973. *The Comic-Stripped American; What Dick Tracy, Blondie, Daddy Warbucks and Charlie Brown Tell Us about Ourselves.* New York: Walker.

Bjordahl, Hans. 1987. Where the Buffalo Roam: The Story. *The Colorado Daily.* The University of Colorado, Boulder. http://www.shadowculture.com/wtbr/story.html (accessed April 27, 2010).

Campbell, T. 2006. *A History of Webcomics.* San Antonio, TX: Antarctic Press.

Fenty, Sean, Trena Houp, and Laurie Taylor. 2005 Webcomics: The Influence and Continuation of the Comix Revolution. ImageTexT: Interdisciplinary Comics

Studies 1 (2). http://www.english.ufl.edu/imagetext/archives/v1_2/group/ (accessed April 27, 2010).

Gordon, Ian. 1998. *Comic Strips and Consumer Culture: 1890–1945*. Washington, DC: Smithsonian Institution Press.

Grossman, Lev. 2007. New Zip for the Old Strip. *Time* 169 (14): 50–51.

Hajdu, David. 2008. *The Ten-Cent Plague*. New York: Farrar, Straus, and Giroux.

Inge, M. Thomas. 1990. *Comics as Culture*. Jackson: University Press of Mississippi.

Jenkins, Henry. 2006. *Fans, Bloggers, and Gamers: Exploring Participatory Culture*. New York: New York University Press.

MacDonald, Heidi. 2005. Web Comics: Page Clickers to Page Turners. *Publishers Weekly*, December 19. http://www.publishersweekly.com/article/CA6291934.html (accessed April 27, 2010).

McCloud, Scott. 2000. *Reinventing Comics*. New York: HarperPerennial.

Nyberg, Amy Kiste. 1998. *Seal of Approval: The History of the Comics Code*. Jackson: University Press of Mississippi.

Pustz, Matthew. 1999. *Comic Book Culture: Fanboys and True Believers*. Jackson: University Press of Mississippi.

Reid, Calvin. 2005. U.S. Graphic Novel Market Hits $200M. *Publishers Weekly*, April 18. http://www.publishersweekly.com/article/CA525426.html (accessed April 27, 2010).

Sabin, Roger. 1993. *Adult Comics: An Introduction*. London: Routledge.

Sabin, Roger. 1996. *Comics, Comix, and Graphic Novels*. London: Phaidon Press Ltd.

Serchay, David. 1998. Comic Book Collectors: The Serials Librarians of the Home. *Serials Review* 24 (1): 57–70.

Strang, Ruth. 1943. Why Children Read the Comics. *Elementary School Journal* 43 (6): 336–342.

Wolk, David. 2003. Comics Shops Sell Books, Too. *Publishers Weekly*, December 22. http://www.publishersweekly.com/article/CA370769.html.

Wright, Bradford W. 2001. *Comic Book Nation: The Transformation of Youth Culture in America*. Baltimore: The Johns Hopkins University Press.

Zorbaugh, Harvey. 1944. The Comics—There They Stand! *Journal of Educational Sociology* 18 (December): 196–203.

10 Information Exchange and Relationship Maintenance among American Youth: The Case of Text Messaging

Arturo Longoria, Gesse Stark-Smith, and Barbara M. Hayes

Introduction

The ways American youth currently exchange information are shaped by the growing number of media, both digital and analog, available to them. Within the past decade alone, we have seen the quick adoption—and, at times, the equally quick abandonment—of digital media that facilitate information exchange among American youth: instant messaging, chat rooms, LiveJournal, MySpace, Blogger, Facebook, and Twitter, to name a few. These applications have augmented but not replaced older media such as the telephone and handwritten notes, which persist as a means of communication for youth. Our goal is to study not only the different media that facilitate information exchange for American youth but also to examine how the simple act of exchanging information helps maintain existing relationships among communication participants. That is, whether the messages exchanged are important or mundane, youth construct their own social worlds and learn how to navigate through these worlds independently through the act of contacting one another using various media.

This is nothing new or exceptional. Before the Internet and the Web, young people used landline telephones, letters, radios, and other media to communicate with one another. Today, American youth accomplish this task with a wide variety of new technologies—some of which may already be outdated at the time of publication, while others do not yet exist as we write. Spaces such as arcades, restaurants and diners, parks, and clubhouses have also facilitated relationship management and information exchange among American teenagers. Examined in isolation, these communication technologies and the conditions in which they flourish may seem trivial, but they often constitute the space for key developmental processes among youth. These technologies are exogenous to relationships, but their adopters quickly shape them to serve fundamental human needs. Our long-term project is to study relationship maintenance and information exchange from the perspective of the young people participating in these exchanges.

This chapter will focus on a single aspect of this larger topic: text messaging. As the authors will explain in this chapter, we see text messaging as a medium whose use straddles the line between written and verbal, synchronous and asynchronous communication. This medium's potential to combine the strengths of different kinds of communication can make it particularly expressive, and, as such, we find it worthy of study. Future work can further inform the study of information exchange and relationship maintenance through a wide variety of communication media.

Text messaging has become increasingly popular throughout much of the developed world. In many places it has caused concern. Parents and educators worry that it is changing the way the younger generation communicates. Critics in the popular press fear that extensive use of text messaging may impair teenagers' writing abilities, negatively affect their personal interactions, and, along with other new media such as instant messaging and email, bring about unfortunate changes in how we use the English language. These concerns may or may not be warranted, but they demonstrate that there is much to be considered in this new medium. What is it about text messaging that has propelled it to the levels of popularity it has attained? Is there anything really new about this form of communication? In order to push past the popular generalizations about text messaging and to discover more substantive answers to questions such as these, researchers from a variety of disciplines have already begun to explore how the medium functions. While much of this work is excellent and illuminating, it falls short of providing a comprehensive view of text messaging in the United States.

A variety of specific aspects of the medium have already been studied, including the use of stylized language in text messages; the content discussed via text message; and the implications of making oneself available constantly. By and large, however, these studies do not consider text messaging from the perspective of the teenagers and young adults who use it. As yet, we do not have a full account of the development and use of text messaging in the United States that contextualizes the medium within the social worlds of Americans teenagers. To that end, this chapter will examine the work that has already been done on text messaging, seeking not only to explain how rich such a seemingly limiting medium can be but also to increase awareness of research that is left to be done.

In order to give a broad account of text messaging, this chapter is divided into four sections. We begin with a history of the medium itself.

This first section will locate text messaging as a user-driven medium of communication by exploring how the original intents of its creators were subverted by the uses the public made of it. Though the technology was not initially projected to be a popular one, it was quickly and widely adopted in certain areas. We will enumerate some of the motivations people initially may have had to use this seemingly unattractive medium.

Once we have established an historical context in which to view the development of text messaging, the second section will examine text messaging behaviors, particularly among teenagers and young adults. We will look at the language people use in text messages—what is known as "text speak"—and its implications. Furthermore, we will address the norms that govern how people text: in particular, the etiquette that shapes the frequency and formality of their text messaging.

Next we will recount the primary social functions of text messaging. This third section will explore how text messaging may serve to help maintain relationships within one's social network. We will also touch on how text messaging has facilitated a more up-to-the-minute approach to coordinating social plans. For example, being in constant contact enables a user to alert a friend instantly if, say, a restaurant is full or a car has broken down.

The fourth and final section of this chapter will take a step back and discuss text messaging more abstractly. We will consider what effects text messaging appears to have on users of this medium, again with a focus on teenagers and young adults. What differences in social structure and social interaction seem to follow from text messaging? Are there aspects of text messaging that change the way people who use it interact? This section will focus on how teenagers use the medium both as a way to build independence from the family sphere and to signal inclusion and exclusion within their social circles. Finally, we end our discussion by analyzing the possible social ramifications of being constantly available to others.

Before our discussion begins, let us briefly explain our interest in this field and the approach we took in looking at text messaging. Our inquiry was spurred by a news article published in the *New York Times* (Steinhauer and Holson 2008) that discussed the physical dangers text messaging creates when its users are distracted during attention-critical tasks. To get a better impression of text messaging itself, we then reviewed other articles in the popular press and industry publications, but nothing we found was especially illuminating. We were hoping to find something more substantive that would speak to the way text messaging functions as a part of social interaction. As we approached the scholarly literature about text messaging, we also referred to broader theoretical literature in fields such as sociolinguistics, history, and communication studies to help frame our research. Our approach has been to focus on text messaging inasmuch as it relates to interpersonal communication in social circles and peer groups and so we do not treat commercial uses of the medium (e.g., voting on reality television shows) or uses of the medium that are not direct, one-to-one exchanges (e.g., mass text messaging or tweeting via text). We feel that our literature review draws attention to the gaps in the existing body of research, particularly in the story of text messaging among American youth. These gaps are to be expected when examining such a recent communication medium. We hope that the synthesis presented here prompts further research and is a step toward a more well-rounded understanding of text messaging as a medium for interpersonal communication.

History of Text Messaging

The idea to equip cell phones with a short messaging service (SMS, i.e., text messaging) emerged in Europe in the 1980s as telecommunication operators sought to formalize a pan-European mobile technology. In 1987, fifteen telecommunication operators from thirteen European nations signed a memorandum of understanding in which they committed to provide interoperability and standards of use via network protocols set up by the Global System for Mobile Communication (GSM, originally known as the *Groupe Spécial Mobile*). Four working parties were established within the GSM to work toward creating an interoperable mobile technology. The fourth of these—initially known as the Implementation of Data and Telematic Services Experts Group but later renamed simply WP4, Working Party 4—appointed Finn Trosby as the chairman of a drafting group responsible for the specifications of a short messaging data service.

In his account of the design and development of this messaging service, Trosby (2004) makes it plain that the medium was not intended to be commercially significant nor did its designers expect or foresee the popularity it has now attained. Years before, Trosby chaired the group that essentially designed text messaging. Telenor, a telecommunications company based in Norway, had conducted some of the initial GSM research into a messaging service and concluded that short messages might be useful for alerting subscribers of pending messages or other service alerts since cell phone users may frequently be out of their coverage area or have their phones turned off.

Trosby outlines the technical details that led to the specification of a 160-character limit for each message—a key feature of text messaging that we discuss in more detail later in this chapter. The drafting group chaired by Trosby considered a couple of already established GSM signaling systems that could be used to implement a short messaging service. The system that the drafting group ultimately chose had been designed to transport small amounts of information. Upon analyzing and optimizing the programming code that handled data transfers within the signaling system, Trosby's team found that there was enough space for a few more than 160 characters per message. A decision was made to round down to 160 characters for the sake of having a simpler number.

As both Gerard Goggin and Trosby discuss, though the technical aspects were in place for text messaging to be a service offered by all GSM-compliant telecommunication operators, it was still unclear how the service would be used. In fact, "the decidedly unsexy SMS" was of no real concern to the telecommunications industry, which was committed to being viewed as highly modern and futuristic (Taylor and Vincent 2005, 79). Because the industry was not expecting the text messaging service to become popular or for it to be lucrative, the prepay or pay-as-you-go packages offered by European cell phone companies did not make any mention of text message use in their billing plans. This created a loophole that essentially allowed pay-as-you-go subscribers to send text messages for free.

At this time in Europe (the early 1990s), teenagers were the widespread adopters of pay-as-you-go cell phone subscriptions (Ling 2004, 9). Their subsequent exploitation of the free text messaging service can be seen as the impetus for the initial popularity of text messaging in Europe and explains the demographic of its primary users. Companies were quick to realize what was happening, though; they closed the loophole and began to charge customers a fee to send a text message. Despite the introduction of a billing system, the charges for sending a text message remained lower than the cost of a phone call, which partially explains why younger cell phone users on a limited budget continued to use text messaging (Kasesniemi 2003, 21).

It is important to note, though, that the development of text messaging in Europe was facilitated by one important factor not present in the United States: the concerted effort by telecommunication operators to create an interoperable messaging system. The United States was not involved in the GSM's initial efforts to create a short messaging service nor was there a move by cell phone companies to create an interoperable network. The American cell phone market was not influenced by the European models created by the GSM such as the "calling party pays" billing system (Ling 2004, 15). The cost of a phone call or a text message in the United States is shared between the parties involved and a text message cannot be rejected like a phone call. By contrast, in Europe the initiator of the cell phone call or text message is fully responsible for the cost. These major differences between the two telecommunication industries demonstrate that factors driving American adoption of text messaging are not parallel to those driving adoption in Europe. Nevertheless, we feel that text messaging culture in Europe can inform the use of text messaging in the United States because, although these differences may have resulted in a delayed and slower uptake of the medium in the United States, its usage levels continue to rise.

The underlying question we seek to address is why, despite these differences, the level of text messaging in the United States is approaching, if it has not already reached, the level of use present in Europe. We argue that there may be something inherent in the medium that appeals to its users. Text messaging must meet certain interpersonal communication needs if its adoption is growing at such impressive rates *in the absence of* a significant economic incentive. Perhaps adopters of text messaging prefer to periodically and quickly touch base with others without the need for a telephone call that could prolong the length of the interaction. In fact, some researchers have credited text messaging with recreating the brief, passing communication exchanges that were possible and prevalent in smaller, preindustrial communities (Fox 2001). With this in mind, we can see texters using the medium not only to transmit pertinent information but also to briefly connect with members of their social circles in ways that other media have not allowed—quickly and with no restrictions on mobility. (Others may argue that email or instant messaging already allow for these quick bursts of communication, but these technologies require that one be near a computer whereas a cell phone allows mobility as long as the texter is in his or her specific coverage area.) Assuming there is a desire to touch base or have brief exchanges with others asynchronously (or synchronously, depending on one's use of text messaging, an idea explored later in this chapter), we see text messaging as an ideal medium for this type of interpersonal communication. It is an exchange limited in length that can be accomplished from nearly any location.

Text Messaging Behavior

One of the most commonly remarked upon features of text messaging is the language that is used in the messages. This language is often marked by a terse and direct style, which bypasses social niceties, progressing directly to the topic at hand (Crystal 2008). In addition to the overall style of the exchange, the language of text messaging is often characterized by distinctive atypical orthography, which is commonly referred to as "text speak." Text speak is most often associated with a combination of abbreviations, acronyms, and emoticons, which is similar to language popularly used in instant messaging and email.

Some criticize this sort of language as incapable of expressing complex emotions or conveying the nuances of meaning that are so crucial to interpersonal connection. There is worry that the frequent use of such language will erode people's ability to use formal or more traditional language in other realms of their lives. Many writers in both the popular and academic press seem quick to jump to these dire conclusions about the language of text messaging. However, studies that delve deeper into the communication practice of teenagers who use text messaging tell a different story.

Before exploring the possibilities this style affords, let us first look at how and why it may have developed. At first glance, it seems that the stylistic elements of text message language may be a direct byproduct of the confines of the medium. Within 160 characters, one does not have room for pleasantries or for traditional and cumbersome spellings, and the limitations of the medium may be responsible for the initial development of this style. However, as Crispin Thurlow notes in his article "Generation Txt?," explaining the use of language in text messaging is significantly more complex than that: "The linguistic and communicative practices of text messages emerge from a particular combination of technological affordances, contextual variables and interpersonal priorities" (Thurlow 2003a). This reflects the idea that the use of text speak may be driven by additional factors beyond the technological limitations of the medium.

Thurlow suggests that the language of text messaging follows three main imperatives or "sociolinguistic maxims": "brevity and speed," "paralinguistic restitution" and "phonological approximation." These maxims illustrate the way text messaging has developed into a mode of expression that can be seen as straddling the line between written and spoken communication (Goggin 2006). Though asynchronous in theory, text messaging often takes place as part of a prolonged exchange between participants who reply almost instantaneously. As Eija-Liisa Kasesniemi notes in her seminal study of texting in Finland, *Mobile Messages: Young People and a New Communication Culture*, text messaging functions more like speech than email does because in text messaging, "the sender may receive an answer in less than a minute" (Kasesniemi 2003, 172). Kasesniemi likens text messaging to instant messaging, noting, however, that some of her participants prefer text messaging since it is more portable. "One undeniable strength of text messaging is the constant availability of the device and the immediacy of communication enabled by this" (Ibid., 171). In this view, text messaging may be as close to speech as two parties can get without actually speaking.

This similarity to spoken language is the root of Thurlow's maxims for the style of text messaging. Brevity facilitates the normal back and forth in conversation. The strikingly synchronous nature of text messaging supports this characteristic of verbal exchanges. The other two elements similarly infuse text messages with much of the flavor of speech. Texters use paralinguistic restitution by simulating expressive features of speech—like tone—with orthographic features (e.g., emoticons, ellipses, capitalization) and phonological approximation to directly echo forms of speech, often to indicate a certain level of informality or to use a regional variation (e.g., gonna, wanna).

As Crystal (2008) argues in his book *Txtng: The Gr8 Db8*, the use of text speak may also reflect the impulse people naturally have to play with language. He traces many of the techniques used in text messaging back to older systems of shorthand and, notably, to linguistic puzzles like rebuses. He proposes this ludic impulse as an explanation for why people sometimes opt to use text speak in situations where it may not be the easiest (or fastest) choice. Crystal also makes the important point that text messaging does not compel users to employ text speak or to adopt other conventions of the form. In fact, in Thurlow's analysis of the text messages of Welsh university students, he found that atypical orthography and typology only accounted for roughly 20 percent of the overall message contents.

When texters do decide to use the linguistic trappings that are often identified with the medium, it may be not only to enrich the conversation with speech-like conveyers of meaning but also to signal a connection to a social group or a particular person. This can happen when we are face to face or during interactions in which we are not present. As Mary Chayko explains, "When we form close relationships, our thoughts and actions can become synchronized. This can take the form of our unconscious mimicking of another's facial expressions and movements or of synchronizing our speech patterns to another's. Online, a kind of textual synchronicity can develop whereby people gravitate toward a similar syntax" (Chayko 2008, 28). Chayko is speaking of online communication, but there is no obvious reason why such an idea would not hold true for text messaging. She goes on to point out that this type of synchronicity helps people feel closer to one another in terms of shared emotion rather than physical proximity.

In addition to fostering or cementing a one-on-one relationship, language choices in text messages may signal membership in a wider social group. Work in sociolinguistics on style and variation among adolescent peer groups is particularly pertinent to this concept (e.g., Eckert 2000, 2005; Irvine 2001). As Penelope Eckert notes, "A key role of language in the creation and maintenance of peer groups is in creating and supporting differentiation. This is done both in conversation and through the construction of linguistic style—a way of deploying linguistic resources that sets the community off from others, and that is designed to represent the community's character" (2005, 106). This may be especially true for youth as "adolescents lead all other age groups in linguistic change" (Chambers 1995, quoted in Eckert 2005).

Another important aspect of the way people text is the etiquette that governs the quality and frequency of the exchange. As previously mentioned, Kasesniemi found that many of her research participants reply immediately to messages, treating the medium primarily as a synchronous one. As Ditte Laursen explores in her article "Please Reply! The Replying Norm in Adolescent SMS Communication," when the receiver of a message does not reply quickly, the lag is usually interpreted as having a definite meaning, for example indicating an intentional distance between the sender and the receiver or perhaps merely signaling a transmission error (Laursen 2005). Kasesniemi also notes that in addition to a quick reply, senders of messages expect to "be replied to using the same medium, with a text message" (Kasesniemi 2003, 197). Text messaging is, in this way, considered distinctive. Though not unrelated to other media, text messaging is distinguished by the language used within messages as well as in the etiquette that has been developed around the sending and receiving of messages. These features of the medium are ripe for meaning making because their consistency affords opportunities for specific interpretations (i.e., a break in etiquette is assigned a particular meaning).

Looking at Internet forums helps us understand how the etiquette of text messaging is being expressed. Discussions in these forums reflect a clear understanding of what constitutes proper and improper behavior. Derek Rake, a writer for *EzineArticles*, provides the following advice for effective text messaging in dating:

Principle #1 "First impression" SMS counts! OK, if you have met this girl and successfully gotten her number, then it is crucial to make the first SMS count. You will need to come off in the SMS as *full of character* . . .

Principle #2   Flirt! Flirt! Flirt! The SMS is designed to do one thing well—which is
to *flirt!* Be suggestive in your texts and get her to guess your intentions . . .
Principle #3   Go for the kill. Once you have developed the rapport using the two
principles above, then casually ask her out. (Rake 2009)

(Note that it is still necessary to secure a telephone number before using
text messaging as a means of increasing contact.)

Though the etiquette of text messaging may still be evolving, certain
parameters have been clearly defined. The complaint that "He/she texts
me constantly!" is found frequently on Internet forums and blogs, illustrat-
ing that there is a clear etiquette being breached. The existence of such
rules allows for intentional breaks in etiquette as conveyors of meaning,
such as when someone does not respond to a text in order to intentionally
signal the cooling of a relationship.

The behavior of texters illustrates the balance that is struck between the
technological influences and social influences that affect how a medium
of communication is used. The language that has developed around the
use of texting is not solely determined by the medium itself, nor is it purely
a function of the social world of texters (see Ling 2004 for a more detailed
discussion of social determinism and technological determinism). It is easy
to overestimate the role that technology plays in shaping how a medium
is used. As Alex S. Taylor and Richard Harper note in their article, "The
Gift of the Gab?," overly deterministic views "ignore the fact that when
technologies are adopted they become part of an already established social
context and this context often shapes the use of technologies" (Taylor and
Harper 2003, 267). The use of text messaging demonstrates the interplay
between the technology implicit within the medium and the social worlds
of those who use it.

Critics of text messaging cite the behavior of its users (from quick reply
norms to nonstandard orthography) as evidence that the medium cannot
express complicated ideas and is, instead, restricted to simple exchanges
for social coordination. This perception may be caused by a variety of
factors, but one particularly strong motivator for such beliefs is the general
distrust and dismissal of adolescent culture. As Eckert puts it, "Both adolescents and their language are commonly viewed as sloppy, rebellious
and irresponsible" (2005, 93). Teenagers are frequently regarded as difficult
to understand. It is easy for adults, who view teenage culture from the
outside, to reject it as incapable of producing meaningful interaction. As
Sherry Turkle puts it in her exploration of the ways new media are affecting communication, "Although the culture that grows up around the cell
is a talk culture . . . it is not necessarily a culture in which talk contributes

to self-reflection" (Turkle 2006, 225). Turkle bases this assessment both on the frequency of communication via cell phone (which she feels limits an individual's time for processing his or her feelings independently) and on the language used in text messages, such as emoticons, which she believes are "meant to communicate an emotional state quickly. They are not meant to open a dialogue about complexity of feeling" (Ibid.). It is unclear what the basis for such claims is, aside from Turkle's own personal observations of text message communication.

Those authors who do attempt to look at text messaging among youth from the inside, which is to say with an ethnographic approach that often relies heavily on interviews, paint a much more expansive picture of what is possible within the confines of this medium. As Kasesniemi reports, "Teens' messages quite frequently deal with the more fundamental aspects of life such as grief, the illness of a family member or death" (Kasesniemi 2003, 167). In many ways, strategies employed in text speak such as paralinguistic restitution may provide an enhanced ability to convey sensitive or complex topics by allowing for the nuances and subtlety commonly found in speech. These features of spoken communication may combine with attractive features of writing such as the ability to edit (if only within a short window of time) and the easy retention of a somewhat permanent record of what was said.

Furthermore, when actually asked, teenagers do not consider communication a simplistic act. As Thurlow notes, summarizing work done on this subject, "young people seem consistently to construe communication in complex and varied terms, recognizing its basis in relational-contextual variables such as motivation, power inequalities and empathic insight" (Thurlow 2003b, 54). Stereotypes about teenagers may influence the way we view their communication and, specifically, the way we view text messaging. As researchers, we strive to find ways to understand the phenomenological experience of those we are researching. At this point, text messaging has inspired more popular commentary than actual research on user experience. As Thurlow puts it, "however much may be said by adults about communication in adolescence, there has been very little research that has sought to understand communication from the perspective of young people themselves" (Ibid., 53). As academic work on text messaging moves forward, it will be important to try to change this. For example, the authors of *Hanging Out, Messing Around, Geeking Out: Living and Learning with New Media* (Mizuko et al. 2009), state in their introduction that they "take youth seriously as actors in their own social worlds." Studies of text messaging that carefully consider the reasons why youth have adopted this technology are needed. There is little data available on the ways it may affect communication. This, coupled with the fact that this medium is most popular among the younger generation, makes it all too easy to speculate without grounding the resultant ideas either in statistical data or in a rich, qualitative anthropological approach.

Functions of Text Messaging

Looking at the practice of text messaging from a slightly different angle, we can continue to explore the role it plays in teenagers' social lives. Many of the studies about text messaging categorize compiled messages based on social function. The categorization schemes differ between the studies, but by looking at those of Thurlow and Kasesniemi, we can achieve a basic sense of how this technique works and what it reveals. Thurlow creates a detailed set of divisions, which essentially forms a spectrum ranging from those messages primarily oriented toward information exchange ("Informational Practical" and "Practical Arrangement") to those that serve the purpose of relationship maintenance ("Friendship Maintenance," "Romantic," "Salutary"). This distinction is reflected in the broad division that Kasesniemi makes between those messages that can be seen as "complying with everyday information needs" and those that are "initiating and maintaining social relationships" (Kasesniemi 2003, 165).

Kasesniemi also presents a third category, "pure pastime and entertainment," though we suggest that this category can largely be subsumed under relationship maintenance as it would be difficult to make a case that messages sent as pastime are not also serving to strengthen relationships. Further, Kasesniemi does not robustly establish the basis for this category. As she notes, it can be difficult to separate messages along these lines as many of them may serve dual functions. For example, expressing a desire to coordinate socially may strengthen a relationship. Acknowledging that there may be overlap, we can proceed to explore how text messaging supports these functions. Broadly speaking, a historical approach to an emergent phenomenon always seeks to locate this new trend within the wider pattern of what has come before it. This approach is skeptical of exceptionalism and therefore wary of the idea that a new technology can fundamentally change the way people communicate. With that said, if text messaging can claim a part in the creation of a solid change in communication patterns, it is due to the ad hoc coordination it makes possible. As Rich Ling puts it, "Arguably, the greatest social consequence arising from the adoption of the mobile telephone is that it challenges mechanical timekeeping as a way of coordinating everyday activities" (Ling 2004, 69). Here, Ling is speaking of mobile communication in general, but the point may be even sharper for the use of text messaging, which is often seen as requiring less attention than a phone call and can be easily used to "progressively refine an activity" (Ibid., 72). This capacity to target and refine allows for flexibility in emerging situations (e.g., traffic jams that might cause a late arrival) without a loss of courtesy. As Kasesniemi puts it, "many Finns are openly indulging in the luxury of perpetual connectedness" (Kasesniemi 2003, 25). The implications for social relationships of this ability for easy, up-to-the-minute communication will be explored in the next section.

In addition to ad hoc coordination, the immediacy of texting lends itself to the maintenance of social relationships. Much evidence suggests that teenagers mostly text those who are already in their social group (Kasesniemi 2003). As Richard Harper explains, "teens constrain their social worlds to those who have a right to contact them and exclude those who don't" (Harper 2005, 110). This preference for using text messaging to interact with those already a part of their world situates the medium as a useful tool for relationship maintenance and development.

The interviews that Kasesniemi documents with the Finnish teenagers in her study are full of references to this type of communication and give particular emphasis to another aspect of text messaging that makes it attractive for them: the privacy it affords. As Goggin mentions, "as texting is clandestine by nature, it enables secret dialogue away from parental eyes" (Elwood-Clayton, quoted in Goggin 2006, 76). This brings to the fore a feature of text messaging that is perhaps unique in the world of mobile communication. Although a call via cell phone may afford the advantages associated with mobility, the call can always be overheard. Not only does text messaging allow for the privacy necessary for intimacy, but also the relative silence of the medium lends itself to situations where communication to a nonpresent friend is not socially acceptable. (See, for example, Kasesniemi's account of a teenager text messaging while in church.)

The initiation and support of primary intimate relationships is a major subset of text messaging and a driving force behind a great deal of texting activity by teenagers and young adults. Historically, young Americans have always had some control over their romantic relationships. Texting has emerged as an important tool for supporting and maintaining intimate relationships in contemporary America.

David Newman and Liz Grauerholz, in their book *Sociology of Families*, provide a brief history of courtship and dating in the United States that highlights the ways youth have negotiated intimate relationships apart from the family sphere. They trace the progression of the different ways teenagers have found an independent place to create or maintain a romantic relationship. Even when courtship rituals took place in the home, there was still some amount of freedom from parental supervision. For example, the colonial practice of bundling allowed a young unmarried couple to spend the night together in a bed in her parent's home while still discouraging intimate contact (Newman and Grauerholz 2002, 226–227). As technology and social norms evolved, teenagers were presented with increasing opportunities to enact these rituals free from the constraints of parental supervision.

"By the 1930s dating had pretty much moved out of the home and into the public world. In the process, family surveillance was replaced with peer supervision and judgment. Dating now involved activities and places that were virtually off limit to adults" (Ibid, 228). Since then, parents have been more and more marginalized in the process of courtship. Today, many recognize and worry about the amount of unsupervised communication that texting offers, particularly for those too young to drive. This historical perspective gives context to the ways text messaging functions within romantic relationships among youth.

In addition to intimate relationships, an examination of the main functions of text messaging reveals how this medium is used within the wider social interactions of youth. It also begins to establish the richness and possibilities of this mode of communication. What may seem like gossip or terse social coordination can play an important role in maintaining and strengthening relationships. Teenagers devote a great deal of time to talking. A review of forty-five studies of average daily time use by adolescents found that postindustrial-era, schooled American teenagers spend between two and three hours a day talking. It is not clear if texting supplants some of this time or occurs in addition to these hours (Larsen 2001). As Leo Hendry and Marion Kloep explain, "Being with others and being in communication is in itself a very important leisure activity for young people. Through these interactions, they receive social support, offer feedback on new skills, give advice and comfort and provide security just by being 'there' for friends. Through such apparently meaningless conversations, social bonds are negotiated and strengthened" (Hendry and Kloep 2005, 166). This is true for the conversations that teenagers have over lunch or at night over the phone. We believe that it is also true of their text messaging communication, though the newness and perceived limitations of the medium may make it more difficult to see the part it plays in facilitating the key interactions that sustain social ties.

Social Implications of Text Messaging

Whereas the previous two sections have discussed the decisions one makes when using the medium—linguistic style, tone—and the practical functions it serves—ad hoc coordination, relationship maintenance—the present section takes a step back and examines the observable social implications of the use of text messaging. In other words, how do the medium and its affordances fit into the lives of texters? Does the act of text messaging mean anything in itself, and if so, what role does it play in texters' lives? To this end, we focus our discussion on three interrelated issues—that text messaging is a way by which young people can feel independent and emancipated from their guardians; how with this newfound independence young people are able to signal inclusion and exclusion in their social circles via text messaging; and the effects of being "always on."

In his research into cell phone culture among teenagers and young adults, Ling notes that teenagers are often given the responsibility of paying for some, if not all, of their cell phone charges (Ling 2004, 112). Teenagers and young adults, many of whom may not have had such economic responsibilities before, are forced to prioritize their expenses. Such responsibility adds a dimension of ownership not previously experienced by many teenagers and young adults: their economic responsibilities become a validation and confirmation of their adulthood (Ibid., 118). In these cases, then, it is not difficult to see how teenagers could feel that this medium of communication affords them a personal space in which they can easily exclude their parents or guardians because they are paying for the service themselves. It is important to note here, though, that these observations were made in Europe and the circumstances may not be the same in the United States. Most American cell phone service providers have, within the last decade, begun offering family plans through which a family can easily add multiple cell phones at a more affordable price than opening a separate account for a teenager. Because of this, and the lack of demand for pay-as-you-go cell phones in the United States, many American teenagers and young adults may not experience this economic emancipation from their families via cell phones.

Even though American teenagers may not have the same level of financial responsibility for their cell phone use, they do utilize mobile communication as a platform for emerging independence. This is true in spite of the fact that many teenagers may have initially acquired their phones because of parental concern for their safety. As Ling (among many other researchers) noted in his interviews with parents, many attribute the decision to permit their child to have a cell phone to safety reasons. Parents hope that if anything were to happen while their son or daughter were out, a cell phone would enable the teenager to reach out for help. While this is very valid, it is not reasonable to expect that the teenager will solely use the cell phone for this purpose. In fact, while a cell phone may increase a parent's ability to be more aware of their child's whereabouts, that same cell phone also causes the parent to become more out of touch with the child's social circle (Kopomaa 2005). This is because much of the communication that occurs within the social circle migrates to the cell phone as there is no longer a need for, say, communication via the parentally monitored landline house telephone. Mobile communication and text messaging allow for a more direct and private form of interpersonal communication than is possible with the house telephone. And even within the realm of mobile communication, text messaging affords an even greater "sphere of freedom" (Ibid., 156) as its use cannot be overheard, unlike a cell phone call. Researchers have found that many young texters report that their family would be surprised to know what is being discussed via text message—a clear sign that text messaging allows a space in which a teenager can present a self-image different from the one they present to their family members (Reid and Reid 2005, 116).

This sense of independence, coupled with the ability to communicate frequently and from any location—a topic we cover in more depth at the end of this section—may also affect the structure of teenagers' social groups. Researchers claim that as communication within a teen's social circle migrates to an always-on mode, the social circle becomes more tightly bound. Cell phones are ubiquitous and cell phone users have a habit of always carrying their phones and rarely, if ever, turning them off. These factors lead to a type of interpersonal communication Ling describes as an "anytime-anywhere-for-whatever-reason type of access to other members of the peer group" (Ling 2004, 18). This immediate, pervasive access creates a "perpetual connectedness" (Kasesniemi 2003, 25) within a given social group that transcends copresence. That is, teenagers whose lives may still be tightly scheduled or supervised by parental figures can feel that they are never truly away from their peers irrespective of physical proximity. And in this environment wherein peers are merely a few thumb taps away, a tightly bound social group dynamic is created. Teenagers are able to express themselves as they wish in this private space, which not only gives them a sense of emancipation and independence but also allows them to negotiate peer relationships privately at any given time.

One of the ways teenagers negotiate peer relationships is by deciding with whom they maintain contact. That is, the ability to decide who is and who is not allowed in one's social circle comes with being able to function independently in this mobile sphere of freedom. Within peer groups, young people use text messaging to signal inclusion and exclusion in their immediate social circles. Researchers have noted that there exists a strict correlation between the members of a teenager or young adult's social group and the entries in his or her cell phone address book (Kasesniemi 2003; Harper 2005; Elwood-Clayton 2005). Harper's idea that teenagers "constrain their social worlds to those who have a right to contact them and exclude those who don't" (Harper 2005, 110) manifests itself through the cell phone address book. In short, if a person is listed in a young person's address book, they are included in the social group; otherwise, the person is considered an outsider. As romantic relationships or friendships end, there may be a ritualized process of removing the person from the address book in order to symbolically mark the end of the relationship (Elwood-Clayton 2005, 216). This primarily accomplishes two things. First, it is a signal that the other person has officially been excluded from the young adult's social group. Second, by deleting the other person's address book entry, the other person is, in essence, "anonymized." Most cell phones will display the name under which a person's phone number is saved in the address book when a call or text message is received. If no entry exists, the actual phone number is displayed, and in a culture where phone numbers are not memorized, this provides no method of identifying a caller or texter. These "summons from an anonymous person" (Harper 2005, 109–110) are no longer viewed as a message from a social peer but rather an attempt at communication from an anonymous person who is clearly not part of the social circle. Were the teenager to respond with a "who r u??" text message, for example, this could be construed as a personal insult or slight (Elwood-Clayton 2005, 216).

Of course, these methods of inclusion and exclusion presuppose that all members of the social group have cell phones and are able to send text messages, which is not always the case. We were not able to find any existing research that fully treated the inclusion or exclusion of peers who did not own or have access to cell phones. One study of deaf teenagers in Norway did tangentially touch on this subject, though, by describing how the deaf teenagers interviewed for the study felt included in the coordination of social plans among peers because of text messaging (Bakken 2005). Before they had access to text messaging, these teenagers had been restricted to using specialized landline telephones in order to communicate telephonically. This restriction severely limited their ability to communicate with others because these specialized telephones are not widely available. The introduction of text messaging to their lives allowed the deaf teenagers to communicate through a medium that requires no special equipment. They were freely able to communicate with peers who did not have hearing impairments, and as a result, they reported feeling more included in the arrangement of social plans. Conversely, they reported feeling out of touch or excluded when they did not have access to their cell phones because they were not able to communicate via text message. From this, we can reasonably extrapolate that the lack of access to text messaging or mobile communication within a social group that uses the medium as a primary means of contact correlates to feelings of exclusion for those without the ability to text. Further research into this topic could help to substantiate this claim.

Clearly, though, not all texters are teenagers or young adults. While our work focuses on this age group, as its members are the primary users of text messaging, this demographic is changing as text messaging is adopted by the greater population. We have previously touched on the always-on aspect of the medium in relation to teenagers and their newfound independence, but does being always on create the same sense of connectedness and social bonding in older texters? While text messaging may have softened our conceptions of time and allowed for what Ling refers to as "microcoordination"—a type of ad hoc coordination where loose plans are made and then concretized via mobile communication as the hour nears (Ling 2004)—many adults report feeling overwhelmed because of this constant availability. To a teenager, it may seem completely normal to be "on" despite doing homework or having a family meal or while at a part-time job, but many adults describe this as "mentally and emotionally draining" (Chayko 2008, 134). In her work on mobile and online connectedness, Chayko explains this as being a generational issue: younger people who grew up with mobile communication may not make the same distinctions between "work," "home," and "leisure" that older adults were exposed to growing up. These observations lead us to ask about the future social norms of today's teenagers. As this generation grows up, will they continue to embrace always-on technology because they are so accustomed to it? Or is their comfort with it mostly a function of the typical adolescent desire for near-constant peer interaction? Will they grow up into adults with communication norms similar to our own or will their style of interaction continue to be marked by their use of text messaging and other technologies that facilitate being always on? These questions will not be easily or quickly answered, but we believe that this topic is ripe for research that can help shed light on the broader question of how technologies fit into the communication behaviors of their users.

For the purposes of our current work, we argue that if so many people are adopting text messaging and making themselves always available despite experiencing a certain amount of mental and emotional stress, there must be something inherently worthwhile in the medium. Chayko has studied the always-on factor of mobile connectedness and she presents a simple answer to the question of why we make ourselves available constantly: comfort. There may be a mental and emotional drain to always being available for the older generation, but in the research we have studied, teenagers and young adults rarely, if ever, voice this complaint. Perhaps feeling overwhelmed is part of the learning curve when adopting a new, always-on technology. Adults may be willing to endure this because the comfort one gets from his or her social circle surpasses the annoyance of being always on. When you find yourself in a stressful or hostile environment, it may be a relief to know that your social peers are only a few thumb taps away.

In making yourself always available, you also accept that your peers are doing the same. Chayko explains that this helps confirm that "certain relationships persist" (Chayko 2008, 117). In fact, many of the people Chayko interviewed reported feeling that bonds between family members strengthened because of constant mobile communication. Chayko explains that this is because humans are comforted in knowing that others are around and there for them despite not being physically present.

As Fox further posits, it is not simply knowing that your social circle is never more than a text message away: communicating via text message itself signals a return to the type of interpersonal communication that existed in smaller, more tightly-knit communities in preindustrial times (Fox 2001). The fleeting preindustrial "village green conversations" Fox describes are not so much about content but rather about a quick connection with a member of one's network that reinforces a social bond. In other words, it is not so much the content of the message that impacts the participants, but, rather, the act of sending and receiving a message that helps reinforce the bonds within a social circle. When viewed in this light, it is not hard to understand why teenagers and young adults have been quick to adopt text messaging as a form of interpersonal communication.

This is not to say that unlimited use of text messaging is always a positive force in teenagers' lives. Some researchers suggest that a constant connection can actually hinder a teenager's independence because they never really leave home—they are always tapped into and can ask for advice from a parent (Turkle 2006). Text messaging has only recently achieved popularity in the United States, and so it is still unclear whether such claims are founded in truth. In-depth, interdisciplinary research into text messaging and any possible developmental effects it may have on teenagers who rely on it excessively is needed. At this point in the story of text messaging, we, along with many other researchers studying new media, simply aim to show that sensationalistic concerns in the popular press about text messaging are currently unsubstantiated. Rather than condemning teenagers and young adults for being tethered to their cell phones, we must examine why this is taking place and study it in its own context. As danah boyd reminds us (2006), we must be careful not to fall prey to unwarranted moral panics. Instead, we should seek to actually understand what teenagers are doing and why they do it.

Conclusion

After examining the behavior associated with text messaging, the functions the medium can serve, and the potential long-term social effects of these uses, a picture of it has emerged as a medium that provides, on the one hand, opportunities for meaning-rich communications within particular social groups. On the other hand, many questions about the effects of text messaging remain and further research, particularly into the use of text messaging in the United States, can flesh out the picture we currently have. More information is necessary to make determinations about the impact this medium will have on our society. In his popular book *Bowling Alone*, about the social changes our country has gone through in the last century, Robert Putnam writes, "When the history of the twentieth century is written with greater perspective than we now enjoy, the impact of technology on communications and leisure will almost surely be a major theme" (Putnam 2000, 216). Putnam may very well be right about this, but just what this impact will look like is open for debate.

Beyond questions about the quality of the impact of technology, it seems premature to present it as either purely positive or purely negative. It is unclear how influential it will really be. Although some researchers, like Turkle, feel that the use of text messaging and mobile communication is significantly changing the way we communicate, others like Fox see this phenomenon as a throwback, returning us to earlier styles of communication. Does technology always push us in new directions, or, by enabling us to communicate casually with distant friends and family, can it in some ways bring us back to a preindustrial form of socialization?

Clearly, this is not a question that can be answered definitively, but we believe that research into text messaging and other new forms of communication can help to illuminate the situation. In this chapter, we have argued that such research must consider the medium at hand from the perspective of its users. As Thurlow and others have posited, an outside perspective often fails to grasp the essentials of an experience for people inside the community of use and this may be particularly true when the community is populated by adolescents, a group easily dismissed by those of us who no longer are adolescents ourselves.

Furthermore, given our focus on interpersonal communication and our belief that text messaging may be particularly suited for this type of communication, it is crucial to consider the particular circumstances of this social interaction. Meaning in social interactions is often determined by the particular relationship between the actors. Nikolas Coupland explores this point in his article "Language, Situation, and the Relational Self" as a part of his critique of the overly general categorizations of sociolinguistics: "It is at the level of the person—an individual's personal and social identity—that our social judgments of speech styles reside" (Coupland 2001, 199). Coupland writes this in reference to features of language determined by regional and dialectical differences that may vary in meaning based on the context of their use, but his point can be extended to stylistic choices like when and to what extent one employs text speak. The use of abbreviated or non-standard language may have a very different meaning within a particular exchange (between two adult friends who have developed an ironic use of text speak, for example) than it would within a different exchange (between a teenager and her mother, say). The use of an ethnographic approach could help to add information about this medium's use from the perspective of insiders and also to illuminate the particular factors that come into play in specific instances of such use. A respect for the importance of the individual context of a situation can provide specific knowledge to complement a general picture abstracted from many cases, allowing for a more meaningful account of text messaging as a medium for communication.

Text messaging is a relatively recent form of communication, and though it has reached a notable level of popularity, it is difficult to know if it will persist at this level for some time to come. As we have noted throughout this chapter, text messaging is not necessarily the easiest or most logical communication medium. Though it appears to be an integral part of the fabric of communication within particular social groups, it remains to be seen whether it will continue to be so prominent. Even if the story of text messaging is nearing its end, the insight that we can gain from studying it will continue to have value. We see text messaging as a fascinating medium in and of itself, with particular metacommunicative properties that are worthy of examination, but its chief value lies in the insights it can offer into the larger story of the way people relate with communication technologies.

A further exploration of other communication media commonly used by American youth would complement the understanding of text messaging that we have offered in this chapter. Investigating the similarities and differences in how youth use distinct media to share information and maintain relationships can help illuminate the interplay between technology and interpersonal communication. As we have stressed throughout this chapter, it is important that further research examine communication from the perspective of the people involved, thereby creating a richer portrait of the way people interact with communication technologies.

Bibliography

Bakken, Frøydis. 2005. SMS Use Among Deaf Teens and Young Adults in Norway. In *The Inside Text: Social, Cultural and Design Perspectives on SMS*, ed. R. Harper, L. Palen, and A. Taylor, 161–174. Norwell, MA: Springer.

boyd, danah. 2006. Identity Production in a Networked Culture: Why Youth Heart MySpace. Talk, American Association for the Advancement of Science, St. Louis, MO, February 19. http://www.danah.org/papers/AAAS2006.html (accessed June 28, 2009).

Chambers, J. K. 1995. *Sociolinguistic Theory: Linguistic Variation and Its Social Significance*. Oxford, UK: Blackwell Publishers.

Chayko, Mary. 2008. *Portable Communities: The Social Dynamics of Online and Mobile Connectedness*. Albany: State University of New York Press.

Coupland, Nikolas. 2001. Language, Situation, and the Relational Self: Theorizing Dialect-style in Sociolinguistics. In *Style and Sociolinguistic Variation*, ed. Penelope Eckert and John R. Rickford, 185–210. New York: Cambridge University Press.

Crystal, David. 2008. *Txtng: The Gr8 Db8*. New York: Oxford University Press.

Eckert, Penelope. 2000. *Linguistic Variation as Social Practice*. Malden, MA: Blackwell Publishers.

Eckert, Penelope. 2005. Stylistic Practice and the Adolescent Social Order. In *Talking Adolescence: Perspectives on Communication in the Teenage Years*, ed. Angie

Williams and Crispin Thurlow, vol. 3 of Language as Social Action, 93–110. New York: Peter Lang.

Elwood-Clayton, Bella. 2005. Desire and Loathing in the Cyber Philippines. In *The Inside Text: Social, Cultural and Design Perspectives on SMS*, ed. R. Harper, L. Palen, and A. Taylor, 195–219. Norwell, MA: Springer.

Fox, Kate. 2001. Evolution, Alienation and Gossip: The Role of Mobile Telecommunications in the 21st Century, Social Issues Research Centre. http://www.sirc.org/publik/gossip.shtml (accessed June 28, 2009).

Goggin, Gerard. 2006. *Cell Phone Culture: Mobile Technology in Everyday Life*. New York: Routledge.

Harper, Richard. 2005. From Teenage Life to Victorian Morals and Back: Technological Change and Teenage Life. In *Thumb Culture: The Meaning of Mobile Phones for Society*, ed. Peter Glotz, Stefan Bertschi, and Chris Locke, 101–116. New Brunswick, NJ: Transaction Publishers.

Hendry, Leo, and Marion Kloep. 2005. "Talkin', Doin' and Bein' with Friends": Leisure and Communication in Adolescence. In *Talking Adolescence: Perspectives on Communication in the Teenage Years*, ed. Angie Williams and Crispin Thurlow, vol. 3 of Language as Social Action, 163–184. New York: Peter Lang.

Irvine, Judith T. 2001. Style as Distinctiveness: The Culture and Ideology of Linguistic Differentiation. In *Style and Sociolinguistic Variation*, ed. Penelope Eckert and John R. Rickford, 21–43. New York: Cambridge University Press.

Ito, Mizuko, et al. 2009. *Hanging Out, Messing Around, Geeking Out: Living and Learning with New Media*. http://digitalyouth.ischool.berkeley.edu/report (accessed June 28, 2009).

Kasesniemi, Eija-Liisa. 2003. *Mobile Messages: Young People and a New Communication Culture*. Tampere, Finland: Tampere University Press.

Kopomaa, Timo. 2005. The Breakthrough of Text Messaging in Finland. In *The Inside Text: Social, Cultural and Design Perspectives on SMS*, ed. R. Harper, L. Palen, and A. Taylor, 147–159. Norwell, MA: Springer.

Larsen, Reed W. 2001. How U.S. Children and Adolescents Spend Time: What It Does (and Doesn't) Tell Us about Their Development. *Current Directions in Psychological Science* 10 (5) (October): 160–164.

Laursen, Ditte. 2005. Please Reply! The Replying Norm in Adolescent SMS Communication. In *The Inside Text: Social, Cultural and Design Perspectives on SMS*, ed. R. Harper, L. Palen, and A. Taylor, 53–73. New York: Springer.

Ling, Rich. 2004. *The Mobile Communication: The Cell Phone's Impact on Society*. San Francisco: Morgan Kaufmann Publishers.

Ling, Rich. 2005. Mobile Communication vis-à-vis Teen Emancipation, Peer Group Integration and Deviance. In *The Inside Text: Social, Cultural and Design Perspectives on SMS*, ed. R. Harper, L. Palen, and A. Taylor, 175–193. Norwell, MA: Springer.

Newman, David M., and Liz Grauerholz. 2002. *Sociology of Families*. Thousand Oaks, CA: SAGE Publications.

Putnam, Robert D. 2000. *Bowling Alone*. New York: Simon & Schuster.

Rake, Derek. 2009. How to Seduce a Girl Via Texting—Three Principles. EzineArticles, March 9. http://ezinearticles.com/?How-to-Seduce-a-Girl-Via-Texting—Three-Principles&id=2079028 (accessed November 18, 2009).

Reid, Donna J., and Fraser J. M. Reid. 2005. Textmates and Text Circles: Insights into the Social Ecology of SMS Text Messaging. In *Mobile World: Past, Present and Future*, ed. Lynne Hamill and Amparo Lasen, 105–118. New York: Springer.

Steinhauer, Jennifer, and Laura M. Holson. 2008. As Text Messages Fly, Danger Lurks. *New York Times*, September 20.

Taylor, Alex S., and Richard Harper. 2003. The Gift of the Gab?: A Design Oriented Sociology of Young People's Use of Mobiles. *Computer Supported Cooperative Work* 12: 267–296.

Taylor, Alex S., and Jane Vincent. 2005. An SMS History. In *Mobile World: Past, Present and Future*, ed. Lynne Hamill and Amparo Lasen, 75–92. New York: Springer.

Thurlow, Crispin. 2003a. Generation Txt?: Exposing the Sociolinguistics of Young People's Text-Messaging. *Discourse Analysis Online* 1 (1). http://extra.shu.ac.uk/daol/articles/v1/n1/a3/thurlow2002003-paper.html (accessed December 10, 2008).

Thurlow, Crispin. 2003b. Teenagers in Communication, Teenagers on Communication. *Journal of Language and Social Psychology* 22: 50–57.

Thurlow, Crispin. 2005. Deconstructing Adolescent Communication. In *Talking Adolescence: Perspectives on Communication in the Teenage Years*, ed. Angie Williams and Crispin Thurlow, vol. 3 of Language as Social Action, 1–20. New York: Peter Lan.

Thurlow, Crispin, and Angie Williams. 2005. From Apprehension to Awareness: Toward More Critical Understandings of Young People's Communication Experiences. In *Talking Adolescence: Perspectives on Communication in the Teenage Years*, ed. Angie Williams and Crispin Thurlow, vol. 3 of Language as Social Action, 53–71. New York: Peter Lang.

Trosby, Finn. 2004. SMS, The Strange Duckling of GSM. *Telektronikk* 3: 187–194.

Turkle, Sherry. 2006. Tethering. In *Sensorium: Embodied Experience, Technology, and Contemporary Art*, ed. Caroline A. Jones, 220–226. Cambridge, MA: List Visual Art Center.

11 Conclusion

William Aspray and Barbara M. Hayes

This book has explored nine common activities in everyday American life. They can be grouped in four categories: the purchase of goods and services (cars, airline travel), the pursuit of hobbies (sports, genealogy, comics reading, and cooking), participation in various sectors of society (government information, philanthropy), and communication and relationship maintenance (text messaging). These activities illustrate just a few of the many information behaviors important in everyday American life.

Each of these activities has its own set of exogenous and endogenous forces that shape the information sought and sources used over time. Nevertheless, there are some common shaping forces—particularly exogenous ones—that emerge as significant across several of the case studies. One of the major forces has been government regulation. Some examples include the Freedom of Information Act, which has mandated that more government-generated information be made available to ordinary citizens; government regulations about safety and environmental impact, which have dictated elements of the manufacturing of automobiles since the late 1960s; deregulation of the airline industry, which has radically changed the travel options available to American citizens; and the decade census and visa systems, which have provided valuable information to individuals tracing their genealogies.

Another exogenous force that emerged as critical to several of the everyday information activities discussed here is the advent of modern media. Over time, the information environment for all of the activities examined in this volume has become increasingly varied and rich. We have seen how each new media technology—from radio to television to the Internet—was explored almost immediately for commercial purposes. Examples are advertising automobiles and delivering sports information. Each new medium has extended the information available for particular hobbies and made them much more popular. Examples are the impact of the *Roots* television series on genealogical research and *The French Chef* hosted by Julia Child on gourmet cooking. The widespread adoption of Web-based technologies has made government information available to citizens at any time of the day or night. User-driven new digital media applications such as texting have also influenced the ways in which American youth keep in touch with one another and develop their personal identities.

War, and the concomitant surge of patriotism, has had a dramatic impact on information and information sources in everyday American life. Examples include suspension of car production and most domestic airline travel during World War II, which forced car owners to gather information about how to keep older cars running and obtain sufficient fuel coupons; the patriotism in the aftermath of that war that shaped the content of comics reading; and the ways in which every American war has caused Americans to change their normal patterns of charitable giving and dig deeper to assist individuals and families who have suffered the effects of war. Religion has also been a significant shaping force. American religious institutions have served as powerful brokers of information about philanthropy. They encourage their congregations to engage in philanthropy and filter information, helping givers select worthy targets. Religious requirements were also an influence on the rise of genealogy as a hobby in America.

Many additional exogenous forces have shaped American information-seeking behaviors in everyday life. Terrorism and its impact from 9/11 onward required Americans to gather new information on airline travel and necessitated the development of new venues for gathering government information on the threat and official responses to it. The increasing mobility of Americans has impacted the use of cars as well as the ability to trace one's ancestors. Economic downturns such as the Depression of the 1930s curtailed many discretionary purchases and activities and expanded the need for philanthropy. Other exogenous forces noted include the influence of extremely wealthy businessmen-turned-philanthropists on the practice of philanthropy by the rest of the population and major national celebrations such as America's Bicentennial in 1976, which led to a craze for genealogy.

Claims are often made for the revolutionary character of the Internet. Indeed, the Internet has caused many changes in information-seeking behavior in everyday American life. First and foremost, it has served as an information enhancer, giving sharply increased amounts of information. Instant, extensive information abounds about sports statistics, the costs of cars, the financial situation of nonprofit organizations, recipes, cooking instruction, and genealogical information. Searches sometimes supply too much information, necessitating the development of new skills with which Internet users can prioritize information and judge its quality and trustworthiness.

The Internet has removed some of the information asymmetry between buyers and sellers. Americans who have Internet access and are willing to spend the time can now approach purchases armed with pricing information derived from Web sites that equalize the balance of power in financial transactions. Car buyers can begin a negotiation with the "blue book" price in hand as well as detailed information about available promotions, discounts, and incentives. Air travelers can search Web sites for the lowest possible fares and even receive information on whether those fares are likely to change in ensuing weeks. The Internet now provides tools that enable donors to research the trustworthiness, efficiency, and effectiveness of charities seeking donations.

In addition, the Internet serves as a highly modifiable delivery vehicle for information and entertainment. Games and comics are now available in many different online formats. Social networking sites and their associated applications, available via cell phone, instant messaging, and texting, have become major communication devices especially for young people. These sites, adopted quickly and somewhat unquestioningly by the young, have caused their parents to seek new sources of information about the impact of these technologies on their children.

The Internet has transformed passive activities into participatory activities, such as comic book reading or fantasy sports. Civic-minded individuals can engage in philanthropy in more active, direct ways by utilizing common, inexpensive Internet tools to gather support for their favorite charities or even start new ones. Amateur genealogists can explore and add to online archives. Aspiring cooks can watch streaming videos that provide step-by-step instruction.

The Internet has also helped to change basic business models that touch people every day. It has been a factor in the disintermediation of the travel industry. Airline travelers can book flights directly, bypassing traditional travel agencies. This change, however, has shifted time and cost burdens to the traveler and reduced the availability of face-to-face help for such tasks as booking and advice on hotels. Often, Americans using the Internet to gather information must parse what they find there without the help of an intermediary.

In some ways, however, the Internet has not made much difference: it simply provides new ways of delivering the same information. For example, while the tools of genealogy are different and are augmented by the Internet, the basic information sought has not changed. Although customization has increased choices, car buyers today are asking many of the same questions that they asked eighty years ago. Other questions raised by car buyers, such as fuel economy or financing options, are prompted by other societal changes and not by the Internet. Although the Internet provides a powerful new tool in gathering information, people still ask their friends and family or local garage mechanic for advice about car buying and they continue to read the *Consumer Reports* reviews as they have for fifty years.

The investigations in this volume are probably most pertinent to middle-class, white America. However, many middle-class American everyday activities are similar, regardless of race or ethnicity. Blacks were much less likely to pursue genealogy until the *Roots* miniseries aired on television, but today blacks are about as active in this hobby as whites. Youth who use text messaging do not vary markedly by race, ethnicity, family income, or geographical location. Philanthropy is greater in middle- and upper-class families, but it reaches into every economic sector. Lower income Americans may choose volunteering over cash donations. Elfreda Chatman (1992, 1996) developed a theory of information poverty that examines the information needs and information-seeking behavior of marginalized populations. If we were to expand our study and differentiate the activities of different demographic types, this theory might be useful in studying the poor, those living in rural communities, and the aged, all of whom might be considered information outsiders.

Some Final Methodological Considerations

The method employed in this book is useful for identifying change in the questions asked and the sources used over time. It suggests but does not truly answer the issues about information quality (e.g., what one learns at the car dealership or from advertisers) or user satisfaction with the information that one obtains. There are many questions about the information process that are at best partly or indirectly addressed by this approach. These include, for example, task definition, the relative importance of the different information sources, the relative trustworthiness of different information sources, the relationship between particular questions and particular information sources, the mental models and cognitive styles with which a person filters and processes the various information sources, the reason why individuals seek out particular sources among others, the degree to which users unconsciously gain information that supports a particular everyday activity in the course of their everyday life, the temporal sequencing of information questions or information sources in the course of carrying out some particular everyday life activity, the synthesis of information from various sources, or any of the other psychological aspects of information-seeking behavior (e.g., affect or emotional state). Fortunately, there are many other theories and methods of information-seeking behavior that can be used in conjunction with the method employed in this book.

We want to return here to several important scholarly works that we mentioned in the introduction. As we indicated there, one of the most important scholars of everyday life is Michel de Certeau. His work on the practice of everyday life (de Certeau 1984) focuses on its informal and mundane activities and positions them as a contest between tactics of individuals and groups to deal with the strategies of powerful institutions such as those of business, government, and formal education. Certeau did not himself discuss information-seeking behavior in everyday life, but a number of scholars have attempted to build on his work to do so, as is reviewed in Spink and Cole 2001.

Another of the scholarly movements mentioned in the introduction is reader response theory. It was developed in the 1970s and 1980s in opposition to the new criticism that had been dominant in literary studies before this. Reader response theory emphasizes the agency of the reader, that is, the role the reader plays in giving meaning to the text instead of having the meaning reside in the text itself. (See, e.g., Suleiman and Crosman 1980.) This movement was similar in character to the Birmingham School of media studies that occurred at about the same time, and has been followed in recent years by the interest in fandom, bloggers, and participation of fans as producers as well as consumers of media that is studied by Henry Jenkins and others (Schulman 1993; Jenkins 2006). These theories fit well with our chapter on comics reading and its evolution from passive reading to active engagement.

Four other general theories of information behavior are closely related to the work in this book, and it would be profitable to use each of them in addition to the method used here. The first is Reijo Savolainen's theory of everyday life (Savolainen 1995, 1999, 2002). It addresses how social and cultural factors affect people's preferences and uses of information in everyday-life activities. The work is based in Pierre Bordieu's notion of *habitus*, which is a systematic way of thinking that is internalized by individuals, and considers such issues as how people budget their time, consume goods, and pursue hobbies (Bordieu 1984).

A second example of a relevant general theory is Diane Sonnenwald's concept of the information horizon, which gives one a way of thinking about the sources that an individual uses to gather information in carrying out an everyday activity (Sonnenwald 1999; Sonnenwald, Wildemuth, and Harmon 2001). According to this theory, for each individual and each activity there is an information horizon within which one seeks information. The information horizon includes one's social network, subject matter experts, formal information paper and electronic sources, and knowledge of the world through experimentation and observation.

Allen Foster's theory of nonlinear information seeking is a third example of a theory that fits well with the everyday-life information seeking as described in this book (Foster 2004). Many theories of information-seeking behavior frame the search as a discrete activity, with a beginning and an end, and with clearly identifiable sequenced stages in between. This nonlinear theory does not fix the beginning or end, and it admits repeats of searches, concurrent searches, and introduction of new searches at any time, until the user is satisfied with the results. The theory aims to give an understanding of how searches are opened, how the user orients herself in the midst of a search, and how information that is gathered is consolidated.

Finally, there is institutional ethnography, which is a method developed by Dorothy Smith out of feminist thought and elaborated by DeVault and McCoy (Smith 1987; DeVault and McCoy 2002). This method is similar to the rubric used in our study, except for the fact that it examines a single point in time using an ethnographic rather than a historical approach. According to this approach, one identifies an experience, identifies institutional processes that shape it, and examines how these institutional processes undergird the experience.

Three theories about search strategies and other ways of gaining information are particularly appropriate in the study of everyday life. The first is Williamson's ecological theory of human information behavior, which is useful in understanding how people gather information in everyday-life activities not only through purposeful searching but also through simply being aware (Williamson 1998). For example, car buyers often know a lot about the system of dealerships, model years, and manufacturers and their products through passive knowledge.

A second theory about search that is relevant here is optimal foraging, which was introduced by anthropologists and adopted by information scientists to understand how information seekers decide how to use their resources (Winterhalder and Smith 1981; Cronin and Hert 1995;

Sandstrom 1994). The theory involves an information seeker who can choose among alternative search methods, a currency (such as information value) that enables the seeker to conduct a cost–benefit analysis, constraints that limit the search behavior, and a strategy for carrying out the search. The theory has also been applied in understanding information-seeking tools, and one can see how it can be used to understand the development of online tools for booking airline reservations in an era when there is not an affordable intermediary to help sift through the numerous travel options.

The last of these relevant search theories is Marcia Bates's concept of cherrypicking, which was developed to look at the ways in which nonlibrarian users carry out searches that are not the single structured query that had been postulated in the early years of information science, but included an iterative and emergent approach that drew upon various databases and interfaces that involve such actions as looking through journal runs or searching the materials on a library shelf adjacent to the material identified by the structured search (Bates 1989). The process used in everyday-life domains clearly has this iterative and emergent character, but it draws on sources that include but go well beyond those found in a library or in an online database. For example, the everyday user may well scan the other books or magazines near the specific one sought in a bookstore; or might read part of a book and then formulate a question for a friend, family member, or fellow hobbyist. Similarly, the user might explore the first Web site suggested by a search engine query, but also several of the secondary (adjacent) sites returned.

One of the most important exogenous forces shaping information gathering is changes in the media. Three general theories have direct bearing on the influence of the media, and in particular the influence of the Internet, on everyday information-seeking behavior. The first is Everett Rogers's diffusion theory (Rogers 2003). It is a useful way of thinking about the role of media (newspapers, radio, television, the Internet) as sources of information for everyday-life activities. Rogers considers the suitability of information gained through these sources by considering their relative advantage (over the information provided in other sources), compatibility (with other sources), complexity (of the information provided, how readily it is understood), trialability (whether we can test out information before incorporating it), and observability (the degree to which the value of the information is readily apparent).

The second is Savolainen's concept of network competence. Although his theory goes beyond the topics covered in this book toward a theory of social cognition, the concept of network competence helps us to understand the reasons why youth are so enamored with interactive and online platforms for reading comics, maintaining identity through text messaging, and buying airline tickets online. This theory defines competence as "the mastery of four major areas: knowledge of information resources available on the Internet, skilled use of the ICT tools to access information, judgment of the relevance of information, and communication" (Savolainen 2002, 211, as quoted in Fisher, Erdelez, and McKenchie 2005, 55).

Finally, there is George Zipf's "Principle of Least Effort," which states that people are more likely to seek information sources that are accessible and easy to use even though there may be ones of higher quality. Zipf's principle explains some of the attractiveness of using the Internet (Zipf 1949).

There are several general theories of information-seeking behavior that inform the material in specific chapters. We identify those here, in the order in which the chapters appear in the book. Chapter 3 (philanthropy) applies Elfreda Chatman's concept of "small world" to philanthropy (Chatman 1999). A small world is a group of people, associated in some way, who share a common worldview. This is a part of Chatman's larger theory of "Life in the Round," an approach to information behavior in everyday life, which considers social norms, social types, and worldviews as well as small worlds. Fisher's concept of information grounds can also be applied to chapter 3 (Pettigrew 1999, 2000). This concept is a useful one for understanding how people exchange information when getting together in a place for an instrumental purpose that is other than information sharing, for example at a church bazaar to raise funds for the local animal shelter. The purpose of the meeting is a social activity, but information flows as a byproduct of informal interaction. Thus one might learn at the church bazaar about the nature of various charities that are not part of the purpose of the bazaar.

The concept of communities of practice and situated learning are useful concepts for chapters 5 and 6. According to these theories, knowledge is created and retained in communities of practice, and individuals learn by being participants in these communities (Lave and Wenger 1991; Wenger 1998). Hobbyists, for example, genealogists or knitters or sports fans, learn from practicing their hobbies with other hobbyists.

Chapter 5 (genealogy) and especially chapter 7 (gourmet cooking) are informed by the theoretical work on serious leisure. Founded by Robert Stebbins in 1982 and pursued more recently by the author of our gourmet cooking chapter, Jenna Hartel, serious leisure examines the skills, knowledge, and experience that one dedicates oneself to acquiring in order to pursue this activity in one's leisure time. Although the study is interdisciplinary, the main method is ethnographic (Stebbins 2001).

The work of Chapman and Newell in chapter 8 on government information seeking has close connection to the work of Marcella and Baxter on information interchange (Marcella and Baxter 1999; Marcella, Baxter, and Moore 2002). This latter work examines the dichotomy between the perspectives of the government information provider and the user. In fact, Marcella and Baxter have applied this work to freedom of information legislation in the United Kingdom.

Finally, chapter 10 (text messaging) is closely aligned with social positioning theory. This is a theory that was introduced by the psychologist W. Hollway in 1982 and has been developed by several scholars (van Langenhove and Harre 1999). It is a postmodern theory that contends that identity is socially constructed, contextual, and highly individualized. The theory argues that individuals are actively involved in developing their own identities.

This volume has examined changes over time in both information questions and information sources used in the prosecution of everyday life activities in the United States. The authors have attempted to give a more intimate, everyday perspective to information-seeking behavior, reaching into American history and American homes in an effort to understand its social context. This more intimate perspective and study is just beginning to appear in a new and evolving literature that focuses on information activities at home. The authors hope these modest attempts will spur additional contributions to this complex topic, which is deserving of further study. Everyday activities whose information-seeking behaviors seem particularly worthy of further study include food, clothing, and shelter; getting an education and finding a job; obtaining health care and financial advice; and finding a mate.

Bibliography

Bates, M. J. 1989. The Design of Browsing and Berrypicking Techniques for the Online Search Interface. *Online Review* 13 (5): 407–424. http://www.gseis.ucla.edu/faculty/bates/berrypicking.html (accessed May 3, 2010).

Bordieu, P. 1984. *Distinction: A Social Critique of the Judgment of Taste*. London: Routledge.

de Certeau, M. 1984. *The Practice of Everyday Life*. Berkeley: University of California Press.

Chatman, E. A. 1992. *The Information World of Aging Women.* Westport, CT: Greenwood.

Chatman, E. A. 1996. The Impoverished Life-World of Outsiders. *Journal of the American Society for Information Science* 47: 193–206.

Chatman, E. A. 1999. A Theory of Life in the Round. *Journal of the American Society for Information Science* 50: 207–217.

Cronin, B., and C. A. Hert. 1995. Scholarly Foraging and Network Discovery Tools. *Journal of Documentation* 51: 388–403.

DeVault, M. L., and L. McCoy. 2002. Institutional Ethnography: Using Interviews to Investigate Ruling Relations. In *Handbook of Interview Research: Context and Methods*, ed. J. F. Gubrium and J. A. Holstein, 751–776. Thousand Oaks, CA: SAGE.

Fisher, Karen E., Sandra Erdelez, and Lynne McKenchie, eds. 2005. *Theories of Information Behavior.* ASIS&T Monograph Series. Medford, NJ: Information Today.

Foster, A. E. 2004. A Non-linear Model of Information Seeking Behavior. *Journal of the American Society for Information Science and Technology* 55: 228–237.

Jenkins, H. 2006. *Fans, Bloggers, and Gamers.* New York: New York University Press.

Lave, J., and E. Wenger. 1991. *Situated Learning: Legitimate Peripheral Participation.* Cambridge, UK: Cambridge University Press.

Marcella, R., and G. Baxter. 1999. The Information Needs and the Information Seeking Behaviour of a National Sample of the Population in the United Kingdom. *Journal of Documentation* 55 (2): 159–183.

Marcella, R., G. Baxter, and N. Moore. 2002. Theoretical and Methodological Approaches to the Study of Information Need in the Context of the Impact of New Information and Communication Technologies on the Communication of Parliamentary Information. *Journal of Documentation* 58 (2): 185–210.

Pettigrew, K. E. (now Fisher). 1999. Waiting for Chiropody: Contextual Results from an Ethnographic Study of the Information Behavior Among Attendees at Community Clinics. *Information Processing & Management* 35: 801–817.

Pettigrew, K. E. 2000. Lay Information Provision in Community Settings: How Community Health Nurses Disseminate Human Services Information to the Elderly. *Library Quarterly* 70: 47–85.

Rogers, E. M. 2003. *Diffusion of Innovations.* 5th ed. New York: Free Press.

Sandstrom, P. E. 1994. An Optimal Foraging Approach to Information Seeking and Use. *Library Quarterly* 64: 414–449.

Savolainen, R. 1995. Everyday Life Information Seeking: Approaching Information Seeking in the 'Way of Life.' *Library & Information Science Research* 17: 259–294.

Savolainen, R. 1999. Seeking and Using Information from the Internet: The Context of Non-Work Use. In *Exploring the Contexts of Information Behaviour*, ed. T. D. Wilson and K. Allen, 356–370. London: Taylor Graham.

Savolainen, R. 2002. Network Competence and Information Seeking on the Internet: From Definitions towards a Social Cognitive Model. *Journal of Documentation* 58: 211–226.

Schulman, Norman. 1993. Conditions of Their Own Making: An Intellectual History of the Centre for Contemporary Cultural Studies at the University of Birmingham. *Canadian Journal of Communication* 18 (1): 51–73.

Smith, D. E. 1987. *The Everyday World as Problematic: A Feminist Sociology*. Boston: Northeastern University Press.

Sonnenwald, D. H. 1999. Evolving Perspectives of Human Information Behavior: Contexts, Situations, Social Networks and Information Horizons. In *Exploring the Contexts of Information Behaviour*, ed. T. D. Wilson and D. K. Allen, 176–190. London: Taylor Graham.

Sonnenwald, D. H., B. M. Wildemuth, and G. Harmon. 2001. A Research Method Using the Concept of Information Horizons: An Example from a Study of Lower Socio-economic Students' Information Behavior. *The New Review of Information Behaviour Research* 2: 65–86.

Spink, A., and C. Cole. 2001. Introduction to the Special Issue: Everyday Life Information Seeking Research. *Library & Information Science Research* 23 (4): 301–304.

Stebbins, R. A. 2001. *New Directions in the Theory and Research of Serious Leisure*. Lewiston, NY: Edwin Mellen Press.

Suleiman, S. R., and I. Crosman, eds. 1980. *The Reader in the Text: Essays on Audience and Interpretation*. Princeton: Princeton University Press.

van Langenhove, L., and R. Harre, eds. 1999. *Positioning Theory: Moral Contexts of Intentional Action*. Oxford: Blackwell.

Wenger, E. 1998. *Communities of Practice: Learning, Meaning, and Identity*. Cambridge, UK: Cambridge University Press.

Williamson, K. 1998. Discovered by Chance: The Role of Incidental Information Acquisition in an Ecological Model of Information Use. *Library & Information Science Research* 20 (1): 23–40.

Winterhalder, B., and E. A. Smith, eds. 1981. *Hunter-Gatherer Foraging Strategies*. Chicago: University of Chicago Press.

Zipf, G. K. 1949. *Human Behavior and the Principle of Least Effort*. Cambridge, MA: Addison-Wesley.

Editors and Contributors

William Aspray and Barbara M. Hayes are editing their second book together. Their first book, *Health Informatics: A Patient-Centered Approach to Diabetes*, is scheduled for publication by MIT Press in 2010. Aspray is the Bill and Lewis Suit Professor of Information Technologies in the School of Information at the University of Texas at Austin. He has formerly taught at Harvard University, Indiana University at Bloomington, University of Pennsylvania, Virginia Tech, and Williams College. He has also held leadership positions in the Charles Babbage Institute, IEEE Center for the History of Electrical Engineering, and Computing Research Association. Author or editor of approximately twenty books, his two most recently published works have each won a Choice Award for best academic books of the year: *Women and Information Technology* (edited with Joanne Cohoon, MIT Press, 2006) and *The Internet and American Business* (edited with Paul Ceruzzi, MIT Press, 2008).

Barbara M. Hayes is associate dean for administration and planning at the Indiana University School of Informatics, Indianapolis campus. She earned an MS in informatics and also holds an MSW in social work. She teaches organizational informatics, social informatics, IT project management, and a service learning course Computing for a Cause. Hayes has pursued information and communications studies throughout her career. Beginning as a professional journalist, she worked as an editorial writer for a large city newspaper. After earning her MSW, she provided patient-physician communication and liaison services in nonprofit medical settings. She pursued an informatics degree to explore ways in which digital technology might be used to deliver information to medical professionals and their patients. Her career in informatics has focused on the social aspects of computing.

Gary Chapman is director of The 21st Century Project at the Lyndon B. Johnson School of Public Affairs at the University of Texas at Austin, a project dedicated to expanding public participation in the development of new goals for science and technology policy in the post-Cold War era and specializing in the social implications and trends of new developments in information technologies and telecommunications. Chapman was formerly the executive director of the national public interest organization Computer Professionals for Social Responsibility. He writes extensively, and has published in such venues as *The New York Times*, *The New Republic*, *Technology Review*, *The Los Angeles Times*, *The Washington Post*, and *The Bulletin of the Atomic Scientists*, among others.

James W. Cortada   is a member of the IBM Institute for Business Value. He conducts research on the history and management of information technology, with particular emphasis on its effects in business and American society at large. Recent publications include *Making the Information Society* (Pearson, 2002), which discusses the role of information in American society; a three-volume study of how computing was used in over forty industries over the past half century, *The Digital Hand* (Oxford University Press, 2004–2007); and most recently a study of how Americans embrace information technology, *How Societies Embrace Information Technology* (Computer Society, 2009).

Nathan Ensmenger   is an assistant professor in history and sociology of science at the University of Pennsylvania. His current research interests are aimed at reintegrating the history of the "information revolution"—very broadly defined to encompass a wide range of nineteenth- and twentieth-century scientific, technological, and social developments—into mainstream American social and cultural history. In addition to his work on the social and cultural history of software and software workers, he has studied the disciplinary history of artificial intelligence and artificial life; the formation of a distinctive computing subculture and programming "aesthetic"; and the crucial and often misunderstood role of women in computing. He has also developed and taught courses on the computer and Internet "revolutions," and on the relationship between technological innovation and social change. His book *The Computer Boys* will soon appear with the MIT Press.

Jenna Hartel   is an assistant professor in the Faculty of Information at the University of Toronto. Her PhD dissertation was an ethnographic study of the information aspects of gourmet cooking as a hobby. She is a theorist, methodologist, and historian of library and information studies. Her empirically based work is organized around the question: What is the nature of information in the pleasures of life?, which she investigates through the concatenated study of *serious leisure* realms, which are crossroads of information and enjoyment.

Rachel D. Little received a masters degree in information studies in 2010 at the University of Texas at Austin School of Information, where she was a Senior Tarlton Fellow at Tarlton Law Library (University of Texas School of Law). She received her JD from the University of Houston Law Center in 2005 and practiced immigration law for two and a half years before returning to school. Before law school Little worked for five years with immigrant rights and domestic violence prevention organizations in Washington, DC, and El Paso, Texas, including Ayuda, Inc., the NOW Legal Defense and Education Fund (now known as Legal Momentum), and Las Americas Immigrant Advocacy Center. Little graduated from the University of Puget Sound in 1997 with a bachelor's degree in international political economy.

Arturo Longoria received a masters degree in information studies in 2010 from the University of Texas at Austin. He received his undergraduate degree from Columbia University. He has served as an intern in the reference department at the University of Texas at Austin main library, has interned at SUNY Albany's libraries, focusing on public services and reference, and worked as a public services assistant at Huston-Tillotson University in Austin. Before graduate school Longoria worked in the information technology field as a programmer analyst. He plans to pursue a career as a reference librarian in an academic library.

Sara Metz is currently working as a technology consultant for an educational software company, developing instructional resources and pursuing research into future technologies. She received her BA from Trinity University and her MS in Information Science from the University of Texas at Austin. She is active in several professional organizations, including the American Library Association and the Special Libraries Association.

Beth Nettels received her MA in classics from the University of Toronto in 2008 and her MS in information studies at the University of Texas at Austin. She has experience working with medieval manuscripts and early printed books. She has worked with Congressional papers in the archives of the Robert J. Dole Institute for Politics and in the Archives of American Mathematics at the Briscoe Center for American History.

Angela Newell is a PhD candidate in public policy at the Lyndon B. Johnson School of Public Affairs at the University of Texas at Austin. She received her MS from Carnegie Mellon. Her research focuses on government information systems, small decision-making groups, and the implementation of interactive Internet tools. Prior to graduate school, Newell was a scientific researcher contracted to the Department of Energy. She is also the co-creator of an e-budget system for the Pittsburgh School District and researcher of CEMINA: Internet Radio Razil for the World Bank. Newell served as a two-year AmeriCorps member and as a National Service Fellow. She received two national awards, including the Josten's Our Town Award, for her work in community development.

Jameson Otto received a MS from the School of Information at the University of Texas at Austin in 2010 and a BA in history and English from Southwestern University in 2007. Since 2007 he has worked as a legal assistant for the intellectual property firm Meyertons, Hood, Kivlin, Kowert & Goetzel, P.C. While at MHKKG he has managed several large digitization and information organization projects. He also maintains the firm's law library. He was the 2006 Division III Texas State Champion in cross-country.

George Royer received his JD from the University of Alabama School of Law in 2005 and his MS in information studies at the University of Texas School of Information in 2010. He began doctoral study at the University of Texas in 2010. After receiving his law license, Royer practiced in the areas of mass torts and class actions with the Birmingham, Alabama-based firm of Whatley Drake & Kallas, LLC. Royer has also worked extensively with media archives, including the Texas Archive of the Moving Image in Austin and the Academy for Motion Picture Arts and Sciences Film Archive in Hollywood, California. He is researching issues related to digital media, the media industry, and participatory culture.

Gesse Stark-Smith received the MS in information studies program at the University of Texas at Austin and her bachelor's degree from Macalester College. She has served as a teaching and research assistant with Professor William Aspray and an intern in the reference department of the main library of the University of Texas at Austin. She has previously worked in library circulation. Before beginning her master's candidacy, she spent a year teaching English in northern Spain. Stark-Smith hopes to pursue a career in academic librarianship, focusing on reference and library instruction. She has also published on privacy in times of American war with William Aspray and Craig Blaha.

Cecilia D. Williams combines her interest in history and technology through her archival work at the Houston Public Library. She recently reevaluated the Norma Meldrum Juvenile Special Collections and worked on the Digital Archives photograph collections. Her previous work was in public libraries and Web design. She received her MS in information studies from the School of Information at the University of Texas at Austin in 2009.

Jeffrey R. Yost is associate director of the Charles Babbage Institute at the University of Minnesota. He holds a PhD in history of technology and science from the Case Western Reserve University. He is editor in chief of *IEEE Annals of the History of Computing*. His recent projects include a synthetic book on the history of the computer industry, a reference book on the history of scientific computing, and articles and book chapters on the history of medical informatics and privacy, the history of computer security standards, the history of IBM mid-range computer systems, the cultural history of computing and software, and IT historiography. He has written on travel agents in a recent MIT Press publication.

Index

Note: The letter *f* following a page number denotes a figure; the letter *t* following a page number denotes a table.

ABS (anti-lock braking system), 14
ACMP (Association of Comics
 Magazine Publishes), 287–288
Adolescents
 comic book reading and, 284, 283,
  289
 text messaging and (*see* Text
  messaging)
Advertising
 air travel, 135–136, 139–140
 car (*see* Automobile advertising)
African Americans
 discrimination by airline companies,
  126
 interest in genealogical research, 162
 philanthropy and, 85, 114n6
Airline companies
 customer loyalty programs, 127
 government contracts with, 125
 price structure, 142–144, 151n37
Airline Deregulation Act, 132–134
Airline hijackings, domestic, 130
Airline travel, 121–156
 businessmen and, 126
 current information issues, 140–145
 endogenous information-seeking
  factors, 3, 123t
 e-ticketing and, 144

exogenous information-seeking
 factors, 3, 123t
growth in, 121, 129–130, 133
Internet as information source,
 122–123
luxury vs. leisure, 137–138
from mid-1990s to present, 136–140,
 147t
from 1920s–1941, 125–127, 146t
from 1945–1978, 128–132, 146t
from 1978 to mid-1990s, 132–136,
 147t
passengers, information-seeking by,
 121–122
security concerns, 136–137
Air travelers
business, 141
demographic profile of, 140–141
leisure, 141
as strategic buyers, 143
Altruism, 74–75
Alvord, Katie, 19
American Airlines, 127
American Cookery, 230
American Gambling Association,
 209–210
American Red Cross, 102–103, 102t
American Temperance Society (ATS), 81

American Tract Society, 81
Andrews, Peter, 157
Anti-comics movement, 287–292
Anti-lock braking system (ABS), 14
Application programming interfaces
    (APIs), 259
Archie Comics, 285
ARNOVA (Association for Research on
    Nonprofit Organizations and
    Voluntary Action), 92
Aspray, William, 3, 4–5, 9–69, 277–303
Assembly line, 35
Association for Research on Nonprofit
    Organizations and Voluntary Action
    (ARNOVA), 92
Association of Comics Magazine
    Publishes (ACMP), 287–288
Athletes, professional, 208–209
ATS (American Temperance Society), 81
Auburn Automobile Company, 30–31,
    39–40. *See also* Cord automobile
Auto clubs, 43, 63n26
Automobile advertising
    auto shows, 19, 28, 30, 31, 36, 38, 44,
        51, 54, 59, 62n11
    car buying and, 18–23, 60
    of Cord automobile, 30–32, 33*f*–34*f*
    in magazines, 19–23
    in newspaper classified section, 54
Automobile dealerships. *See*
    Dealerships, automobile
Automobile manufacturers. *See also*
    specific automobile manufacturers
    American, in early 1900s, 35
    Big Three, 28, 39, 46, 51
    body styles, 49
    competition between, 37–39, 59–60
    dealer system and, 24–26
    defects and, 15
    foreign competition, 46
    interactive Web sites of, 29
    introductions of new models, 44
    national advertising and, 27–28

postwar, 45–52
prewar innovations, 37
technological innovations, 49–50
World War II and, 40–41
Automobiles. *See* Car buying; Cars
Auto shows, 19, 28, 30, 31, 36, 38, 44,
    51, 54, 59, 62n11

Baseball, 189, 193, 194
Baseball Abstracts, 196
Baseball Prospectus, 204
Bates, Marcia, 335
Batman, 292
Baym, Nancy, 297–298
Benevolent Empire, 80
Better Business Bureau, 109
Bible, 157–158
Bill James Baseball Abstract, 201, 202,
    203
Bjordahl, Hans, 298–299
Blogs
    cook-through, 242
    sports-related, 206–207
Bon Appétit, 233t
Books
    comic (*see* Comic books)
    genealogical, 168–169, 176
Boxing events, 191–192
Brace, Charles Loring, 94
Brin, Sergey, 97
Burkert, Herbert, 254
Bush, President George, 261–262
Business models, Internet effects on, 331

CAB (Civil Aeronautics Board), 132
Cantor, Eddie, 99
Car buying, 3, 9–69
    advertising and (*see* Automobile
        advertising)
    advertising role in, 18–23
    dealership role in (*see* Dealerships,
        automobile)
    exogenous forces, 60

financial issues, 16, 36–37, 58–59
financing options, 36–37, 58–59
foreign competition and, 46
historical issues, 10, 12f–13f
information issues, 11, 14–16, 12f–13f
information sources, 16–18, 17f, 54–57
Internet information and, 331
Internet tools for, 332
postwar issues, 51–54
prewar issues, 41–44
segmentation of market, 58
women and, 22–23
CarFax, 55
Car loans, 37
Carnegie, Andrew, 95
Car races, 44
Cars. *See also specific makes/models*
before 1945, 35–42
after 1945, 45–52
air-conditioned, 49
compact, 47
concept, 51
emission standards for, 48
environmentally friendly, 48
fuel-efficient, 49
manufacturers of (*see* Automobile
  manufacturers)
ownership experience from
  1917–1934, 41–42
postwar, 58
prewar, 57
purchasing (*see* Car buying)
safety features, 47–48, 50
steam-powered, 40–41
technical innovations, 57–58
Case, Steve, 97
Catholic Church records, 177
Celebrities, air travel and, 126
Cell phones
cooking-related tools on, 242
pay-as-you-go subscriptions, 309
short messaging service (*see* Text
  messaging)

social consequences of, 316–317
sports score programs, 197
Center for Effective Philanthropy, 108
Certeau, Michel de, 333
CFC (Combined Federal Campaign), 91
Chadwick, Henry, 190
Chapman, Gary, 4, 249–274, 337
Charitable contributions
group affiliation and, 76
statistics on, 73–74
tax law and, 75
Charitable giving, 330
Charity, vs. philanthropy, 72
Charity Navigator, 109
Charity Organization Societies (COS),
  89–90
CharityWatch, 109
Chatman, Elfreda, 332, 336
Chayko, Mary, 312–313
Chefs Line, 242–243
Cherrypicking concept (Bates), 335
Chevrolet, 9, 18, 22–24, 27, 38, 39,
  45–46, 49, 52
Child, Julia, 242
Children, comic book reading and,
  284, 283, 289
Chrysler, 21, 26, 28, 30, 39, 45, 49, 51,
  64n32
Church of Jesus Christ of Latter-day
  Saints, 159, 172
Citizen philanthropists, 94–98
Civic engagement, 271–272
Civil Aeronautics Board (CAB), 132
Civil War, American philanthropy and,
  94, 85
Clarke, Sally, 31, 63n26
Clean Air Act (1970), 48
Clinton, President Bill, 261
Colonial America
government documents in, 252
religion and philanthropy in, 79–80,
  111, 114n5
Combined Federal Campaign (CFC), 91

Comic Book Nation: The Transformation of Youth Culture in America (Wright), 277
Comic book publishers, 290
Comic books
fan culture of, 285–287
golden age of, 281–287
readership of, 283–285
Comic book shops, 294–297, 301, 302n4
Comics, reading, 4–5, 277–303
anti-comics movement, 287–292
in digital age, 297–300
endogenous forces, 301
exogenous forces, 301
webcomics, 298–300
Comics Code, 290–291, 301
Comics industry, direct market, 295
Comics Magazine Association of America (CMAA), 290–291
Comic strips, history of, 278–281
Common Sense (Paine), 253
Communications
online, 312–313
philanthropy and, 98–101
text messaging (*see* Text messaging)
Communities of practice, 336
Community Chest movement, 86, 91, 112
Comparison, in car buying, 15
Computers
computer reservation systems (CRS), 131, 134–135, 139, 150n18
genealogical research and, 170–176, 173t, 174t
Consumer movement, 59
Consumer Reports (Consumer Union Report), 10, 16–18, 43, 53, 61n5
Cookbooks, 229–231, 245n4
Cooking
down-home, 220–221
gourmet (*see* Gourmet cooking)
teaching, 227

Cooking classes, 235–236
Cook's Illustrated, 232, 234, 233t
Cord automobile, 30–32, 33f–34f, 62nn16–20
Cortada, James W., 3–4, 157–184
COS (Charity Organization Societies), 89–90
CounterActive, 243
Courtship, 318
Craig's List, 54
Creative capitalism, 97
Crime comic books, 286, 287, 288
Crime Does Not Pay, 288
Culinary serials, reading, 232, 234, 233t
Culinary technology, 243

Dating, 318
DC Comics, 282–283, 291
DC-3 airplane, 125–126
Dealerships, automobile
business models, 15
for Cord, 30–32, 33f–34f
high-pressure sales, 63–64n31
networks, 59
special role of, 24–30
"Death of distance," 195
Dell Publishing Company, 267, 281–282, 288
Depression of 1930s. *See* Great Depression
Devine, Edward, 90
Diffusion theory, 335
"Digital divide," 272
Digital media, 330
Digital restoration and preservation services, online, 167t
Dix, Dorothea, 94
DNA testing, for genealogical research, 179
Doctor Fun, 299

Donors, philanthropic
fatigue of, 106–107
protection from fraud, 108–109,
  112–113
tax issues, 109–110

Earl, Harley, 20, 46, 51
eBay, 54–55
EC Comics, 286–287, 288–289, 290
Eckert, Penelope, 313
Economic downturns, 330
E-government evolution, 260–264
Emission standards, for automobiles,
  48
Enlightenment, philosophy of,
  252–253
Ensmenger, Nathan, 4, 185–216
Entrepreneurs
philanthropic, 94–95
social, 97–98
Environmental Protection Agency
  (EPA), 265–266
Equal Credit Opportunity Act, 63n29
Ethnic groups, participation in
  genealogical research, 160

Facebook, 98, 263, 264, 266, 270
Family history
genealogy and, 157–159
written, 166–167
Family Tree Maker, 166
Fan-Addict Club Bulletin, 287
Fan culture
of comic books, 285–287, 291–294,
  297–298
in sports, 4, 187–189
Fantasy sports
baseball, 199
broadcasting companies and, 196–197
fans and, 187–188
football, 185–186
information gathering in, 201–203
Internet and, 197–198

players, information-gathering
strategies for, 203–206
popularity of, 197–198
Rotisserie baseball, 199–201
statistics and, 196
Fantasy sports companies, 203
Federal Aid Highway Acts, 50
Federal Funding Accountability and
  Transparency Act of 2006, 265, 269
Federal Register, 255
Federal Trade Commission (FTC), 109
Feinberg, Lotte E., 257
Financing, automobile, 37–38, 58
Fine Cooking, 233t
Fisher's concept of information
  grounds, 336
Flink, James, 26
FOIA. See Freedom of Information Act
Food cultures, regional, adoption of,
  224–225
Food Network, 234
Food & Wine, 233t
Football, 185, 192
Ford, Henry, 10, 25, 35–36, 268
Ford Motor Company, 25–26, 57
Model A, 59
Model T, 10, 35, 36, 37, 38, 57, 58,
  61n2
postwar era, 45–47, 49, 51
prewar era, 35–40
retail distribution practices, 29
Thunderbird, 20
Foreign cuisines, experiencing,
  223–224
Foster's theory of nonlinear
  information seeking, 334
Franklin, Benjamin, 93, 254
Free agency, 208–209
Freedom of Information Act (FOIA)
in Sweden, 253, 254
in United States, 254–258, 270–271,
  329
FTC (Federal Trade Commission), 109

Fuel/gasoline, 48–49
  availability, 48–49
  consumption taxes, 48
  higher-octane, 50
  prices, car buying and, 14–15
Fundraising activities, 103, 105–106,
    104*t*
  differences in online giving, 107–108
  donor fatigue and, 106–107
  fraud potential, 108–109
  professional, 91–92
  social networking and, 107
Funnies, The, 282

Gaines, William, 288–289, 290
Gallaudet, Thomas, 94
Gambling, on sports, 209–210
Gamson, Bill, 199
Garrison, William Lloyd, 94
Gasoline. *See* Fuel/gasoline
Gastronomy, reading, 231–232
Gates Foundation, 96–97
GDS (global distribution systems),
    150n18
Gender, comic book reading and,
    284–285
Genealogical guides, 168–169
Genealogical research, 3–4, 157–184
  after arrival of computers and
    Internet, 170–176, 173*t*, 174*t*
  Bible and, 157–158
  data not on Internet, 177–178,
    178*t*
  endogenous factors, 160–161
  exogenous factors, 161
  family history and, 157–159
  information needs, 163–167
  information sources, 163–167, 164*t*,
    165*t*
  Internet tools for, 332
  misinformation, spread of, 176–177
  motivations for pursuing, 160–163
  before personal computers, 167–170

profile of persons engaged in research
    on, 159–160
  using DNA testing, 179
  Web sites, 159
Genealogical societies, 168
General Motors (GM), 10, 18–21, 26,
    27, 35, 38–39, 45–46, 51
General Motors Acceptance
    Corporation (GMAC), 37, 58, 62n15
Geodata.gov portal, 265–266
Geospatial One-Stop, 261–262
Girard, Stephen, 94
Girl Scouts, 105–106
*Giving and Volunteering in the United
    States 2001*, 73
Gleason, Lev, 288
Global distribution systems (GDS),
    150n18
Global System for Mobile
    Communication (GSM), 308, 309
GM (General Motors), 10, 18–21, 26,
    27, 35, 38–39, 45–46, 51
GMAC (General Motors Acceptance
    Corporation), 37, 58, 62n15
Goggin, Gerard, 309
Good Housekeeping, 232
Goods and services, purchase of. *See*
    Airline travel; Car buying
Google Inc.
  philanthrophy and, 97
  search engine, using to locate recipes,
    240
Gourmet cooking, 4, 217–248
  consulting projects, 226–227
  culinary serials, reading, 232, 234,
    233*t*
  eating out as a study activity, 223
  experiencing foreign cuisines,
    223–224
  expressing culinary expertise, 222*t*,
    226, 228
  having "foodie" friends and, 225–226
  as hobby, 218–221

launching a cooking episode, 222*t*,
236–237, 241
as lifestyle activity, 221, 223, 222*t*,
226
online forums, 227–228
reading cookbooks on, 229–231
reading gastronomy, 231–232
recipe searches, 237–240, 239*t*
resources, 229
staying informed/inspired, 222*t*,
228–229, 236
taking classes in, 235–236
teaching, 227
visiting ethnic neighborhoods/
markets, 225
Gourmet lifestyle, living, 228, 243–244
Gourmet magazine, 219, 231, 232, 233*t*,
245n2
Government, 249–274
agencies, adoption of information
access rules, 259–260
citizen interaction with, 265–267
data portals, 261–263
evolution of e-government, 260–264
geographic data, 261–262
informed citizens and, 252–254
interactive Web-based tools for,
261–264
Internet and, 250–251
intervention for economic crisis,
86–88
online user-generated content and,
251
provider vs. user perspectives, 337
public access to (*see* Public
information)
public support for access to, 271–272
regulations, 47–48, 329
secrecy, national security and, 254
sensitive, withholding of, 249–250
transparency, 257–259, 272
Graphic novels, 297
Great Awakening, the, 79

Great Depression
airline travel and, 125
car sales and, 21, 31, 41, 58
large philanthropic gifts during, 99
philanthropy and, 85–88
GSM (Global System for Mobile
Communication), 308, 309
GuideStar, 107, 108

Habitus concept, 333
Harper, Richard, 314, 317
Harper's Bazaar, 232
Hartel, Jenna, 4, 217–248, 336–337
Hayes, Barbara M., 3, 5, 71–119,
305–328
HealthReform.gov, 268–269
Hearst, Randolph, 278
Hendrickson, Robert, 290
Higdon, Hal, 9
Highway system, 40, 50
HIV/AIDS philanthropy, 100–101,
114n8
Hogan's Alley (Outcault), 278
Hollway, W., 337
Hoover, President Herbert, 86
Horror comics, 286, 289
Horseless Age magazine, 43, 59
Howe, Samuel Gridley, 94
Hulk, the, 292–293

IdeaStorm, 267
"Impure altruism," 74–75
Income, of professional athletes, 208
Independent sector, 71, 113n1
Information exchange, 336
Information horizon, 334
Information interchange, 337
Information-seeking behavior
for everyday activities, 1–2 (*see also
under specific activities*)
exogenous forces, 329–330
methodological considerations,
332–337

Information sources, 332–333
Infrastructure, for American driving,
    40, 50
Instant messaging, vs. text messaging,
    311–312
Institutional ethnography, 334
Internal Revenue Service (IRS), tax
    issues for donors, 109–110
International Society for Third Sector
    Research (ISTR), 92
Internet
  airline industry and, 136–140
  air travel and, 148n2
  car buying, 54–55
  civic engagement and, 271
  data mining searches, 175
  "digital divide" and, 272
  effect on information seeking,
    330–332
  in everyday life, 5–6
  everyday-life literature and, 55–56
  food- and cooking-themed Web sites,
    235
  fundraising on, 100–101
  genealogical research and, 170–176,
    173t, 174t, 178–179
  government information and, 250–251
  government transparency and,
    258–359
  information, quality/authenticity of,
    142
  as information source for car buying,
    19
  philanthropy and, 88, 113
  price scamming operations, 143–144
  recipe searches, 237
  sports and, 195–196, 210
  user-generated content, 264
  Web sites, for car buying, 28–29
IRS (Internal Revenue Service), tax
    issues for donors, 109–110
ISTR (International Society for Third
    Sector Research), 92

Jefferson, Thomas, 253
Jenkins, Henry, 297
Jets, 131–132
John D. Rockefeller Foundation, 95–96
Johnson, Lyndon B., 255

Kasesniemi, Eija-Liisa, 311–312, 316
Kefauver committee, 289–290

Language, of text messaging, 310–313
Larousse Gastronomique, 231–232
Laursen, Ditte, 313
Learning to Give, 110–111
Lee, Stan, 292–294
Levitz, Paul, 297
Liberty Bonds, 86
Libraries, genealogical research and,
    168–169
Library of Congress, 254–255
Life activities, 221
Life cycle, in car buying, 15
Life in the Round theory (Chatman),
    336
Lifeworlds concept (Chatman), 77, 88
Lindbergh, Charles, 125
Lindeman, Eduard C., 96
Ling, Rich, 316–317
Little, Rachel D., 3, 121–156
LIVESTRONG Foundation, 106
Living Cookbook, 243
Lofton, Kenny, 208
Longoria, Arturo, 5, 305–328
"Long Tail" of philanthropy, 98,
    114n7
Loyalty, of sports fans, 187, 208–209

Madison, James, 253
Magazines. See also specific magazines
  advertising on air travel, 127, 131
  auto/car enthusiast, 43, 53, 61nn6, 7,
    63n25
  car advertising in, 18, 20–24, 53–54
  gourmet cooking, 219, 232, 234, 233t

postwar car buying and, 53–54
prewar car buying and, 43–44
sports, 202
Major League Baseball teams, 194,
  202–205
Manufacturers
  of automobiles (*see* Automobile
  manufacturers)
  use of comic books as premiums, 282
Maps
  environmental, 265–266
  government information for, 261–262
"March Madness," 209
March of Dimes, 99–100
Marvel Comics, 292–294, 295
Mastering the Art of French Cooking
  (Child), 242
Mather, Cotton, 79
McCloud, Scott, 296
Media. *See also specific types of media*
  information gathering and, 335
  philanthropy and, 88
  religious reform organizations and,
    80–81
  sports, historical perspective of,
    188–196
  technology, informational effects of,
    329–330
  use by religious organizations, 82–83
Message boards, sports-related,
  206–207
Metz, Sara, 4, 185–216
Michener, James, 192
Miller, Ellen, 258
Model A, 36–37
Money magazine, 18–19, 54, 61n8
*Morley v. Consolidated Manufacturing
  Co.*, 28
Morreale, Herb, 298–299
Motel chains, 50–51
Motoramas, 51
Motor Vehicle and Air Pollution Act
  (1965), 48

Movies/theater productions,
  philanthropy and, 100
Musicians, philanthropically active,
  106
Mutt and Jeff, 280, 281

NADA (National Association of Auto
  Dealers), 25
Nader, Ralph, 17, 47
NARA (National Archives and Records
  Administration), 254
National Archives and Records
  Administration (NARA), 254
National Association of Auto Dealers
  (NADA), 25
National Collegiate Athletic
  Association (NCAA), 209
National Highway Traffic Safety
  Administration, 47, 63n30
National Tuberculosis Association, 105
NCAA (National Collegiate Athletic
  Association), 209
Nettels, Beth, 4–5, 277–303
Network competence, 335–336
Newell, Angela, 4, 249–274, 337
Newspapers
  airline travel advertising, 131
  comic strips in, 278–281
  sports industry and, 189–191
New York Times, 307
Nonprofit Almanac 2008, The, 71
Nonprofit organizations
  philanthropic (*see* Philanthropic
  organizations)
  U.S. gross domestic product and, 71
  watchdogs organizations for, 109

OAG (*Official Aviation Guide of the
  Airways*), 131, 134, 150n19
Obama, Barack, 87–88, 249–250,
  257–258, 265, 267–269, 271
Office of Management and Budget
  (OMB), 269–270

Official Aviation Guide of the Airways
  (OAG), 131, 134, 150n19
Official Baseball Register, 196
Okrent, Daniel, 199
OMB (Office of Management and
  Budget), 269–270
Omidyar, Pierre, 97
Omidyar-Tufts Microfinance Fund, 97
O'Neil, Shaquille, 208
Online forums, for gourmet cooks,
  227–228
Online shopping, for airline fares,
  141–142
OPEC (Organization of Petroleum
  Exporting Countries), 48–49
Open government approach,
  267–268
Organization of Petroleum
  Exporting Countries (OPEC),
  48–49
Orman, Suze, 110
Otto, Jameson, 185–216
Outcault, Richard Felton, 278–279

Page, Larry, 97
Paine, Thomas, 253
Patriotism, informational effects, 330
Penny Arcade, 299
Philanthropic organizations
  number of, 101–102
  tax issues, 109–110
  Web interactivity and, 102–103,
    102t, 104t
Philanthropy, 3, 71–119, 330
  vs. charity, 72
  citizen philanthropists, 94–98
  communications media, 98–101
  definition of, 112
  depression and, 83–88
  differences in online giving,
    107–108
  donor fatigue, 106–107
  endogenous factors, 3

  in everyday life, 71–74
  historical forces, 77–88
  information overload and, 102–103
  Internet tools and, 331
  large-scale, 95–96
  motives for, 74–77
  professionalization of, 89–92
  reformers and citizen philanthropists,
    93–98
  relationship building and, 103,
    105–106
  religion and, 78–83
  "scientific," 90, 95–96
  as "third sector," 71
  value, educating children on,
    110–111
  war and, 83–88
  Web-enabled approaches, 102–103,
    102t, 104t
Pick 5 for the Environment site, 266
Pierce, Lyman L., 86
Planned obsolescence strategy, 20, 39,
  58
Polio, fundraising efforts for, 99–100
Poor Richard's Almanac, 93
Principle of Least Effort, 336
Public information, 249–275
  Freedom of Information act and,
    254–258
  informed citizens and, 252–254
  interactive Web-based tools for,
    261–264
  national security and, 256–257
  post-FOIA era, 258–260
  promotion of transparency,
    collaboration, and participation,
    267–270
  requests for, 255–256
  support for access to, 271–272
Publishers, of comic books, 282–283,
  287–288
Pulitzer, Joseph, 278
PvP, 299

Race/ethnicity, information seeking
and, 332
Radio
car advertising, 23–24
fundraising efforts and, 99
as information source for car
buying, 19
sports coverage, 191–193
Reader response theory, 333
Reading
comics (*see* Comics, reading)
cookbooks, 229–231
of graphic novels, 297
webcomics, 298–300
Reagan, Ronald, 192
Real, Michael, 195
Recipes, 245n5
adapting/changing of, 240–241
asking friends/families for, 240
googling for, 240
home-based collections, 237–238
Internet searches for, 237
online databases, 238–240, 239*t*
traditional forms, 241
"Reciprocal altruism," 75
Red Cross, 86
Religion, philanthropy and, 78–83,
112, 330
Restaurants
fast food, 51
gourmet, eating at, 223
Revolutionary War, American
philanthropy and, 85
Richmond, Mary, 90
Rockefeller, John D., 95, 102
Rogers, Everett, 335
Romance comics, 284–285
Roosevelt, Franklin Delano, 87,
99
Roots (Haley), 162
Rotisserie baseball, fantasy,
199–201
Royer, George, 4–5, 277–303

SABR (Society for American Baseball
Research), 204
Safety features, automobile, 14, 37, 47
Saturday Evening Post, The, 17*f*, 31, 36,
43, 52, 53, 56
Saveur, 233*t*
Savolainen, Reijo, 333, 335
Scholarship, on everyday life, 5–6
Search strategies, theories of, 334–335
Second Great Awakening, the, 79
Seduction of the Innocent, The
(Wertham), 289
September 11, 2001, air travel and,
136–138
Sheen, Fulton John, 82
Situated learning, 336
Slater Fund, 96
Sloan, Alfred, 38–39, 51
Small world concept, 336
Smog, 48
Social entrepreneurship, 97–98
Social networking sites, 331
Social positioning theory, 337
Social workers, early, 90
Society for American Baseball Research
(SABR), 204
Sociology of Families (Newman and
Grauerholz), 317–318
Software for genealogical research,
165*t*, 166, 171–172
Sonnenwald, Diane, 334
Spider-Man, 292–293, 298
Sports
American identity and, 187
gambling and, 209–210
Internet and, 195–196
message boards/blogs, 206–207
newspaper coverage of, 189–191
as social currency, 187–188
Sports Blog Nation, 207
Sports fans, 4, 187–189
Sports video games, 188
Stanford, Leland, 95

Stark-Smith, Gesse, 5, 305–328
Steam-powered cars, 40–41
Stebbins, Robert, 336
Stevenson, Matthew, 9
Strat-O-Matic baseball, 198
Super Bowl, 209
Superheros, in comics, 282–283,
    284–285, 292
Superman, 292
Sweden, freedom of information law,
    253, 254
Syndication, of comic strips, 280

Tabletop baseball, 198
Tax law, charitable contributions and,
    75
Taylor, Alex S., 314
Teenagers. See Adolescents
Telenor, 308
Television shows, 329–330
  car advertising on, 19, 24, 62n9
  from Crystal Cathedral, 81
  food and cooking information on,
    234–235
  football viewership, 185
  fundraising efforts and, 100
  informational effects, 329–330
  philanthropy and, 88
  sports coverage, 193–194, 210
Terrorists, 330
Text messaging, 5, 305–328
  behavior of texters, 307, 310–316
  calling party pays billing system,
    309–310
  criticisms of, 306, 314–315
  dangers of, 307
  etiquette of, 313–314
  in Europe, 309–310
  functions of, 307, 316–319
  history of, 307, 308–310
  vs. instant messaging, 311–312
  language of, 310–313
  popularity of, 306

social implications of, 319–324
social interactions and, 317–318
of sports scores/statistics, 197
Text speak, 310–313
Theories of information-seeking,
    333–334
Theory of information poverty, 332
Thurlow, Chrispin, 311, 312, 316
Train travel
  with air travel, 129
  vs. air travel, 126–127
Travel agents, 122, 131, 134–136, 138,
    150n29, 151nn30, 31
Trosby, Finn, 308–309
Turkle, Sherry, 314–315
Twitter, 242, 263, 264, 266, 270
Txting: The Gr8 Db8 (Crystal), 312

United States Geological Survey
    (USGS), 265
United Way (United Fund), 91–92, 112
USAspending.gov, 270
U.S. Census, 158, 172
U.S. Department of Defense, 265
USGS (United States Geological
    Survey), 265
U.S. Postal Service, 125, 254

Vanderbilt, Cornelius, 95
Video games, sports-related, 188
Volunteerism, decline in, 103, 105–106

Walker, Sam, 202
War. See also World War I; World War II
  genealogical research and, 161–162
  informational effects, 330
Ward & Hill Associated fundraising
    firm, 91
Webcomics, 298–300
Web 2.0 Heroes: Interviews with 20 Web
    2.0 Influencers (Jones), 267
Web 2.0 information tools, 265–267
Web material, fan-generated, 298

Web sites. *See also specific Web sites*
air travel-related, 142
food- and cooking-themed, 235
genealogical, 159, 165*t*, 163, 174–175
gourmet cooking, 239*t*
government, 249–250
online fantasy sports, 202
sports-related, 203–205
with sports simulations, 198
travel-related, 136, 138, 139, 144–145
White House, 263–264
Wertham, Frederick, 289
Whitefield, George, 79
WhiteHouse.gov, 263–264
Williams, Cecilia D., 3, 121–156
Women
air travel and, 130, 149, 148
comic book shops and, 295–296
volunteer activities of, 80
Wonder Woman, 296
World Vision, 107
World War I
fundraising efforts, 86
newspaper sports coverage after, 190
World War II
air travel and, 128
automobile manufacturers and, 40–41
fundraising efforts, 87
soldiers, comic book reading and, 284
WP4 (Working Party 4), 308
Wright, Frank Lloyd, 29

Yahoo! Sports Web site, 200, 202
The Yellow Kid (Outcault), 278–280
YMCA, 86
Yost, Jeffrey R., 3, 121–156
Young Romance, 285
YouTube, 263, 264

Zipf, George, 336